Analytical Chemistry

Séamus Higson
Professor of Bio- and Electroanalysis
Cranfield University

OXFORD
UNIVERSITY PRESS

OXFORD
UNIVERSITY PRESS
Great Clarendon Street, Oxford OX2 6DP

Oxford University Press is a department of the University of Oxford.
It furthers the University's objective of excellence in research, scholarship,
and education by publishing worldwide in

Oxford New York

Auckland Bangkok Buenos Aires Cape Town Chennai
Dar es Salaam Delhi Hong Kong Istanbul Karachi Kolkata
Kuala Lumpur Madrid Melbourne Mexico City Mumbai Nairobi
São Paulo Shanghai Taipei Tokyo Toronto

Oxford is a registered trade mark of Oxford University Press
in the UK and in certain other countries

Published in the United States
by Oxford University Press Inc., New York

First published 2004

A Catalogue record of this book is available from British Library

Library of Congress Cataloging in Publication Data
Data available

ISBN 0-19-850289-3

Typeset by Newgen Imaging Systems (P) Ltd, Chennai, India
Printed in Great Britain by
Antony Rowe, Chippenham, Wiltshire

Contents

Part 4 Analytical chemistry in practice: contemporary analytical science

Preface

Introduction to the Book

Analytical Chemistry is intended to be an unintimidating, student-friendly text covering all of the mainstream analytical techniques encountered both within undergraduate chemistry programmes and masters degrees. It is hoped that the text will also be of benefit to students embarking on research careers through PhD or other research degree programmes who might want an introduction to techniques they are unfamiliar with.

The motivation for writing this book arose from watching many bewildered students trying, or rather struggling, to find information from within the many well written but simply mammoth analytical chemistry textbooks on the library bookshelves. It struck me that there was a real need for a text that allowed students to grasp information quickly, in a user friendly format. I tried hard to find such a text but failed, and so finally decided to write this book!

The need for a text of this type has become more apparent as time has progressed, since undergraduate syllabi both in Europe and the US have given progressively more importance to analytical chemistry as a subject in its own right. This trend has been mirrored in degrees such as biochemistry, forensic science, and many others in which analytical chemistry plays a key role.

Analytical chemistry clearly encompasses other areas of chemistry: physical chemistry, for example, is especially relevant to the way in which many measurements are made, while aspects of the subject directly relate to the analysis of inorganic or organic compounds. By its very nature, analytical chemistry is a *generic* discipline, and overlaps the realms of biotechnology, forensic science, physics, materials science and many other subject areas.

This book introduces the basis of many contemporary analytical techniques by means of an integrated approach. Even if, as a student, you have no intention of entering the field of analytical chemistry, it is hoped that this book will help you understand the nature of analytical chemistry and how it is performed. It is also hoped the material may even help make more sense of, and be of relevance to, some of the physical, inorganic and organic chemistry that you may already be familiar with.

It is often overlooked that more chemistry students end up working in the field of analytical chemistry than in any other branch of the subject. It is a truly dynamic and exciting area of science, and one that impacts upon all of our lives, every day. Above all I hope that this book inspires budding analysts to be enthused about the subject. If this happens then I will have achieved what I set out to do.

Principal Features and Structure of the Book

- ***The text is fully illustrated throughout*** to help explain both concepts and the principles of instrumentation; in all cases I have tried to make diagrams as clear and as informative as possible without being too complicated.

- To help learning, numerous ***fully worked examples*** are included within shaded boxes throughout the text. Problems are also included at the end of each chapter to help you as a student test understanding. Model answers for these problems are provided on the book's ***companion web site*** at **www.oup.com/uk/booksites/chemistry.**
- Many of the ***chapters are cross-referenced*** to help the reader understand how techniques are related to, or may be used in conjunction with, each other.
- ***Key points, definitions and highlights of scientific principles are highlighted*** within the margins as comments to help either further explain some of the more complicated concepts or to help emphasise the importance of a particular point or principle.

As a final point ***the sixteen chapters of this book are arranged into four sections*** to help guide students along a directed learning pathway. These are:

1) The scope of analytical chemistry: ground rules and fundamentals
2) Chemical analysis: key principles and processes
3) The key analytical techniques, and,
4) Analytical chemistry in practice: contemporary analytical science

How to Use the Book from a Learning Perspective

From the outset I wanted the text to be easy to dip in to and for this reason each topic is described within a self-contained section, although many sections are cross-referenced where techniques are either related to, or may be used in conjunction with, each other.

I have tried to compile as comprehensive an index as possible to help the reader access information quickly and easily as and when required.

Throughout the book subject material is designed to used as either as an independent learning resource or as an accompaniment to lecture courses. Highlighted boxes and worked examples can be skipped on first reading to help acquire an understanding of a given topic quickly and easily. The same topic can then be revisited in greater depth by working through examples and also the problems at the end of each chapter.

In some cases, suggestions for further reading are also provided if the reader wishes to access further information on a particular subject area.

Companion web site

Analytical Chemistry is accompanied by a free-to-view **companion web site at www.oup.com/uk/booksites/chemistry.**

The site includes:

- **Model answers** to the problems presented at the end of each chapter
- **Figures** from the book, available to download.

Acknowledgements

The writing of this book has been greatly helped by many friends and colleagues whose input has been truly invaluable. I would firstly like to thank the three commissioning editors, Melissa Levitt, Howard

Stanbury and particularly, in the later stages of the book's preparation, Jonathan Crowe, as well as the rest of the editorial team at Oxford University Press, without whose help this book would never have appeared.

I would like to extend special thanks to Mrs Sally Creveul of the Institute of Bioscience and Technology here at Cranfield, who has done such a marvellous job in preparing the figures for this book. I would also like to thank Mrs Linda Chapman and Mrs Liz Wade, also of my department, for their help with many stages of the book's production.

Many people have helped proof read and edit the book but, in particular, my friend and former colleague Dr Paul Monk has contributed a sterling effort by reading the entire manuscript and coming up with many useful suggestions to help improve the final text.

My thanks too to the following academics for providing many constructive comments on draft material: Tom McCreedy, University of Hull, UK; Mark Lovell, University of Kentucky, USA; John Lowry, National University of Ireland, Maynooth, Republic of Ireland; Carsten Müller, University of Cardiff, UK; David Littlejohn, University of Strathclyde, UK; and Philippe Buhlmann, University of Minnesota, USA.

My research group namely: Dr Stuart Collyer, Dr Frank Davis, Dr Karen Law, Andrew Barton, Davinia Gornall, Emma Lawrence and Daniel Mills, have also been particularly helpful in shaping the book to meet the requirements of students.

This book is dedicated to my late parents Shelia and Jack Higson who supported and encouraged me through many years of education; I would in the same context like to thank my sister Ethne who through my student days gave much support, both emotionally and financially.

Final thanks must be given to my wife, Josephine, who also has read the entire manuscript (several times), and my two daughters Rachael and Sarah who have put up with many evenings with me in front of the computer.

After all this help any errors or omissions must sadly remain the responsibility of the author. I hope students find this book useful and that their understanding of the subject is helped!

Séamus Higson
Professor of Bio- & Electroanalysis
Cranfield University at Silsoe
August 2003

Part I

The scope of analytical chemistry: ground rules and fundamentals

The scope of analytical chemistry and the nature of analytical measurements

1

Skills and concepts

This chapter will help you to understand:

- What is meant by analytical chemistry.
- The scope of analytical chemistry and its applications.
- The importance of high-quality data and the implications of poor quality or erroneous data.
- How to differentiate between a qualitative and quantitative test.
- What is meant by replicate measurements.
- The concepts of specificity, sensitivity, and accuracy of a test.
- What is meant by the terms analyte, interferent, and aliquot.

1.1 When and where analytical chemistry is used?

1.1.1 What is analytical chemistry?

We might not realize it, but many of us unwittingly carry out analytical chemistry on a daily basis. It is a common scene: wandering into the kitchen first thing in the morning, bleary-eyed, to make a cup of coffee to kick-start the day. While the kettle is boiling, we go to the fridge and take out the carton of milk. The carton has been sitting there for a few days and so we are not sure whether it is still fit to use, so we open it up, lift it to our nose, and gingerly sniff it. In so doing, and whether we realize it or not, we are carrying out an analytical test. In this case, our nose is evaluating the products of a variety of bacteria such as *Salmonella typhimurium*. If the milk is too old, and the bacteria have had a chance to multiply, then the action of the chemicals that they produce whilst multiplying will make the milk unfit to drink. We will know whether the chemicals are there or not by sniffing—and aiming to detect that characteristic smell of 'gone-off' milk.

You might recognize this test as being a chemical analysis more readily if you were to use a so called ***electronic nose***. At the time of writing, electronic noses represent an emerging state-of-the-art technique in the field of electronic chemical sensors even though there is a long way to go before they can sense as wide a range of smells as the human nose.

This example demonstrates that ***Analytical Chemistry encompasses any type of test that provides information relating to the chemical composition of a sample.***

We all benefit from the activity of analytical chemists. We all eat food, live in homes, wear clothes, and many of us drive motor cars. These are all examples that rely on the modern manufacturing chemical industry. This, in turn, is critically dependent on its quality control processes, the responsibility for which lies largely with analytical chemists.

Every one of us acts as a consumer and relies on the analytical chemist to play a major role within the manufacturing process to ensure that the food we eat, the clothes we wear, and the medicine we take, are of a suitable quality. The chemical industry has some input into almost every manufacturing industry, and represents the largest manufacturing sector of most major industrialized countries. Indeed, many economists say a good indicator towards the economic health of a nation can be gained by looking at its chemical industry. It follows that the role of the analytical chemist is a truly fundamental one!

If you are reading this text as an undergraduate chemist, and intend using your degree following your university studies, there is a greater than 50% chance that you will be employed in some analytically related role. Many chemists perform analyses as one part of their job even if they do not think of themselves as analytical chemists. For example, the first thing synthetic organic chemists will frequently do having made a new compound is to analyse what they have just produced.

It is clear that the population is becoming ever more demanding of analytical chemistry for ensuring both the quality of the products we consume and how we treat our environment.

The safety of the food we eat is entwined with many issues relating to modern farming methods, the use of agro-chemicals such as preservatives, pesticides, and fertilizers. We are also concerned about issues such as our cholesterol intake, how much fibre a food contains, its vitamin content, and the strength of alcoholic drinks. We demand low or 'acceptable' benzene contents in the petrol we put into our cars and are then concerned with the quantities of CO and CO_2 cars pump into the atmosphere. As the world population increases and our planet becomes ever more crowded, we can be sure that analytical chemists will be called upon to provide ever more information upon which future decisions can be rationally based.

1.1.2 So when and where is analytical chemistry used?

Many people will think of analytical chemistry as involving high-tech instrumentation such as state-of-the-art mass spectrometry, high performance liquid chromatography, and infrared techniques. Perhaps, on the other hand, we remember our school laboratory where we learnt the basics of titimetric analysis and spot chemical tests. While each of these techniques play their role within the arsenal of the modern analytical chemist, we should not forget simpler approaches using, for example, pH meters, litmus paper, and analytical balances; these are often used prior to the more elaborate approaches that sometimes come to mind more readily. The important thing is to view the subject as a *whole*.

Sensors reflect the push towards developing highly simplified analytical tests that may be performed by *non-chemists*. Every time we take our car for a fuel emission test, the mechanic will place a CO gas sensor within the exhaust outlet pipe to determine whether or not the levels of CO exceed a legal threshold. In another context, diabetics may use an electrochemically based sensor to monitor their blood glucose levels, and by this reading, determine the insulin dosage required prior to the next meal. Automated instrumental techniques are also being ever more widely used as more interest is being shown towards environmental chemistry and pollution issues; for example, weather reports often contain some reference to 'air quality'—analytical chemistry is at work here too.

This discussion makes two things clear; first, the use of analytical chemistry touches upon almost every aspect of our lives—and, if anything, our reliance on analytical chemists is set to increase further. Second, the subject is responding to changing needs and is therefore a truly dynamic subject; this is reflected by the vast research effort that is being channelled into this subject.

1.2 The nature of data

In general terms, there are two stages to chemical analysis: ***data collection*** and ***data analysis***—in other words, gathering information and determining what that information is telling us. Broadly speaking, data will come in two forms: ***qualitative*** and ***quantitative***. Likewise, the analysis of this data will either give a qualitative or a quantitative result.

Qualitative analyses are those that give negative/positive, or yes/no types of data; in other words, they say whether or not some substance (the ***analyte***) is present in a sample but do not actually measure the *quantity* of the substance(s) present. A home pregnancy test represents a good example of a qualitative analysis since the result will either indicate the presence (positive result) or lack (negative result) of a pregnancy.

Quantitative analyses determine how *much* of a particular substance there is within a sample rather than just its presence or absence. An example of a quantitative test is the measurement of the pH of an aqueous solution; here the result can range from 0 to 14.

Quantitative data are inherently normally expressed in a numerical format; the sign (negative or positive) and the magnitude both give meaningful information. The accurate use of units when quoting numerical data is also of paramount importance, yet often overlooked by students. Even the yes/no or negative/positive types of qualitative data can be mathematically handled by statistics. So almost all data demand some mathematical treatment, even if at a rudimentary level.

1.2.1 The limitations of data

Both qualitative and quantitative data analyses face limitations. In the case of a qualitative test, there may be a *threshold* below which the test may not be able to identify the presence of the substance. For example, a pregnancy test may give a false negative result if performed too early during the pregnancy if the level of the hormone ***human chrorionic gonadotrophin*** **(hCG)** within the urine is at too low a concentration to cause a colour change within the test strip. Therefore, even though the mother will be pregnant, the test will fail to detect this, and will give a false negative result. The important point to be noted here is that *even a qualitative test has a lower limit of detection* below which it will fail to detect the presence of the analyte.

Absolute accuracy is also impossible with quantitative data—there will always be a margin of error that must be accounted for. For example, no two pH meters will give *absolutely* the same measurement of pH; no two electronic balances will measure *absolutely* the same mass of substance being weighed. The treatment of quantitative data demands that ***error limits*** be determined. In this way a data point or set of data may be quoted to within a **known range** of possible error. A tap water supply may be quoted as containing 100 ± 10 parts per million (ppm) Pb. It is therefore possible to fix the concentration range to be between 90 ppm at the lower limit to 110 ppm at the upper limit. Information of this kind may often be highly useful for ensuring that the *correct and most appropriate* information may be derived from an analysis. More on the determination of error limits is given in Chapter 2.

1.3 Should a qualitative or quantitative test be chosen?

Analytical chemistry should always be performed for a purpose; this may sound obvious yet this fact is often forgotten. It has been estimated that up to 10% of tests performed each year world-wide are unnecessary.

Apart from the waste of money (which at the time of writing is estimated to run to as much as 5% of gross national product for most industrialised nations), there is clearly a huge wastage in human effort and resourcing. What is often lacking is a clear focus as to *why* tests are being performed—and indeed what useful information may be obtained from them.

One obvious point that should always be kept in mind is who the end recipient of the information is and what information is actually required—that is, why the test is to be performed and what useful purpose it may fulfil this. The most common reason for any project failing is due to a lack of planning at the outset and this is just as true for chemical analyses. The important message here is to ***plan your analysis carefully and appropriately***.

In many situations, you may only want a qualitative determination to be performed. For example, you may wish to only know whether or not a pollutant is present above a reasonable threshold but not need to know the quantities at which it is present. In many other circumstances a qualitative test may be performed as a first filtering process—and if the result is positive then a more complicated analysis may be performed in order to ***quantify*** the measurement. The contamination of water samples with lead is a good case study to illustrate this point. The lead iodide test (see Chapter 3) provides a simple positive negative result for the presence of lead above a concentration of approximately 0.2 $g\,dm^{-3}$. This technique provides a simple wet chemical approach which may be carried out at, for example, the side of a river bank using only rudimentary equipment. If a sample proves positive then the analyst may wish to quantify how much lead is actually present. Another approach such as the lead dithiazone test might be chosen to perform a quantitative analysis. This test provides a colour change which is proportional to the content of lead present—the more lead that is present, the deeper red in colour the solution will become. The colour of the lead dithiazone solution can be measured using a spectrophotometer to actually ***quantify*** the amount of lead present. (Since the colour of the solution will be proportional to the amount of lead present, measuring the intensity of the colour will give a direct indication of the quantity of lead in the solution.) Analyses of this kind are described in Chapters 5 and 6. Even in this situation, ***erroneous*** or incorrect results may be caused by the presence of other heavy metal ions, and if further specificity and or sensitivity is required, then the analyst may use atomic absorption spectroscopy (see Chapter 7).

Above all else it should be remembered that the correct test for any particular situation is the one which best meets the requirements of the end user. A prospective mother will be carrying out a pregnancy test to get a positive/negative result—she will not require a test to give the exact concentration of hCG present. Using the same argument, many pollutants have limits specified by legislation, which are deemed to be acceptable or not. In these situations, the water company (or regulator) is not just interested in a qualitative result (i.e. whether the pollutant is present or not), but in the actual concentrations in which it is present—and it is this figure that will determine whether or not the water is deemed to be of an acceptable quality.

1.4 Data handling and terminology

Data handling involves using a number of terms and it is important that these are clearly defined to avoid confusion.

Before data may be handled it must be collected or ***collated***. A sample is often taken from a larger volume and this sample is often known as an ***aliquot*** if in the form of a solution.

The test is often repeated with two or more samples to evaluate reproducibility and measurements of this kind are known as ***replicate*** measurements. Precision is the term used to describe the reproducibility of two or more replicate measurements that have been performed in the same way. There are several ways in which precision can be expressed and these are discussed in Chapter 2.

The substance to be analysed within the sample is known as the ***analyte***, and substances which may cause incorrect or erroneous results are known ***as chemical interferents***. (If an analysis monitors the concentration of a heavy metal ion, a different metal ion may have very similar chemical properties and therefore may interfere with the analysis and acts as a chemical interferent.)

The lowest concentration below which the test will fail to recognize the presence of an analyte is known as the ***lower limit of detection***.

The ***specificity*** of the test defines how the test may respond to the presence of a particular analyte, so if a test is totally specific then it will only respond to the analyte of interest and in this case no chemical interferents will interfere with the analysis.

The ***sensitivity*** of a test meanwhile describes how close or similar in magnitude two readings may be, and still be distinguished from each other. If a particular technique has a sensitivity of 1 ppm for Pb^+, then two determinations for 220 and 222 ppm may be taken as being two distinguishably different readings. By contrast two readings of 220.1 and 220.9 ppm Pb^+ *may not* be differentiated from each other. Data should never be quoted beyond the level of sensitivity and/or accuracy which may be appropriate for the test or instrument. Indeed, inappropriate numbers after decimal points may imply a level of sensitivity that is in fact meaningless and can in fact be totally misleading.

Accuracy describes how close the measured value is to the *true* value, which may in reality be very hard to determine. Certified reference materials (Chapter 2) are often used to help estimate the levels of experimental error which might be expected to be associated with a particular analytical technique.

Once collected, the data will normally contain replicate measurements for each data point. Sufficient information should also be collected to allow an estimation of the uncertainty or errors associated with the method. Only in this way may the experimental error of the technique be quantified.

The data may be collected by human observation and written down by hand, or be collected by some form of automatic sampling technique. An instrument may sometimes collect and process the data directly (e.g., by an autotitrator—Chapter 3). It is becoming increasingly common to use computers to assimilate and process the data. In each situation, however, the data must be evaluated or processed for their quality and reproducibility, and finally of course for its meaning. Statistical methods are frequently used for data handling and processing, and these will be described in Chapter 2.

1.5 The quality of analytical data

Reproducibility and accuracy are normally the most important criteria for the end user of a test. If a blood sample is analysed for its alcohol level in connection with a possible drink-driving conviction, it is crucially important that two differing laboratories would come to the same—and correct—conclusion. In a similar manner, the fuel emission testing equipment used by differing gaseous road-worthy testing stations should be able to give concordant results to within specified limits. There will always be experimental error associated with any test (see Chapter 2), however, the uncertainty of the result should be ***clearly quantified***, if any reliable judgements are to be made from the data. The scrutiny and assessment of the quality of the data may be carried out by some form of ***data validation*** process; much effort, time, and money is expended in statistical analysis of data and validation processes, both of which are discussed in Chapter 2.

Numerous studies have shown how hard it is to attain *concordant* analytical information from different laboratories. ***Poor or unreliable data are at best useless.*** If poor data cause the wrong decision to be made, the result may be very costly indeed. A plant manager, for example, may dispose of a batch of some product believing it to be contaminated when in fact it was perfectly acceptable. The mistaken action may even be dangerous or life threatening, if for example, a clinician administers an inappropriate drug dosage, due to an erroneous pathology laboratory test.

1.5.1 You as the analytical chemist

The quality of data and their interpretation are of paramount importance to any analytical chemist. Data are frequently numerical in nature and their handling and interpretation involves some simple statistics. You may well find the mention of mathematics and statistics offputting, yet the numerical handling of data is intended to add clarity to complex issues and lies at the very heart of analytical chemistry. If approached slowly and gently, none of the maths required to become truly confident in any aspect of data handling should be too problematic! These skills are required from the point at which the subject is first studied; so Chapter 2 of this book covers the statistical handling of numerical analytical data. This and the following chapter have been written to *gently* aid learning in these areas and to make the learning experience a non-traumatic and possibly even a pleasant and enlightening one! It is hard to study or use chemistry without resource to some form of analysis and so it is worth getting the basics firmly established at an early stage. By the time you have worked your way through Chapter 2, and the worked examples it contains, you should have at your disposal all of the principal mathematical skills you need to be able to fully grasp all of the material contained within every other chapter of the book.

Exercises and problems

1.1. Discuss how modern society is dependent upon analytical chemistry.

1.2. What is meant by: (i) a qualitative analysis; and (ii) a quantitative analysis? Give two examples of each.

1.3. Explain the difference between what is meant by the specificity, accuracy, and sensitivity of a technique.

1.4. What is meant by the term 'lower limit of detection' for a technique and how does this differ from the 'sensitivity'?

1.5. What is meant by the term 'replicate measurements'? Why are replicate measurements desirable when performing analyses?

1.6. Explain what is meant by an interferent. How might an interferent affect an analysis? Give two examples.

1.7. What is meant by a data validation process? How might a data validation process be performed?

1.8. What is meant by an aliquot?

1.9. What are meant by error limits?

Summary

1. Analytical chemistry encompases any type of test that provides information relating to the chemical composition of a sample.

2. Qualitative analyses are those that provide information relating to the presence of an analyte.

3. Quantitative analyses are those that allow the concentration of an analyte to be determined.

4. All data contain errors—and these should be estimated—normally by statistical means.

5. An interferent is a substance that may erroneously affect analytical measurements.

6. Replicate measurements are multiple measurements upon the same sample.

7. The specificity of an analytical test describes how selective the test is towards a given analyte.

8. The sensitivity of an analytical test describes how close in magnitude two readings may be and still be distinguished from each other.

9. The accuracy of a test describes how close a measured value is to the true value.

10. Data validation processes are vital if confidence is to be assigned to data.

11. Poor or unreliable data are at best useless and at worst may be dangerous or costly.

Further reading

Anand, S. C. and Kumar, R. (2002). *Dictionary of analytical chemistry*. Anmol Publications.

Kennedy, J. H. (1990). *Analytical chemistry practice*. Thomson Learning.

Analytical quality assurance and statistics

2

Skills and concepts

This chapter will help you to understand:

- How to quote data with the correct number of significant figures.
- What is meant by the spread of data.
- How to calculate the mean and median of a data set.
- How to quantify experimental error by calculation of the standard deviation, variance, and coefficient of variance.
- The difference between an indeterminate and a determinate error and also understand how these can occur and be identified.
- How to calculate the relative and population standard deviations for data sets.
- What is meant by an outlier.
- How to calculate confidence levels for the exclusion of outliers using either the Q-test or T-test.
- How to calculate the equation for a best-fit straight line by means of a least-squares fit approach.
- What is meant by a certified reference material and appreciate how these may be used for accrediting analytical protocols.
- How to use quality control charts.
- How to construct a calibration profile using either a step-by-step or a standard addition approach.

2.1 The consideration of experimental errors: an introduction

All data contain some degree of uncertainty, inaccuracy, and associated errors. It is, therefore, imperative that these are estimated so that they may be either accounted for, or, alternatively, if the errors are deemed unacceptable, the data may be rejected and the measurements taken again.

It should never be forgotten that data with unknown reliability are at best useless and at worst may give an incorrect answer to an important question. Errors or inaccuracies may only ever be estimated but if we have some idea of the source of the error, then some steps may be taken to either account for the problem(s) or even overcome them.

The principal methods for quantifying and handling errors involve the use of simple statistics. We shall, therefore, first consider a few simple definitions and methods for handling experimental data, before considering the sources of errors together with specific case studies and examples.

2.2 Significant figures and the reporting of numerical data

Calculations may use any number of figures for intermediary steps throughout the calibration, but the final value should again be reported with the correct number of significant figures. The correct number of significant figures should be the same as the least number used within the calculation.

There are a number of rules that should always be used when either (a) reporting numerical data or (b) quoting a value that has been calculated from the original data.

Data should be collected with the correct number of significant figures. Significant figures include all of the digits that are known with certainty plus one further estimated digit.

EXAMPLE 2.1

A burette reading, for example, may be reported as 23.76 cm^3. The 23.7 may be taken directly from the burette graduations; the ._6 is estimated by eye. In this example the burette reading is quoted to four significant figures.

2.3 Replicate measurements

A good chemist will always acknowledge that errors occur within any *data set*, however carefully they perform the analysis. It is, therefore, good practice to perform an analysis, if possible, several times, in order that some certainty may be given that the test gives a true and valid reading. If one or more analyses gives a reading that looks suspect when compared with the rest of the data, it may be wise to take further readings before rejecting the suspect data. The suspect data may in this case be useful in highlighting some process that may lead to incorrect readings. If the data set has a wide spread of values with little correlation to each other, the validity of the whole analytical procedure may be brought into question. In each case, the consideration of the collective data set may prove very useful.

The practice of taking multiple readings is known as obtaining ***replicate measurements***. Processes designed to monitor the quality and reliability of the data are known as ***quality assurance techniques***.

2.4 The spread of data, the mean, and the median

2.4.1 The spread, or range, of data

The ***spread or range*** of data is the arithmetic difference between the greatest and the smallest data points for a set of measurements. The data must first be arithmetically arranged in ascending order and the smallest value subtracted from the greatest.

EXAMPLE 2.2

An analytical determination for Pb in an aqueous solution gives six replicate measurements. Find the spread (or range) of the data.
ppm Pb^{2+}

(a) 20.1

(b) 19.5

(c) 20.3

(d) 19.7

(e) 20.0

(f) 19.4

(g) 19.6

Method

The spread of data describes the difference between the highest and lowest data points. The greatest value corresponds to 20.3 ppm Pb^{2+} and the smallest value to 19.4 ppm Pb^{2+}.

The spread is, therefore, 20.3 ppm Pb^{2+} – 19.4 ppm Pb^{2+} = 0.9 ppm Pb^{2+}.

2.4.2 The mean

The ***mean*** of a set of replicate measurements is sometimes also known as the ***arithmetic mean*** or ***average***; they are simply synonyms for the same term.

The mean of a data set is equal to the sum of all the data values divided by the number of measurements included in the data set.

The letter N is normally used to denote the total number of data values or replicate measurements.

i is often used as a subscript to identify each data value, i may range from $i = 1$ to $i = N$. If there are five data values, i may, therefore, be 1, 2, 3, 4, or 5. The Greek capital term for the letter *Sigma* 'Σ', is used to denote the sum of a number of data points. Σ is normally accompanied by sub- and superscripts, which are used to describe the lowest and highest data values for which the data are summed.

It, therefore, follows that $\sum_{i=1}^{N}$ means that data are to be summed from the first (1st) to the last (Nth) data value. There are often multiple data points within a set of data, and in this situation it is normal to label each value to avoid confusion. If the data set, x, is to be summed this may then be written as $\sum_{i=1}^{N} x_i$, which means all of the data points ($i = 1$–N) are to be summed.

It, therefore, follows that the mean $\overline{x}$ of a data set (x) will be equal to:

$$\overline{x} = \frac{\sum_{i=1}^{N} x_i}{N} \tag{2.1}$$

EXAMPLE 2.3

If we take the same data set as for Example 2.2
ppm Pb^{2+}

(a) 20.1
(b) 19.5
(c) 20.3
(d) 19.7
(e) 20.0
(f) 19.4
(g) 19.6

then

$$\sum_{i=1}^{N} x_i$$

may be calculated as follows:

Step 1

$$\sum_{i=1}^{N} x_i = 20.1 + 19.5 + 20.3 + 19.7 + 20.0 + 19.4 + 19.6$$

$$= 138.6 \text{ ppm}$$

Step 2

$$\sum_{i=1}^{N} x_i = 138.6$$

$N = 7$

$$\bar{x} = \frac{\sum_{i=1}^{N} x_i}{N}$$

so

$$\bar{x} = \frac{138.6}{7}$$

$$\bar{x} = 19.8 \text{ ppm } Pb^{2+}$$

Note: The result is quoted in ppm Pb^{2+}—that is, with the same units as the original data. The data are also quoted with the same number of significant figures as the original data (see Section 2.2). Remember that you should never quote data with inappropriate number of figures following the decimal point since they are meaningless. This is especially true when one performs calculations that are related to the quality or reliability of the data.

2.4.3 The median

If a data set contains an odd number of data values then the ***median** is the data value that lies in the middle of the data set when arranged in arithmetic order.*

If, however, a data set contains an even number of data values then the median *is the mean of the two data points that lie in the middle of the data set when arranged in arithmetic order.*

EXAMPLE 2.4

Find the median for the data of Example 2.2.

ppm Pb^{2+}

(a) 20.1

(b) 19.5

(c) 20.3

(d) 19.7

(e) 20.0

(f) 19.4

(g) 19.6

Step 1

Arrange the data (in ppm Pb^{2+}) numerically:

(a) 19.4

(b) 19.5

(c) 19.6

(d) 19.7

(e) 20.0

(f) 20.1

(g) 20.3

The middle point arithmetically is 19.7 ppm Pb^{2+} and in this case: the median = 19.7 ppm Pb^{2+}.

If one has a data set that contains an even number of data values then an additional step must be taken:

EXAMPLE 2.5

Calculate the median for the data given below.

The following data are the same as that for Example 2.3 but with an additional replicate reading. There are eight (and, therefore, an even number of data points):
ppm Pb^{2+}

(a) 20.1

(b) 19.5

(c) 20.3

(d) 19.7

(e) 20.0

(f) 19.4

(g) 19.6

(h) 19.9

Method

1. Arrange the data numerically.
2. Take the mean of the two data values that fall in the middle of the data set to find the median.

Step 1

Arrange the data numerically
ppm Pb^{2+}

(a) 19.4

(b) 19.5

(c) 19.6

(d) *19.7*

(e) *19.9*

(f) 20.0

(g) 20.1

(h) 20.3

Step 2

Take the mean of the two data points that fall in the middle of the data set. The two data points italicized fall numerically in the middle of the data and may then be summed together and divided by 2 to find their mean, and, therefore, the median for the data set:

$$\bar{x} \text{ for } 19.7 \text{ and } 19.9 = \frac{19.7 + 19.9}{2} \text{ ppm}$$

Median = 19.8 ppm Pb^{2+}

2.5 Quantifying experimental errors

Precision and ***accuracy*** are two terms that are often confused. ***Precision*** describes the reproducibility of the results—in other words, how closely the replicate measurements lie to each other. The reproducibility and, therefore, the ***precision*** of the data set is found by looking at the ***spread*** of the readings.

The precision of a set of data may be assessed by:

1. the ***standard deviation***;
2. the ***relative standard deviation*** (sometimes known as the coefficient of variance), or
3. the ***variance***

Each of these terms is a function of the spread of the data, and will be considered in turn throughout this chapter.

The ***accuracy*** of the data by contrast describes how close the data is to the *true or accepted* value for the measurement. The accuracy of a data value may never, of course, be determined exactly since this would assume that the true value was already known with an absolute certainty.

The accuracy of data may be described in terms of the error in the reading.

2.5.1 The absolute error

The absolute error of a system is equal to the difference between the actual reading, x_i, and the true (or accepted) value x_t:

$$E_A = x_i - x_t \tag{2.2}$$

It should be remembered here that the true value x_t may be very hard to determine or even agree upon, which, in turn, makes the use of the absolute error difficult.

2.5.2 The relative error

The ***relative error***, E_r, describes the error in relation to the magnitude of the true value, and may, therefore, be more useful than considering the absolute error in isolation.

The relative error is normally described in terms of a percentage of the true value, or in parts per thousand of the true value.

If the relative error is to be described in terms of a percentage, then E_r may be calculated according to Eqn (2.3):

$$E_r = \frac{x_i - x_t}{x_t} \times 100\% \qquad (2.3)$$

Similarly, if the relative error is to be expressed in terms of parts per thousand of the true value, then E_r may be calculated according to Eqn (2.4):

$$E_r = \frac{x_i - x_t}{x_t} \times 1000 \text{ ppt} \qquad (2.4)$$

EXAMPLE 2.6

Calculate the relative error in percentage terms for an iron analysis that gives a value of 115 ppm Fe content when the true value is, in fact, 110 ppm.

Method

Assign the true value x_t and x_i; then calculate the percentage error in the result:

Step 1: $x_t = 110$ ppm Fe, $x_i = 115$ ppm Fe.

Step 2: The relative error E_r in percentage terms will be equal to

$$E_r = \frac{115 - 110}{110} \times 100\%$$

$$E_r = \frac{5}{110} \times 100\%$$

$$E_r = 4.5\%$$

Note that E_r may be negative if the measured value is smaller than the true value. The negative sign serves to indicate that the reading is low. A positive value for E_r indicates a reading that is larger than the true value.

EXAMPLE 2.7

Again using the same data as for Example 2.6, calculate the relative error in terms of parts per thousand for an analysis that gives 115 ppm Fe and the true value is, in fact, 110 ppm Fe content.

Method

Assign the true value x_t and the measured vallue x_i; then calculate the error in the result in parts per thousand of the true value.

Step 1: $x_t = 110$ ppm Fe, $x_i = 115$ ppm Fe.

Step 2: The relative error E_r will then equal:

$$E_r = \frac{115 - 110}{110} \times 1000 \text{ ppt}$$

$$= \frac{5}{110} \times 1000 \text{ ppt}$$

$$= 45 \text{ ppt}$$

One may compare accuracy and precision by visually thinking of a target at which different sportsmen shoot. If a skilled marksman aims well he might be expected to hit the bulls eye time after time; Fig. 2.1(a). This situation is analogous to *an analytical procedure with both high levels of accuracy and precision*.

In a similar manner, if one considers the performance of an amateur sportsman, his inexperience may lead to a considerable scatter of hits around the centre of the target; Fig. 2.1(b). This scenario is analogous to *a poor level of analytical precision*.

The skilled marksman may, however, not be able to hit the centre of the target if the sights are incorrectly set. His or her skill (precision) will ensure that all shots hit close to each other; they will, however, all be displaced from the centre. This situation is, therefore, analogous to *an analytical procedure that exhibits a high level of precision, but a low level of accuracy*; Fig. 2.1(c).

2.6 Determinate, indeterminate, and gross errors

Any reading will contain some errors however carefully the measurement is taken. Errors may be classified as being either being ***indeterminate*** or ***determinate*** in origin.

Indeterminate errors are those that cause a random distribution of the data around a mean point. Indeterminate errors are sometimes known as

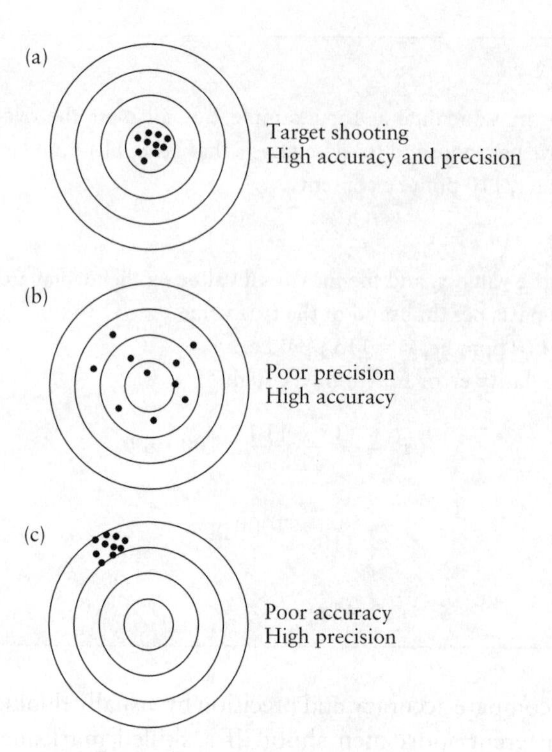

Figure 2.1 Targets to demonstrate the meaning of precision and accuracy.

random errors. Errors of this type are normally associated with the net effect of a number of small unpredictable fluctuations that may not be readily identified or eliminated. Errors of this type lead to poor precision.

Determinate (or systematic) errors, however, cause all of the data to be shifted in one direction. The results are, therefore, typically shifted to values that are either all too low or all too high. Errors of this type lead to poor accuracy.

A third type of error known as a ***Gross*** error can also occur. This type of error is normally large and essentially arises when a significant error has been made with the analytical procedure itself, so rendering the readings invalid. Gross errors lead to ***outliers*** that may under certain circumstances be rejected so that the data set is not distorted.

The influence of indeterminate, determinate, and gross errors may be illustrated as shown in Fig. 2.2. Figure 2.2(a) shows that indeterminate errors simply cause the data to be scattered around a mean point that is often close to the true value. Taking the mean value of a number of replicate measurements usually minimizes the effect of errors of this kind. The magnitude of the indeterminate errors are often a function of the magnitude of the reading but this is not necessarily the case.

(a) Indeterminate or random errors

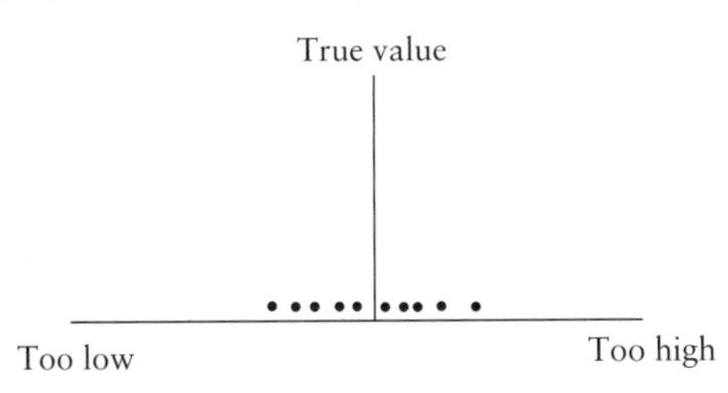

(b) Determinate systematic errors

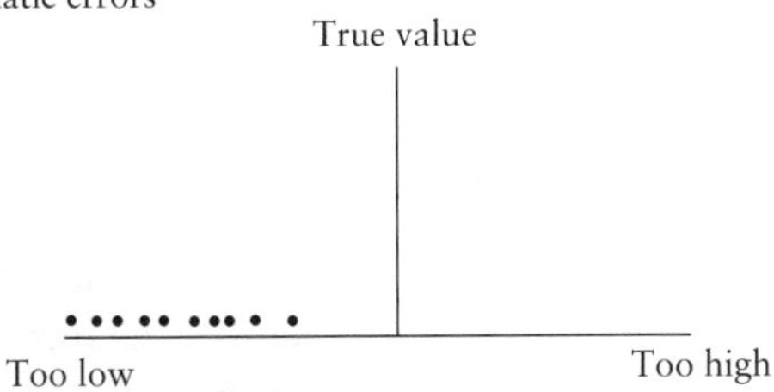

(c) A gross error leading to an outlier

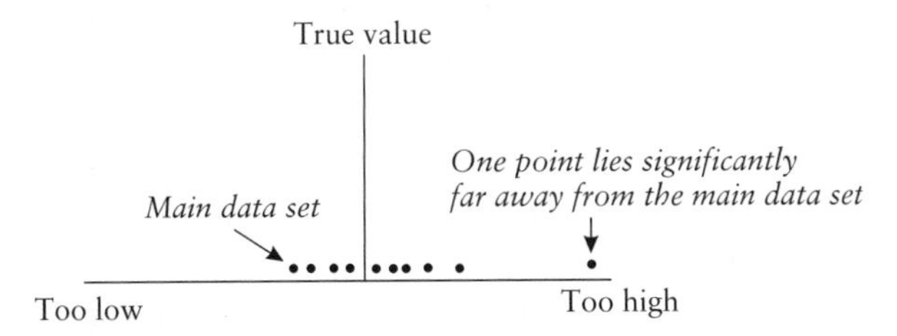

Figure 2.2 Effects of indeterminate and determinate errors.

Determinate errors, Fig. 2.2(b), on the other hand, shift all of the data in one direction and all by the same amount. Determinate errors are, therefore, most significant with smaller data values since the percentage error for the data increases accordingly.

A gross error usually causes one data point to lie significantly away from the rest of the data, and in this way it is often readily identified, Fig. 2.2(c).

2.7 Sources of indeterminate error

Indeterminate or random errors arise from a number of small unpredictable variations. The source of error may be due to many factors such as human error, fluctuations in temperature, or small differences in the quantities of reagents used. Since there are several different sources of error that may sometimes randomly lower or raise the reading, the data are scattered around the true value. In some cases two or more random errors may add together to raise the data value. In other circumstances the data may cause a net decrease in the measured data point. In some cases the net effect may be negligible due to different factors largely cancelling each other out.

Since the nature and magnitude of indeterminate errors are random in their origin, the net effect of these errors is to cause a Gaussian distribution of the data. (Gaussian distributions are discussed in greater detail in Section 2.9.)

2.8 Sources of determinate error

Determinate or systematic errors cause all of the data to be shifted in one direction. The errors are often of very similar magnitude. This behaviour is caused by the same type of error that keeps occurring every time a measurement is made. It is easy to see how an error of this type can occur. Imagine an analytical top-pan balance that is not zeroed or ***tared*** prior to the first measurement, and gives a reading of, for example, of 0.5 g when nothing is placed on the pan. Every mass that is subsequently weighed out will, in fact, be 0.5 g *less* than the value recorded on the balance. It is also clear to see that the error becomes more significant when smaller quantities of reagent are measured out.

There are three main source of determinate error and these are:

1. instrumental error;
2. methodology based errors;
3. personal errors.

2.8.1 Instrumental errors

Errors of this type typically occur as a result of, for example, inadequate maintenance of instruments or lack of calibration with known standards.

2.8.2 Methodology errors

Errors may occur because the actual method being performed is either at fault or is carried out incorrectly. An example here might include the use of a glass pipette that possesses a cracked tip and, therefore, does not allow for the small residual volume of titrant to be retained. The pipette is calibrated to take this volume into account and if this volume is not retained, all the titration equivalence points will be displaced by the same value. In a similar manner, a student may rigorously shake the last drop from the pipette, when good practice dictates it should be retained; once again the titration equivalence point will be distorted.

2.8.3 Personal errors

Personal errors, as the name suggests, are normally linked to errors in personal judgement. Many analyses involve making a judgement. Examples here could include the recording of a titration equivalence point by eye or the estimation of the position of a reading on a scale. Some people, for example, will consistently over-shoot titration end points if they are colour blind, while other people will always tend to round-down or -up the position of the needle to the nearest division. Errors of this kind are hard to eliminate since we all have built in biases however objective we try to be. It is also very easy indeed to have a pre-conceived idea of what result 'should be obtained' before an experiment is actually performed. It is especially important that errors of this type should be guarded against.

2.9 Standard deviations

Indeterminate or random errors may normally be treated by simple statistics. Most of the techniques within the next few sections assume a ***normal*** or ***Gaussian*** distribution of the indeterminate variables that affect the data. Statistical analyses of the data may be used to give some indication of the precision or reproducibility of replicate measurements.

A ***sample standard deviation*** should be calculated for data sets of less than 10 data points; ***population standard deviations*** should be calculated for data sets with greater than 10 data points.

Standard deviations have the same units as the original measurements; if the original analyses give readings in parts per million (ppm) Pb then the standard deviation will also be expressed in ppm Pb. The concept of a standard deviation *assumes* a distribution of data around the mean or true data value; a large standard deviation, therefore, corresponds to a large spread of data.

If the data follows a Gaussian distribution, 68.3% of the data will lie within one standard deviation, 95.5% of the data will lie within two standard deviations, and 99.7% of the data will lie within three standard deviations.

A normal distribution of data is one around a mean value that follows a characteristic pattern. For a Gaussian distribution 68.3% of the data falls within limits of ± one standard deviation around the mean. Similarly 95.5% of the data falls within ± 2 standard deviations, and 99.7% within ± 3 standard deviations. A typical Gaussian distribution is shown in Fig. 2.3.

2.9.1 Sample standard deviation

The ***sample standard deviation***, s, describes the spread of data around the mean data point for a set of replicate measurements.

The sample standard deviation for a set of data is given by Eqn (2.5) and is used to calculate the standard deviation for data sets with 10 or less values:

$$s = \sqrt{\frac{\sum_{i=1}^{N}(x_i - \bar{x})^2}{N-1}} \tag{2.5}$$

A re-arrangement of Eqn (2.5) may lead to an expression for the standard deviation, which is much easier to use in calculations, Eqn (2.6):

$$s = \sqrt{\frac{\sum_{i=1}^{N}x_i^2 - (\sum_{i=1}^{N}x_i)^2/N}{N-1}} \tag{2.6}$$

EXAMPLE 2.8

Replicate measurements for the lead content of a water sample taken from a river gave the following data. Calculate the standard deviation of the data.

ppm Pb content

(a) 19.4

(b) 20.6

(c) 18.7

(d) 19.2

(e) 21.6

(f) 18.9

(g) 19.9

Method

Calculate each term separately and then calculate the standard deviation.

Step 1: Calculate $\sum_{i=1}^{N}x_i^2$

$$\sum_{i=1}^{N}x_i^2 = (19.4)^2 + (20.6)^2 + (18.7)^2 + (19.2)^2 + (21.6)^2 + (18.9)^2 + (19.9)^2$$

$$= 376.36 + 424.36 + 349.69 + 368.64 + 466.56 + 357.21 + 396.01$$

$$= 2738.83$$

	x_i	x_i^2
	19.4	376.36
	20.6	424.36
	18.7	349.69
	19.2	368.64
	21.6	466.56
	18.9	367.21
	19.9	396.01
Totals $\sum$:	138.3	2738.83

Step 2: Calculate $(\sum_{i=1}^{N} x_i)^2$

$$(\textstyle\sum_{i=1}^{N} x_i)^2 = (19.4 + 20.6 + 18.7 + 19.2 + 21.6 + 18.9 + 19.9)^2$$

$$= (138.3)^2$$

$$= 19126.89$$

$N = 7$ since there are seven data points, so

$$s = \sqrt{\frac{2738.83 - 19126.89/7}{7 - 1}}$$

$$= \sqrt{\frac{2738.83 - 2732.41}{6}}$$

$$= 1.03 \text{ ppm Pb}$$

Note: s is in the units of the original analysis.

2.9.2 Population standard deviation

When a data set contains a larger number of data points (typically >10) the expression used to calculate the standard deviation is slightly altered, Eqn (2.7):

$$\sigma = \sqrt{\frac{\sum_{i=1}^{N} x_i^2 - (\sum_{i=1}^{N} x_i^2)/N}{N}} \tag{2.7}$$

σ is now used to denote the standard deviation for a large data set and is called the ***population standard deviation***.

Note that the denominator of Eqns (2.5) and (2.6) $(N - 1)$ is now replaced by N (Eqn 2.7), and the expression is said to be given an ***extra degree of freedom***. It should be emphasized that the two expressions essentially describe the same quality (i.e. they give a measure of the variation of replicate measurements). In practice, although the use of the population standard deviation is recommended for use with larger data sets, it is relatively rare to repeat a measurement in excess of 10 times, and, therefore, Eqn (2.6) is far more widely used for the estimation of the standard deviation, σ.

2.10 Relative standard deviations

In the same way that errors may be quoted in relative terms, that is, as a relative error, the standard deviation may be quoted as a ***Relative Standard Deviation*** or ***RSD***.

The RSD is calculated by dividing the standard deviation, s, or σ, by the mean $\bar{x}$ of the data, Eqn (2.8):

$$\text{RSD} = (s/\bar{x}) \tag{2.8}$$

If the RSD is to be expressed in percentage terms, $(s/\bar{x})$ is multiplied by 100, that is,

$$\text{RSD} = (s/\bar{x}) \times 100 \tag{2.9}$$

The standard deviation when expressed in percentage terms is sometimes known as the ***coefficient of variance***, or CV.

Similarly, if the RSD is to be expressed in terms of parts per thousand, then $(s/\bar{x})$ is multiplied by 1000, Eqn (2.10):

$$\text{RSD} = (s/\bar{x}) \times 1000 \text{ ppt} \tag{2.10}$$

EXAMPLE 2.9

Calculate the relative standard deviation for the lead content for the same water analyses as used in Example 2.8. Express your answer in percentage and parts per thousand (ppt) terms:

(a) 19.4

(b) 20.6

(c) 18.7

(d) 19.2

(e) 21.6

(f) 18.9

(g) 19.9

Method

1. calculate the standard deviation, s, for the data set,
2. calculate the mean, $\overline{x}$, for the data set,
3. calculate the relative standard deviation in terms of percentage and parts per thousand (ppt).

Step 1: $s = 1.03$ Pb (from Example 2.8).

Step 2: The mean $\overline{x} = (19.4 + 20.6 + 18.7 + 19.2 + 21.6 + 18.9 + 19.9)/7$
$= 19.8$ ppm Pb

That is,

$$\frac{1.03}{19.8} \times 100\%$$

Therefore,

RSD = 5.2%

or the RSD in parts per thousand terms = $(1.03 / 19.8) \times 1000$ ppt in parts per thousand terms. Therefore,

RSD = 52 ppt

2.11 The variance

The *Variance* is the square of the standard deviation:

$$\text{The variance} = s^2 \text{ (for data sets <10 values)} \quad (2.11)$$

or

$$\text{The variance} = \sigma^2 \text{ (for data sets >10 values)} \quad (2.12)$$

The variance is an alternative measure also sometimes used as a measure of the reproducibility or precision of a technique. You should note that while the standard deviation, s, has the same units as the original measurement, the variance will have the units of the data squared.

EXAMPLE 2.10

Calculate the variance for the lead content in the water samples of Example 2.8

Method

Square the standard deviation, s.

$$s = 1.03 \text{ ppm}$$

Therefore,

$$\text{variance} = 1.06 \text{ ppm}^2$$

2.12 Outliers and confidence limits

If a data set contains one value that is significantly different from all the other data, there is a strong possibility that it may be erroneous and has arisen as a result of a gross error. The choice must be made whether to retain or reject the data value. If a spurious value is kept, the mean of the data and indeed the standard deviation of the data will be distorted. On the other hand, there is, of course, a chance that the suspect data value is indeed valid and simply unexpected; in this case the precision of the analytical procedure might be lower than expected. *Great care should be taken since if a valid data point is rejected a bias will be introduced to the data.*

Unfortunately, there is no guaranteed method for the rejection or retention of individual data points. There are, however, a number of statistical tests for the rejection of suspect data values that allow confidence limits to be calculated for the rejection of data values. Each method takes into account the spread of the entire data set. The most commonly encountered tests are the ***Q*-test** (Section 2.13) and the ***T*-test** (Section 2.14).

2.13 The *Q*-test

If it is suspected that an outlier exists, the *Q*-test permits a quotient 'Q_{exp}' to be calculated and compared with a table to decide whether or not the value should be rejected or retained. The test will not give a definitive answer, but will give some idea of the confidence that can be associated with rejecting a data point. Q_{exp} may be calculated from Eqn (2.13):

$$Q_{exp} = \frac{d}{w} = \frac{x_q - x_n}{x_h - x_l} \qquad (2.13)$$

where x_q represents the suspect data point, x_n is the nearest neighbouring data value, x_h is the data point with the highest value, and x_l is the data point with the lowest value. $(x_q - x_n)$ represents the difference between the data point and its nearest neighbour and $(x_h - x_l)$ or w represents the ***spread*** of the data values.

The value of Q_{exp} may then be compared with a standard *Q*-test table, such as in Table 2.1.

Having calculated the value of Q_{exp}, we must compare the value with the values in the table applicable for the number of replicate data points measured. If Q_{exp} is less than any of the values in the table, then the data cannot be rejected with the certainty quoted within the table. If Q_{exp} is greater than a Q value within the table, the data point may be rejected

Table 2.1 A Q-test table

No of replicate measurements	Reject with 90% confidence	Reject with 95% confidence	Reject with 99% confidence
3	0.941	0.970	0.994
4	0.765	0.829	0.926
5	0.642	0.710	0.821
6	0.560	0.625	0.740
7	0.507	0.568	0.680
8	0.468	0.526	0.634
9	0.437	0.493	0.598
10	0.412	0.466	0.568

EXAMPLE 2.11

A series of replicate measurements for the water content in a sample of ethanol by the Karl–Fischer approach gave the following data:

(a) 0.71%

(b) 0.65%

(c) 0.68%

(d) 0.72%

(e) 0.91%

With what confidence may data point (e) be rejected if one uses the Q-test?

Method

Calculate Q_{exp} and compare with Table 2.1.

Step 1

$x_q = 0.91\%$—as the suspect data value.
$x_n = 0.72\%$—as the nearest neighbouring value.
$x_h = 0.91\%$ as the highest data value.
$x_l = 0.65\%$—as the lowest data value.

Step 2: Compare Q_{exp} with the Q-test table for the appropriate values corresponding to five data points:

$$Q_{exp} = 0.73$$

The Q-values for five data points are 0.642 if data are to be rejected with a 90% confidence, 0.710 for the rejection of data with a 95% confidence, and 0.821 for a 99% confidence for the rejection of a data point:

$$Q_{exp} = 0.73 > 0.710 \quad \text{but} < 0.821$$

It may, therefore, be concluded that this outlier may be rejected with greater than a 95% confidence but with less than 99% confidence.

(at least) with the certainty associated with the Q-quotient as it appears in the table. A value for Q_{exp} frequently lies in between two values, and in this case the data points may be rejected with a certainty in between the two values quoted.

2.14 The *T*-test

Another test for evaluating whether or not an outlier should be rejected is the American Society for Testing Materials (ASTM) T_n test, which is often simply known as the T-test.

Again a parameter (which in this case is denoted as T_n) is calculated, Eqn (2.14)

$$T_n = (x_q - \bar{x}_n)/s \tag{2.14}$$

where x_q is the suspect data point in question and $\bar{x}_n$ is the nearest neighbouring data value.

The value for T_n is again compared with a standard T-test table for the appropriate number of replicate measurements, Table 2.2.

Please note that the T-test should not be confused with the assignment of confidence limits using t distribution tables (Section 2.15). These are different statistical tools.

EXAMPLE 2.12

Let us use the data in Example 2.11 for replicate measurements for the water content of an organic solvent. With what confidence may data value (e) be rejected?

(a) 0.71%

(b) 0.65%

(c) 0.68%

(d) 0.72%

(e) 0.91%

Method

1. First calculate the standard deviation, s, for the data.
2. Calculate the mean for the data.
3. Calculate the T-value and compare with the T_n test table.

Step 1: $s = 0.10\%\ H_2O$

Step 2: $\bar{x} = 0.73\%\ H_2O$

Step 3: $T_n = (0.91 - 0.73)/0.1$

$= 1.8$ for five data points

1.8 is greater than any value quoted in the T value table for five data points and, therefore, this datum may be rejected with greater than 99% confidence that is an outlier.

Table 2.2 A T-test table

No. of replicate measurements	Reject with 95% confidence	Reject with 97.7% confidence	Reject with 99% confidence
3	1.15	1.16	1.17
4	1.46	1.48	1.49
5	1.67	1.71	1.75
6	1.82	1.89	1.94
7	1.94	2.02	2.10
8	2.03	2.13	2.22
9	2.11	2.21	2.52
10	2.18	2.29	2.41

Notice that the two methods give differing results—the T_n tests suggests that point (e) may be rejected with greater than 99% confidence, while the Q-test suggests that you may only reject this value with between 95% and 99% certainty. This discrepancy only highlights that statistical tests do not offer definitive answers and only point towards decisions that may be taken on reasoned thinking. It should be very clear by now that great care should, indeed, be taken if one is to reject any data, even if a Q-, T-test, or other statistical analysis is used.

2.15 Confidence limits

Confidence limits define a range of values either side of the calculated mean that describes the probability of finding the true mean. Several assumptions are made, the most important of which is that the data follow a normal Gaussian distribution around a mean value, and that the population and sample standard deviations have the same value.

The confidence limit for a set of data is described by Eqn (2.15):

$$\mathrm{CL} = \bar{x} \pm \frac{ts}{\sqrt{N}} \tag{2.15}$$

where $\bar{x}$ is the mean, s is the sample standard deviation, N is the sample size, and t is the 't distribution' and can be found in Table 2.3.

Please note that the use of t distribution tables should not be confused with the T-test (Section 2.14).

Figure 2.3 shows how the data population falls around a mean value and the probability of finding the true mean within a set of confidence limits around the calculated values.

Table 2.3 '*t*' Distribution table

No. of replicate measurements	Probability 90%	95%
2	6.314	12.706
3	2.920	4.303
4	2.235	3.182
5	2.132	2.776
6	2.015	2.571
7	1.943	2.447
8	1.895	2.365
9	1.860	2.306
10	1.833	2.262

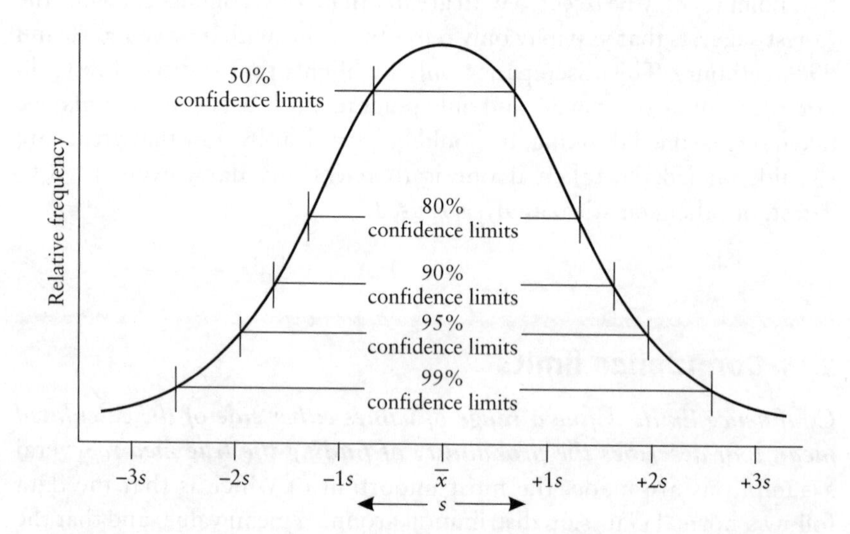

Figure 2.3 Percentage distribution of confidence limits around a mean.

EXAMPLE 2.13

Calculate the 95% and 90% limits for the mean data value of Pb concentration in Example 2.9.

Method

Calculate the confidence limits according to Eqn (2.15).

$$\text{CL } 95\% = 19.8 \pm \frac{2.447 \times 1.1}{\sqrt{7}} = 19.8 \pm 1.0 \text{ ppm Pb}$$

$$\text{CL } 90\% = 19.8 \pm \frac{1.943 \times 1.1}{\sqrt{7}} = 19.8 \pm 0.8 \text{ ppm Pb}$$

2.16 Least-squares fit procedures for calibration plots

For many analytical procedures, we will expect the value of a measured parameter to increase with the value for some key quantity (e.g. the concentration). If the measured signal and the variable of interest increase together linearly, it is possible to construct a calibration profile for two or more experimentally determined data points. Many calibration plots follow a linear profile over a given analyte concentration range. A straight line may be defined as a $y = mx + c$ expression, where y describes the ordinate, x the abscissa, m the slope or gradient, and c the intercept of the ordinate or y-axis, Fig. 2.4.

The $y = mx + c$ profile of a straight line may be used to fit the best straight line through a number of experimentally determined data points, for which the (x) co-ordinate is typically known and the (y) co-ordinate is known experimentally. The (x) values may, for example, be concentration values while the (y) values will be experimental readings.

Since any straight line may be described by $y = mx + c$, it follows that:

$$\bar{y} = m\bar{x} + c \qquad (2.16)$$

For any data set all of the x values and all of the y values will be known. The mean of the y values, $\bar{y}$, and mean of the x values, $\bar{x}$, may, therefore, be readily calculated. If there was some way of calculating the gradient, m, then we can predict every value for Eqn (2.16), except for the intercept, c, which could then be found by simple substitution. Fortunately, there is, indeed, a simple way of finding the best-fit line through the data and hence its gradient. This method is known as the ***Least-Squares Fit*** method for deriving a calibration line.

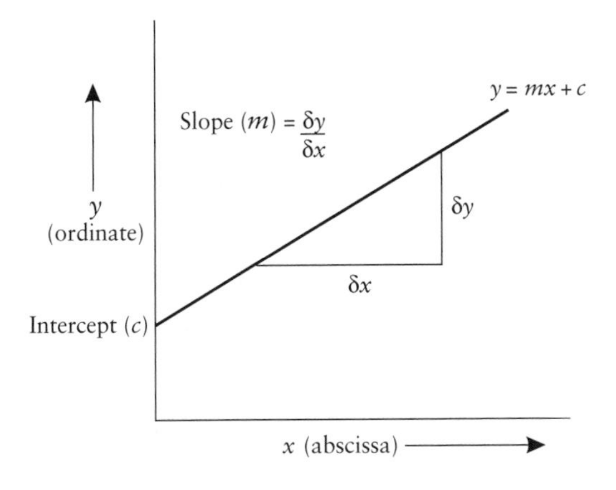

Figure 2.4 The $y = mx + c$ straight line calibration.

The first step is to calculate the gradient of the line, m, which may be found from Eqn (2.17):

$$m = S_{xy}/S_{xx} \tag{2.17}$$

where

$$S_{xy} = \Sigma x_i y_i - \frac{\Sigma x_i \Sigma y_i}{N} \quad \text{and} \quad S_{xx} = \Sigma x_i^2 - \frac{(\Sigma x_i)^2}{N}$$

EXAMPLE 2.14

Determine a best-fit calibration plot for the data below, which correspond to the chromatographic determination of an organic pesticide:

	Pesticide concentration ($\times 10^6$ M) (x)	Chromatographic peak area (arbitrary units) (y)
(a)	6.0	12.4
(b)	9.0	18.9
(c)	12.0	26.0
(d)	15.0	31.2
(e)	18.0	37.1

Method

1. Calculate the best-fit gradient for the calibration line, by calculating each term and substituting each term into Eqn (2.17).
2. Calculate the mean values for x and y and substitute into Eqn (2.16) and substitute to find the intercept c.
3. Plot a best-fit calibration curve.

Step 1:
First, calculate S_{xy}

Assign data to x and y points: $N = 5$ for five data points:

x	y	$x_i y_i$
6.0	12.4	74.4
9.0	18.9	170.1
12.0	26.0	312.0
15.0	31.2	468.0
18.0	37.1	667.8
		$\Sigma x_i y_i = 1692.3$

Next, calculate $[\Sigma x_i \Sigma y_i/N]$

$$\Sigma x_i = 6.0 + 9.0 + 12.0 + 15.0 + 18.0 = 60.0$$
$$\Sigma y_i = 12.4 + 18.9 + 26.0 + 31.2 + 37.1 = 125.6$$
$$[\Sigma x_i \Sigma y_i/N] = 1507.2$$

Next, calculate Σx_i^2

$$\begin{aligned}\Sigma x_i^2 &= (6.0)^2 + (9.0)^2 + (12.0)^2 + (15.0)^2 + (18.0)^2 \\ &= 36.0 + 81.0 + 144.0 + 225.0 + 324.0 \\ &= 810.0\end{aligned}$$

Next, calculate $[(\Sigma x_i)^2/N]$

$$\Sigma x_i = 6.0 + 9.0 + 12.0 + 15.0 + 18.0 = 60.0$$

$$(\Sigma x_i)^2/N = (60)^2/5 = 720$$

Now substitute each value into Eqn (2.17)

$$\begin{aligned}S_{xy} &= 1692.3 - \frac{(60 \times 125.6)}{5} \\ &= 185.1\end{aligned}$$

and

$$\begin{aligned}S_{xx} &= 810.0 - \frac{3600}{5} \\ &= 90\end{aligned}$$

$$m = \frac{185}{90} = 2.06$$

Therefore, $m = 2.06$ arbitrary units per micromolar concentration of pesticide.

Step 2: Calculate the mean values for x and y:

$$\bar{x} = (6.0 + 9.0 + 12.0 + 15.0 + 18.0)/5 = 12.0$$

$$\bar{y} = (12.4 + 18.9 + 26.0 + 31.2 + 37.1)/5 = 25.12$$

Therefore,

$$\begin{aligned}c &= 25.12 - (2.06 \times 12.0) \\ &= 25.12 - 24.72 \\ &= 0.4 \text{ (arbitrary units on the ordinate)}\end{aligned}$$

If one now has the best-fit gradient and the intercept, c, on the y-axis, it is now possible to draw the best-fit calibration plot, Fig. 2.5.

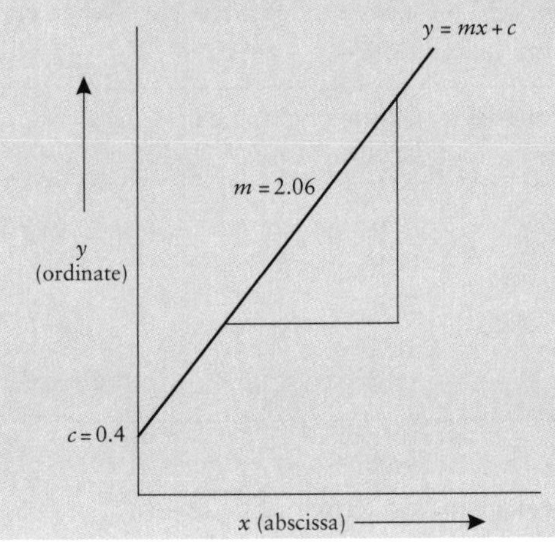

Figure 2.5 Calibration plot for Example 2.14.

Note: You should be very careful to be rigorous with the appropriate units for the calculation of the gradient, m. The gradient corresponds to the ratio of one set of units with another and should therefore also be expressed with the correct units.

2.17 Correlation coefficients

We have already seen that the precision or reproducibility of replicate measurements can be assessed by calculating standard deviation values. In a similar manner, data that should correspond to a straight line calibration plot may be evaluated in terms of ***a correlation coefficient***, which describes how close the data set lies to a perfect straight line. The most commonly used correlation coefficient is the ***Pearson correlation coefficient, r***, which will have a value ranging from 0 to 1.

A value of 1 corresponds to every data point falling on a perfect straight line; a value of 0 means that there is no correlation between the data points whatsoever.

The Pearson correlation coefficient, r, may be calculated from the expression of Eqn (2.18):

$$r = \frac{\Sigma x_i y_i - (\Sigma x_i)(\Sigma y_i)/N}{\sqrt{[\Sigma x_i^2 - (\Sigma x_i)^2/N][\Sigma y_i^2 - (\Sigma y_i)^2/N]}} \tag{2.18}$$

In practice, many experimentally determined data return values for r of >0.9 and it is, therefore, very common to quote r^2. The closer r is to 1, the closer r^2 will remain to 1. By contrast, as r deviates from 1, r^2 will fall away with the square of r. r^2 is a dramatic test of data fit.

EXAMPLE 2.15

Let us use the data of Example 2.14. Calculate the Pearson correlation coefficient (and r^2) for the straight-line calibration.

	Pesticide concentration ($\times 10^6$ M) (x)	Chromatographic peak area (arbitrary units) (y)
(a)	6.0	12.4
(b)	9.0	18.9
(c)	12.0	26.0
(d)	15.0	31.2
(e)	18.0	37.1

Method

1. Assign x and y values and calculate each term for Eqn (2.18).
2. Substitute values and calculate r and r^2.

Step 1: $N = 5$ for five data points.

Calculate $\Sigma x_i y_i$

x	y	$x_i y_i$
6.0	12.4	74.4
9.0	18.9	170.1
12.0	26.0	312.0
15.0	31.2	468.0
18.0	37.1	667.8
		$\Sigma x_i y_i = 1692.3$

Calculate Σx_i and Σy_i:

$$\Sigma x_i = 6.0 + 9.0 + 12.0 + 15.0 + 18.0$$
$$= 60.0$$
$$\Sigma y_i = 12.4 + 18.9 + 26.0 + 31.2 + 37.1$$
$$= 125.6$$
$$\Sigma x_i \Sigma y_i = 60 \times 125.6$$
$$= 7536$$

Calculate $[(\Sigma x_i^2) - (\Sigma x_i)^2/N]$:

$$\Sigma x_i^2 = (6.0)^2 + (9.0)^2 + (12.0)^2 + (15.0)^2 + (18.0)^2$$
$$= 36 + 81 + 144 + 225 + 324$$
$$= 810$$
$$(\Sigma x_i)^2 = 60^2$$
$$= 3600$$
$$(\Sigma x_i)^2/N = 3600/5$$
$$= 720$$
$$[(\Sigma x_i^2) - (\Sigma x_i)^2/N] = 810 - 720 = 90$$

Calculate $[(\Sigma y_i^2) - (\Sigma y_i)^2/N$

$$\Sigma y_i^2 = (12.4)^2 + (18.9)^2 + (26.0)^2 + (31.2)^2 + (37.1)^2$$
$$= 153.76 + 357.21 + 676.0 + 973.44 + 1376.41$$
$$= 3536.82$$
$$(\Sigma y_i)^2 = (12.4 + 18.9 + 26.0 + 31.2 + 37.1)^2$$
$$= (125.6)^2$$
$$= 15775.36$$
$$(\Sigma y_i)^2/N = 15775.36/5$$
$$= 3155.07$$
$$[(\Sigma y_i^2) - (\Sigma y_i)^2/N] = 3536.82 - 3155.07 = 381.75$$

Step 2: $r = \dfrac{1692.3 - (60.0 \times 125.6)/5}{\sqrt{[90][3536.82 - (15775.36/5)]}}$

$$= \frac{1692.3 - 1507.6}{\sqrt{(90)(3536.82 - 3155.07)}}$$

$$= \frac{185.1}{\sqrt{90 \times 381.75}}$$

$$= 0.999$$

Therefore,

$$r^2 = 0.997$$

Note: r^2 is a unitless quantity. In this case, r^2 is still very close to 1, indicating an excellent correlation coefficient and hence a close fit to a perfect straight line.

2.18 Quality control and assurance systems

There is often a great deal of confusion concerning the terms ***Quality control*** and ***Quality assurance*** and so, in practice, they are often used interchangeably and hence incorrectly.

Quality control simply means the *regulation of quality and the mechanism by which it is achieved*. This may be, for example, the rejection of certain analyses (e.g. via the Q- and T_n-tests) and via the use of replicate measurements.

Quality assurance, however, embraces a different concept. A quality assurance system involves a set of procedures put in place to ensure that quality control activities are carried out. A quality assurance system should permit a certain level of confidence to be assigned to results that are obtained from an analytical procedure. An entire analytical laboratory may be the subject of a quality assurance system. A quality assurance system normally involves accreditation with an active outside independent organisation.

Within many countries numerous analytical procedures may now be accredited via, for example, the ***National Measurement Accreditation Service*** (NAMAS) of the United States or the ***Laboratory of the Government Chemist*** (LGC) of Britain in addition to international bodies such as the ***International Standards Organization (ISO)***.

Two key steps within a quality assurance system should involve (a) proficiency testing of the laboratory and (b) the use of certified reference materials.

2.19 Certified reference materials

Certified reference materials are specially prepared samples containing an analyte at a concentration that has been pre-determined by a third party to a high degree of accuracy and precision. These samples are supplied with a ***certificate*** detailing the analyte, and may be used as a reference by which the analytical performance of another laboratory may be judged against.

The use of ***Certified reference materials*** should have a central role in the quality control procedures of almost any analytical laboratory. Indeed, their use will almost certainly be demanded if quality assurance accreditation is to be granted by a third party.

Organizations such as the LGC in Britain and National Institute of Standards and Technology in the United States produce a range of certified reference materials with very accurately known quantities of particular analytes. The samples are often prepared so as to be as similar as possible to the real samples that may be analysed within the laboratory. A certified reference material for atomic absorption spectroscopy might, therefore, contain known concentrations (to within very narrow limits) of, for example, lead.

Certified reference materials of this type are used to ensure the *accuracy* of results so that the results of different laboratories may concur. The use of certified reference materials may also be used to identify and thereby eliminate systematic errors that will not be identified by other means such as verifying that data fall within acceptable limits of standard deviation. (*Remember an unacceptable standard deviation only points to poor precision within an analysis.*)

Certified reference materials may also often be obtained that will contain fixed concentrations of the analyte in question, but *variable* compositions of the remaining composition of the mixture. Samples of this type are extremely useful for identifying possible problems from, for example, interferents.

Certified reference materials are usually supplied with information describing their composition. For accreditation purposes, however, they may sometimes be supplied as '*blind samples*'; the laboratory seeking accreditation must analyse the sample or samples and these results will be compared to that of a third party who will assess the composition of the reference materials from the organization who prepared it.

Samples of this type may be particularly useful if a new technique is being developed that will require accreditation if it is to be accepted by the wider analytical community.

2.20 Quality control charts

A quality control procedure must involve the monitoring of data over some period of time. A quality control manager regulating an individual production process will, for example, often want to monitor several parameters to ensure they stay within pre-defined limits. If some process does start to go wrong then the manager will need to know as soon as possible to allow corrective action to be taken at the earliest opportunity.

A graphical representation often highlights abnormal behaviour most effectively and serves as a record for the behaviour of a process over a period of time. ***Quality control charts*** are often used as part of a quality control process to maintain and record the performance of a system over a period of time. If the process stays within the pre-defined limits then the process is said to be ***under control***. By contrast, if the process goes outside the limits imposed on the system, it is said to be ***out of control***.

There are typically two main types of Quality Control Charts that are typically used and these are known as ***Shewhart*** and ***CUSUM*** charts.

2.20.1 Shewart charts

Shewart charts (named after W.A. Shewart in the 1930s) are designed to graphically plot a process to determine whether or not it remains under control and to help bring in remedial action to get the process back under control if necessary. This requires some appropriate variable to be recorded periodically. If an observation falls outside the outer limits this indicates that the process is out of control and action is required. If two or more consecutive readings fall between the warning limits and the outer limits then action should also be taken.

The inner or warning set of control limits are set to represent $\pm 2\sigma$ or $\pm 2s$. The outer set of control limits is set to represent $\pm 3\sigma$ or s, see Fig. 2.6.

2.20.2 CUSUM charts

An alternative form of quality control chart is the ***Cumulative sum*** or ***CUSUM*** chart. A sequence of sampled analyses is made at regular

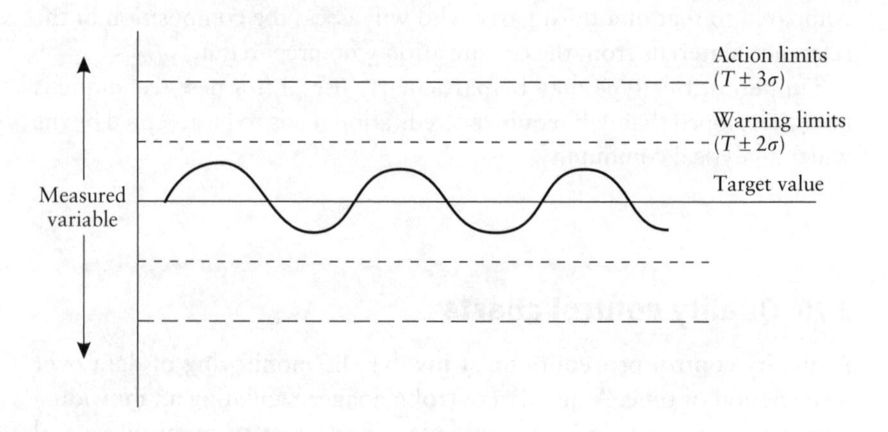

Figure 2.6 Shewart chart.

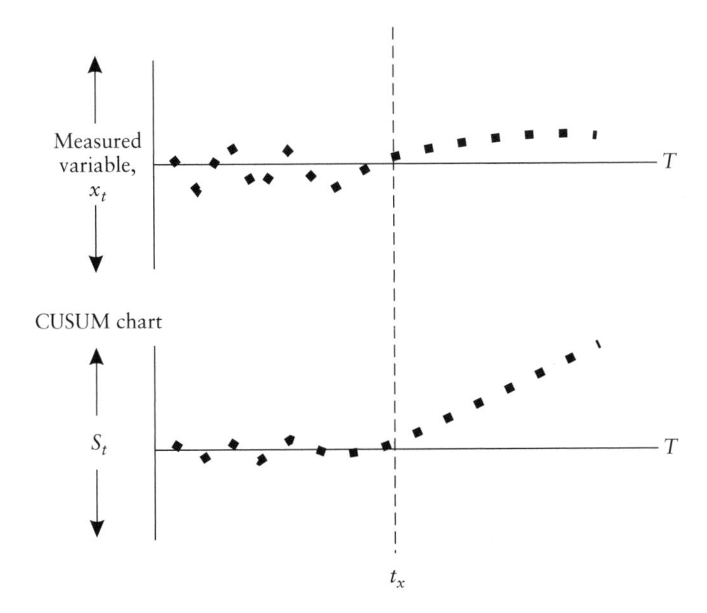

Figure 2.7 Comparison of Shewart (upper) and CUSUM charts (lower) for a process in which all of the measured data values are tending to a value higher than *T*.

intervals. The values are compared with the target for that parameter and the deviations aggregated to provide a cumulative sum that is plotted with respect to time, Fig. 2.7.

If we denote a measured variable, x_t, at time p and the target value T, then at time t, the cumulative sum of deviation about the target T is given by $S_t = \sum_{ts}(x_p - T)$. A CUSUM chart is formed by plotting this cumulative sum against time as shown in Fig. 2.7. For a process that is controlled well, there will be small undulating deviations that will remain close to zero. If the process begins to go out of control, the CUSUM will begin to deviate sharply from the CUSUM zero line.

CUSUM charts offer simplicity since they may be readily plotted with minimal calculations, as each data point is determined. If the data are randomly distributed around T and if little deviation is seen in the value of the cumulative sum of the measured variables tending to a higher or lower value, then the CUSUM chart will show a deviation. A Shewart chart and a CUSUM chart are compared in Fig. 2.7 to illustrate this point. The CUSUM chart in Fig. 2.7 shows a process that starts to show signs of deviation from the target parameters (after t_x).

2.21 Calibration methods

In order for any reliable analytical measurement to be made, the measurement system should itself be calibrated. This may involve the calibration of glassware but will frequently involve the calibration of some instrument.

For an ideal system, the response of the system will be directly proportional to the concentration of the analyte of interest. In this situation, we will obtain a straight-line calibration profile that may be described by a $y = mx + c$ equation, as in Section 2.16. Linear response calibration profiles are rarely seen, however, except perhaps over a limited concentration range. In this case a ***step-by-step*** calibration procedure must be adopted.

Another form of calibration is often used by the addition of a known quantity of analyte to a sample undergoing analysis and this technique is known as the ***standard additions calibration technique***.

2.21.1 Step-by-step calibrations

A step-by-step calibration is often the most reliable method for calibrating an analytical system. The calibration is performed by sequentially analysing a number of prepared samples across the concentration range of interest. In this way a calibration profile may be plotted. The two major disadvantages are that (i) the process may be excessively time consuming, and (ii) the prepared samples do not represent 'real' samples and so may be devoid of interferences that can distort the signal.

Time may be saved in some instances by using a two-point calibration with a best-fit line procedure if previous findings have determined that the system always behaves linearly (at least within the concentration range of interest).

If, however, the effect of any interferences is to be addressed, then a ***standard additions calibration technique*** must be employed.

2.22 Standard additions technique

The ***standard additions technique*** involves the addition of a number of standardized aliquots to a real sample *in order to raise its concentration by a known amount*. The basic principle involves measuring the change in signal in response to a change in analyte concentration.

The signal that is observed before the standard addition must, therefore, be due to:

1. the analyte itself;
2. any interferences;
3. any factors that contribute to a baseline response.

The effects of chemical interferences on the analysis may, therefore, be evaluated (at least to some extent), since the calibration is now performed within a real sample and not in an idealized (and possibly unrealistic) laboratory sample.

Typically, a number of highly concentrated small-volume standard additions are used, so that the overall volume remains essentially constant throughout the calibration. In this way any interferent, volume, concentration, or interferent-related errors can be minimized.

The standard addition technique is most useful when it can be assumed that the system essentially behaves linearly, since it is then possible to assume that an incremental addition will cause an incremental change in response. If the system does not respond linearly then a number of standard additions will have to be made so that the system may be monitored over a defined concentration range.

If a calibration (response versus concentration) plot is prepared via the standard additions method, then we would typically expect to obtain a calibration profile that fails to go through the origin.

In the absence of interferents, the intercept of the x-axis corresponds to the concentration of the analyte in the original sample, although it should always be realized that this value may be distorted if interferents are present.

EXAMPLE 2.16

A flame photometer is used to determine the Ca^{2+} concentration of a water sample. The instrument was calibrated via a standard additions method, and the responses obtained are listed below. Assuming that no interferences are present, determine the Ca^{2+} concentration within the original sample.

Standard addition concentration (mg dm^{-3})	Instrument reading (arbitrary units)
0 (the original sample)	12
3	16
5	27
10	37
15	49
20	61

Method

1. Determine the best-fit line for the data via the least-squares fit method.
2. Plot the calibration graph and determine the x intercept, which corresponds to the concentration of Ca^{2+} within the original sample. (Alternatively, you may mathematically determine the x intercept from the $y = mx + c$ equation, having determined the best fit line.)

Step 1:

The gradient is calculated by the least-squares fit procedure to be 2.49 units per mg dm^{-3} Ca^{+}.

Step 2:

$\bar{x} = (0 + 3 + 5 + 10 + 15 + 20)/6 = 8.83$

$\bar{y} = (12 + 16 + 27 + 37 + 49 + 61)/6 = 33.67$

If $y = mx + c$, then $\bar{y} = 33.67$ and $\bar{x} = 8.83$.

Now, c (intercept of y-axis) may be calculated since:

$$33.67 = (2.49 \times 8.83) + c$$

so

$$\begin{aligned} c &= 33.67 - (2.49 \times 8.83) \\ &= 33.67 - 21.99 \\ &= 11.68 \text{ arbitrary units} \end{aligned}$$

If we re-substitute into the equation

$$y = mx + c \text{ then}$$

$$\begin{aligned} x \text{ intercept} &= (0 - 11.68/2.49) \\ &= -4.69 \text{ mg dm}^{-3} \end{aligned}$$

If the calibration curve is plotted and the best-fit line is estimated by eye, then it can be seen that we arrive at a similar x intercept, which confirms the value obtained for the Ca^{2+} concentration within the original sample (see Fig. 2.8).

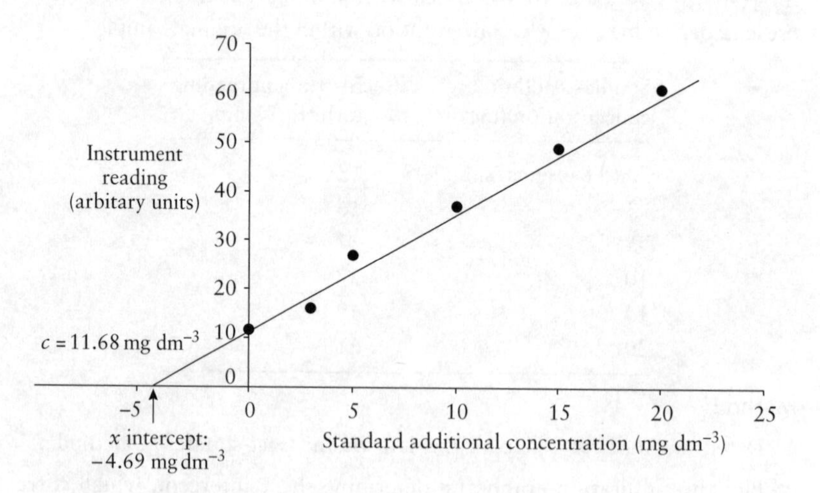

Figure 2.8 Calibration plot for Example 2.16.

Exercises and problems

2.1. Explain what is meant by indeterminate and determinate errors giving examples for each.

2.2. How many significant figures do each of the following data values possess: (a) 7.9×10^5, (b) 300.45, and (c) 5.043×10^{-4}?

2.3. Replicate samples for the iron content of an alloy were determined to contain 94.67, 94.54, 94.62, and 94.93% Fe. Calculate the standard deviation and the relative standard deviation of these analyses.

2.4. Using the data for Problem 2.3, calculate the 90% and 95% confidence limits for the mean of these data.

2.5. Five potassium chromate samples were weighed with the following results: (a) 123.3, (b) 124.2, (c) 121.5, (d) 123.6, and (e) 124.1 g. Calculate the median, mean, and the range of data.

2.6. Two samples of potassium chloride were weighed using an analytical balance and were recorded as being 34.5645 g and 35.5664 g, respectively. Express these figures to four decimal places.

2.7. A burette has calibrated divisions of 0.1 cm^3; when performing a titrimetric analysis titre results are recorded between 10.5 and 10.7 cm^3. To how many significant figures should the titre values be recorded? Explain your reasoning.

2.8. Five samples of soil were weighed prior to analysis. The weights were recorded as:

(a) 23.67 g
(b) 34.53 g
(c) 31.56 g
(d) 26.34 g
(e) 42.19 g

Calculate the mean and median for the weight of these five samples.

2.9. A water sample taken from a lake was analysed for its cadmium content. Six replicate measurements of the cadmium content were recorded as below:

(a) 20.2 ppm
(b) 18.5 ppm
(c) 21.4 ppm
(d) 19.2 ppm
(e) 21.8 ppm
(f) 18.8 ppm

Calculate the spread of the data.

2.10. Calculate the relative standard deviation of the data of Problem 2.9.

2.11. Calculate the relative error in percentage terms for the analysis of a river water sample that gives a value of 15.7 ppm Cu when the true value is, in fact, 18.0 ppm Cu.

2.12. An acid-base titration gives the concentration of an HCl solution as 0.104 M. The true concentration was in fact 0.110 M. Express the relative error for this analysis in parts per thousand terms.

2.13. Calculate the coefficient of variance for the following replicate measurements for the iron content of a water sample:

(a) 34.6
(b) 29.5
(c) 32.2
(d) 33.7
(e) 34.6
(f) 32.4
(g) 35.1

2.14. Calculate the variance for the data of Problem 2.13.

2.15. A series of replicate measurements for the water content in a sample of ethanol by the Karl-Fischer approach gave the following data:

(a) 0.77%
(b) 0.67%
(c) 0.71%
(d) 0.90%
(e) 0.78%

With what confidence can point (d) be rejected if the Q-test is used?

2.16. Six replicate measurements for the concentration of an H_2SO_4 acid concentration are recorded as below:

(a) 0.152 M
(b) 0.153 M
(c) 0.149 M
(d) 0.148 M
(e) 0.151 M

Calculate the 50% and 90% confidence limits for the mean of the data values.

2.17. Calculate the relative standard deviation in percentage terms for the data of Problem 2.15.

2.18. Replicate measurements for the chloride content of a water sample gave the following results: (i) 0.81, (ii) 0.83, (iii) 0.82, and (iv) 0.91 mM. With what confidence can the data point (iv) be rejected as an outlier by (a) the Q-test or (b) the T-test?

2.19. A UV/visible determination for potassium permanganate gave the following results:

Concentration (ppm)	Absorbance
1	0.03
2	0.07
5	0.15
7	0.22
8.5	0.24
10	0.31

(a) Plot the data in the form of a calibration curve.
(b) Using the least-squares fit approach, determine the best-fit line for this calibration plot.
(c) Calculate the Pearson correlation coefficient for this data set.

2.20. A flame photometer is used to determine the Mg^{2+} concentration of a water sample. The instrument was calibrated by a standard additions method; the additions are listed below. Assuming that no interferences are present, determine the Mg^{2+} content.

Standard addition concentration (mg dm^{-3})	Instrument reading (arbitrary units)
0 (blank)	15.6
2.5	22.1
5	35.1
10	48.1
15	63.7
20	79.3

2.21. Calculate the Pearson correlation coefficient for the data of Problem 2.20.

Summary

1. It is important that data are quoted to the correct number of significant figures.

2. The spread or range of data is the arithmetic difference between the largest and the smallest data point for a set of measurements.

3. The mean or average, $\bar{x}$, for a data set is equal to the sum of all the data points, $\sum_{i=1}^{N} x_i$ for a data set divided by the number of data points, that is, $\sum_{i=1}^{N} x_i/N$.

4. The median is the value that lies in the middle of the data set when arranged in arithmetic order.

5. The precision of a data set may be assessed by the standard deviation, the variance, or coefficient of variance.

6. The absolute error of a system is equal to the difference between the actual reading x_i and the true (or accepted value) x_t, that is, $E_A = x_i - x_t$.

7. The relative error, E_r, describes the error in relation to the magnitude of the true value and is equal to $E_r = (x_i - x_t)/x_t \times 100\%$.

8. Indeterminate errors are those that cause random distribution of data around a mean point.

9. Determinate (or systematic errors) cause all of the data to be shifted in one direction (to either higher or lower values).

10. Gross errors lead to outlier data points.

11. The sample standard deviation, s, describes the spread of data around the mean data value for a set of replicate measurements and may be calculated according to:

$$s = \sqrt{\frac{\sum_{i=1}^{N} x_i^2 - (\sum_{i=1}^{N} x_i)^2/N}{N-1}}$$

12. The population standard deviation, s, (for data sets with >10 values) is calculated according to:

$$\sigma = \sqrt{\frac{\sum_{i=1}^{N} x_i^2 - (\sum_{i=1}^{N} x_i^2)/N}{N}}$$

13. Relative standard deviations (RSD) are calculated in terms of either percentage or parts per thousand.
If the RSD is to be expressed in percentage terms, $(s/\bar{x})$ is multiplied by 100, i.e.:

$$\text{RSD} = (s/\bar{x}) \times 100$$

If the RSD is to be expressed in parts per thousand, $(s/\bar{x})$ is multiplied by 1000, i.e.:

$$\text{RSD} = (s/\bar{x}) \times 1000$$

14. The variance is the square of the standard deviation and so the variance is equal to s^2 or σ^2.

15. *Q*- or *T*-tests may be used for rejecting suspect outlier data points with calculated confidences in conjunction with *Q*- or *T*-test tables.

16. Confidence limits define a range of values either side of the mean that describe the probability of finding the true mean. The confidence limit, $\text{CL} = \bar{x} \pm (ts/\sqrt{N})$, where $\bar{x}$ is the mean, s is the sample standard deviation, N is the sample size, and t is the 't statistic'.

17. Best-fit lines may be fitted to $y = mx + c$ straight line plots by means of a least-squares fit approach that involves calculation of the gradient of the line, by $m = S_{xy}/S_{xx}$, where $S_{xy} = \Sigma x_i y_i - (\Sigma x_i \Sigma y_i/N)$ and $S_{xx} = \Sigma x_i^2 - ((\Sigma x_i)^2/N)$.

18. The Pearson correlation coefficient, r, describes how close the data set lies to a straight line and will have a value of 0–1.

19. Certified reference materials are specially prepared samples containing an analyte at a pre-determined concentration and may be used as part of a quality control process.

20. Quality control charts (e.g. Shewart or CUSUM) may be used to monitor a process and maintain it within pre-defined limits.

21. Calibration processes may be performed by a number of approaches including step-by-step calibration standard addition approaches.

Further reading

Anderson, R. (1984). *Statistics for analytical chemists*. Van Nostrand Reinhold, New York.

Meier, P. C. and Zund, R. E. (2000). *Statistical methods in analytical chemistry*. Chemical Analysis Series. Wiley, UK.

Miller, J. C. and Miller, J. N. (1993). *Statistics for analytical chemistry*. Ellis Horwood Series in Analytical Chemistry, Ellis Horwood, New York.

Part II

Chemical analysis: key principles and processes

Standard wet chemical and reagent-based techniques

3

Skills and concepts

This chapter will help you to understand:

- What is meant by the dissociation constant for an acid or base and how to calculate it.
- The concept of the ionic product of water and how to use this within pH calculations.
- How a buffer resists pH changes and how a buffer may be prepared.
- How strong and weak acids interact with each other and how to plot pH profiles as acids and bases are added to each other.
- How to determine the alkalinity of water samples via titration of samples with HCl.
- How to use titrimetric analyses to determine the Ca^{2+} and Mg^{2+} concentrations within aqueous samples.
- How to use a silver nitrate titrimetric approach to determine the chloride content in water samples.
- How to use sodium thiosulphate for the titrimetric determination of, for example, oxygen within aqueous solutions.
- The concept of a back titration and how to calculate the concentration of an analyte within an aqueous sample using a back titration.
- How the water content within an organic sample may be determined using a Karl–Fischer titration.

3.1 Introduction to wet chemical techniques

Wet chemical and reagent-based techniques still form the mainstay of much of modern analytical chemistry. Although modern instrumental and computational techniques undoubtedly have a more exotic image, simpler techniques such as titrimetric analyses or gravimetric analyses are still very widely used.

Wet chemical analyses are often our first school-time encounter with analytical chemistry and will typically comprise pH or litmus based tests, titrimetric analyses and other similar tests. These techniques are still extremely important even if many analyses at some stage demand the use of more sophisticated techniques.

Wet chemical techniques often offer the simplest of approaches since they do not necessitate the use of expensive and complex instrumentation, even if they are inherently more laboratory intensive and generally require a greater level of skill by the analyst than many instrumental approaches. However, wet chemical techniques are usually more difficult to take out of the laboratory into the field than instrumental based monitoring systems. The analysis of a river sample by titrimetry, for example, normally requires the collection and return of samples to a central location since titrimetry can only sensibly be performed within a laboratory setting.

Wet chemical techniques are, nevertheless, still central to modern analytical chemistry, and indeed many wet chemical techniques have now been either partially or fully automated to minimize the labour intensive drudgery often encountered when many similar analyses of multiple samples are required.

3.2 Acid/base equilibria for water, simple acids, and simple bases

3.2.1 The dissociation of water

Many wet chemical techniques are based on acid/base reactions and we shall therefore briefly consider the nature of acid–base interactions and equilibria.

Water in its liquid form largely consists of undissociated (H_2O) molecules; a very small proportion of the water molecules dissociate, however, to form H^+ and OH^- ions, Eqn (3.1):

$$H_2O \rightleftharpoons H^+ + OH^- \tag{3.1}$$

The ***dissociation constant*** K_c of water at 25°C is generally given by Eqn (3.2):

$$K_c = \frac{[H^+][OH^-]}{[H_2O_{(l)}]} \tag{3.2}$$

Since only a tiny proportion of the liquid water dissociates, we may take the concentration of the undissociated water to be constant. Water has a relative molecular mass of 18.015, which equates to a molarity of ~55.55 mol dm^{-3}.

The hydrogen ion concentration $[H^+]$ in pure water is equal to the hydroxyl ion concentration $[OH^-]$, since one molecule of water dissociates to give one proton and one hydroxyl ion. The $[H^+]$ and $[OH^-]$ within pure water may each empirically be shown to be 10^{-7} M, which gives a value for K_c equal to 1.8×10^{-16} mol dm^{-3}.

It is more normal, however, to express the dissociation of water in terms of the ***ionic product*** of water, K_w, where $K_w = K_c\,[H_2O]$. If we re-arrange Eqn (3.2) in terms of K_w then we arrive at Eqn (3.3):

$$K_w = [H^+][OH^-] \tag{3.3}$$

At 25°C K_w is equal to ~10^{-14} mol^2 dm^{-6}.

It should also be noted that the K_w of water increases with temperature, since the dissociation of water is an endothermic process with a $\Delta H =$ ~+58 kJ mol^{-1}.

3.2.2 The pH scale and the pH of aqueous solutions

The pH of an aqueous solution is defined as being equal to the negative $\log_{10}$ of the hydrogen ion concentration, that is,

$$\mathrm{pH} = -\log_{10}[H^+] \tag{3.4}$$

The 'p' within 'pH' comes from the German word *potenz*—power. Since the $[H^+]$ of pure water at 25°C is equal to 10^{-7} M, then the pH of pure water at 25°C should therefore be 7.0.

H^+ and OH^- ions may arise, however, from other sources and since the equilibrium for the dissociation of water will be maintained:

$$H_2O_{(l)} \rightleftharpoons H^+_{(aq)} + OH^-_{(aq)} \tag{3.5}$$

It follows that K_w must also at any particular temperature remain constant.

If H^+ ions are added by the addition of an acid, then according to Eqn (3.5) the concentration of OH^- must fall. In a similar manner, if OH^- ions are added by the addition of a base, then the number of H^+ ions in solution will decrease. As the K_w for water at 25°C, is equal to 1×10^{-14} mol^2 dm^{-6}, then if the $[H^+]$ concentration falls to 10^{-14} mol dm^{-3}, then the $[OH^-]$ must equal 1 mol dm^{-3}. Similarly, if the concentration of H^+ rises to 1 mol dm^{-3} then the $[OH^-]$ must fall to 1×10^{-14} mol dm^{-3}. These values correspond to pH values of 0 and 14, respectively, which is taken as the normal pH range. The concentration of H^+ and OH^- ions are shown in Fig. 3.1.

Note that a change in pH of one unit corresponds to a 10-fold change in the H^+ concentration.

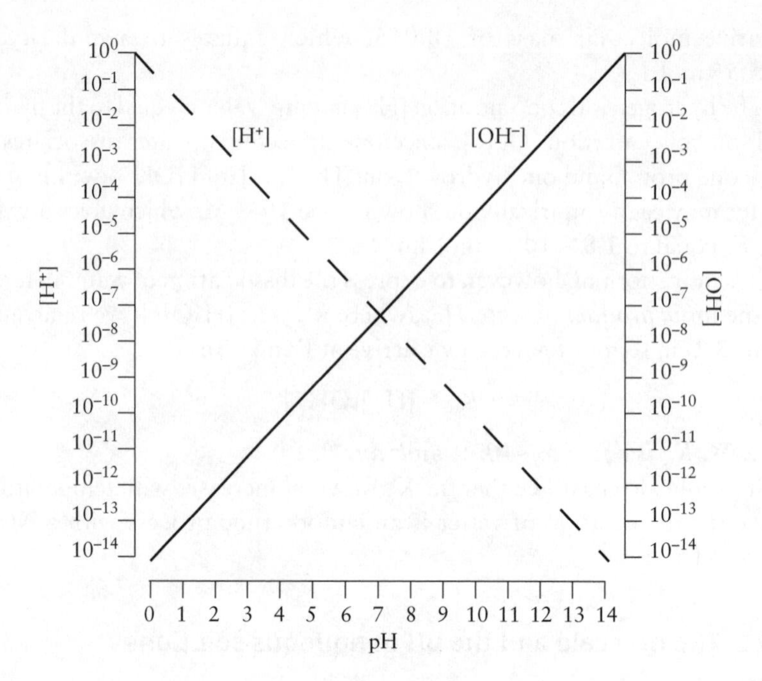

Figure 3.1 H^+ and OH^- concentrations with varying pH.

3.2.3 The calculation of pH values

Strong acids

The pH of a solution depends on: (a) the concentration of the acid or base; and (b) its degree of dissociation.

Let us first consider the pH of a strong acid such as HCl that can be assumed to fully dissociate, Eqn (3.6):

$$HCl_{(aq)} \rightleftharpoons H^+{}_{(aq)} + Cl^-{}_{(aq)} \tag{3.6}$$

If the acid fully dissociates then the concentration of the H^+ can be assumed to be equal to the molarity of the HCl solution.

EXAMPLE 3.1 Calculate the pH of an aqueous 0.1 M HCl solution.

Method

1. Calculate the H^+ ion concentration.
2. Find the $-\log_{10}$ of the H^+ concentration.

Step 1: Assuming that all of the HCl dissociates, a 0.1 M HCl solution will contain:

$$0.1 \text{ M } H^+$$

Step 2: pH of the solution $= -\log_{10} 10^{-1} = 1$

$$\log_{10} 0.1 = -1.0$$

so pH of 0.1 M of a HCl solution = 1.0.

EXAMPLE 3.2 Calculate the pH of a 0.1 M sulphuric acid (H_2SO_4) solution

Method

1. Once again calculate the $[H^+]$.
2. Take the $-\log_{10}$ of the $[H^+]$.

Step 1: H_2SO_4 is once again a strong acid, which can be considered to fully dissociate. Note, however, that it is a di-basic acid—and therefore gives two H^+ ions for every H_2SO_4 which dissociates:

$$H_2SO_4 \rightleftharpoons 2H^+_{(aq)} + SO^{2-}_{4(aq)}$$

0.1 M H_2SO_4 therefore gives $(0.1 \times 2) = 0.2$ M $[H^+]$.

Step 2: $-\log_{10} 0.2 = \sim 0.7$

so pH of 0.1 M H_2SO_4 solution = 0.7.

Strong bases

The calculation of pH is simple for strong acids, but how do we go about calculating the pH of a strongly basic solution? The method is only slightly more complicated, but involves considering the ionic product of water.

A strong base may once again be considered to fully dissociate, but in this case yields hydroxyl ions.

A good example of a strong mono-acidic base is sodium hydroxide, NaOH, which dissolves in water according to Eqn (3.7):

$$NaOH_{(s)} \rightleftharpoons Na^+{}_{(aq)} + OH^-{}_{(aq)} \tag{3.7}$$

By knowing the $[OH^-]$ and the ionic product of water, K_w, one may calculate the $[H^+]$ and so the pH.

EXAMPLE 3.3 Calculate the pH of an aqueous 0.1 M NaOH solution.

Method

1. Calculate the concentration of OH^- ions.
2. Calculate the concentration of H^+ ions by considering the ionic product of water.
3. Calculate the pH of the solution.

Step 1: A 0.1 M NaOH solution may be assumed to have a $[OH^-]$ of 0.1 M.

Step 2: $K_w = 1 \times 10^{-14} = [H^+]\,[OH^-]$

Now if $[OH^-] = 0.1$, then:

$1 \times 10^{-14} = [H^+] \times 0.1$

so
$$[H^+] = \frac{1 \times 10^{-14}}{0.1} = 1 \times 10^{-13}\ \text{M}$$

Step 3: Calculate the pH

$$[H^+] = 1 \times 10^{-13}\ \text{M}$$
$$\text{pH} = -\log_{10}(1 \times 10^{-13})$$

so pH = 13.

Weak acids and bases

Not all acids (or indeed bases) fully dissociate on dissolution in water. Some of the acid molecules ionize, while some of the molecules remain as neutral molecules in solution.

Le Chatelier's principle states that any chemical system will try to resist a change superimposed upon it. Imposing an excess of either H^+ ions (in the case of an acid) or OH^- ions (in the case of a base) clearly acts as an external change—and since we also know that the ionic product of water stays constant, this will have an effect on the concentration of the other ions in solution. Some acids are not as strong as HCl or NaOH—that is, they do not fully dissociate upon dissolution in water. In this situation only a percentage of the acid or base ionizes. The remaining acid or base will in these cases remain as a neutral unionized solute.

The dissociation constant of a weak acid K_a or a weak base K_b allows us to calculate the concentration of H^+ ion or OH^- ions in solution, and we can predict the pH of the solution too.

A weak acid will dissociate according to Eqn (3.8):

$$HA \rightleftharpoons H^+_{(aq)} + A^-_{(aq)} \tag{3.8}$$

The dissociation constant K_a is defined according to Eqn (3.9):

$$K_a = \frac{[H^+][OH^-]}{[HA]} \tag{3.9}$$

In a similar manner, the weak base may dissociate according to Eqn (3.10):

$$B + H_2O \rightleftharpoons BH^+_{(aq)} + OH^-_{(aq)} \tag{3.10}$$

In this case, the dissociation constant K_b is defined according to Eqn (3.11):

$$K_b = \frac{[OH^-][BH^+]}{[B]} \tag{3.11}$$

Strictly, we should include a term for H_2O within Eqn (3.11), although this is often discounted since its concentration is essentially constant and in large excess.

The dissociation constants for some common weak acids and bases are given in Tables 3.1 and 3.2, respectively.

Some acids possess more than one moiety that may dissociate to produce an H^+ ion. Each moiety has its own K_a. An example of such an acid is carbonic acid, H_2CO_3. Similarly, some bases have more than one moiety that can dissociate to give an OH^- ion. Again each dissociation is described by a K_b. An example of such a base is ethylenediamine.

Table 3.1 Values of dissociation constant, K_a, for a number of weak acids

Acid	Molecular formula	Dissociation constant, K_a	
		$K_a(1)$	$K_a(2)$
Acetic acid	CH_3COOH	1.75×10^{-5}	
Benzoic acid	C_6H_5COOH	6.14×10^{-5}	
Carbonic acid	H_2CO_3	4.45×10^{-7}	4.69×10^{-11}
Citric acid	$HOOC(OH)C(CH_2COOH)_2$	7.45×10^{-4}	
Formic acid	$HCOOH$	1.77×10^{-4}	
Lactic acid	$CH_3CHOHCOOH$	1.37×10^{-4}	
Nitrous acid	HNO_2	7.1×10^{-4}	
Oxalic acid	$HOOCCOOH$	5.36×10^{-2}	
Phenol	C_6H_5OH	1.00×10^{-10}	

Table 3.2 Values of dissociation constant, K_b, for a number of weak bases

Base	Molecular formula	Dissociation constant, K_b	
		$K_b(1)$	$K_b(2)$
Ammonia	NH_3	1.76×10^{-5}	
Aniline	$C_6H_5NH_2$	3.94×10^{-10}	
Dimethylamine	$(CH_3)_2NH$	5.9×10^{-4}	
Ethanolamine	$HOC_2H_4NH_2$	3.18×10^{-5}	
Ethylamine	$CH_3CH_2NH_2$	4.28×10^{-4}	
Ethylenediamine	$NH_2C_2H_4NH_2$	8.6×10^{-5}	7.1×10^{-8}
Hydrazine	H_2NNH_2	1.3×10^{-6}	
Hydroxylamine	$HONH_2$	1.07×10^{-8}	
Pyridine	C_5H_5N	1.7×10^{-9}	
Trimethylamine	$(CH_3)_3N$	6.25×10^{-5}	

EXAMPLE 3.4 The dissociation constant, K_a, for acetic acid is 1.75×10^{-5}. Calculate the pH of a 0.1 M CH_3COOH solution.

Method

1. Calculate the concentration of $[H^+]$ using the dissociation constant.
2. Calculate the pH of the solution.

Step 1: Acetic acid dissociates according to the stoichiometry:

$$CH_3COOH_{(aq)} \rightleftharpoons CH_3COO^-_{(aq)} + H^+_{(aq)}$$

Let x = the number of moles of CH_3COO^- (or H^+)
By stoichiometry, we also have x moles of H^+
If α is the number of moles of CH_3COOH then for this weak acid:

$$(\alpha - x) \leftrightarrows x + x$$

Now: $$K_a = \frac{[CH_3COO^-][H^+]}{[CH_3COOH]}$$

So $$K_a = \frac{x \times x}{\alpha - x}$$

$K_a = 1.75 \times 10^{-5}$, and $\alpha = 0.1$ M.
Substituting values for K_a and x gives:

$$1.75 \times 10^{-5} = \frac{[x][x]}{0.1 - x}$$

Rearranging gives:

$$1.75 \times 10^{-5}(0.1 - x) = x^2$$
$$1.75 \times 10^{-6} - 1.75 \times 10^{-5}x = x^2$$
$$0 = x^2 + 1.75 \times 10^{-5}x - 1.75 \times 10^{-6}$$

This equation requires solving as a quadratic equation and so if:

$$x = \frac{-b \pm \sqrt{b^2 - 4ac}}{2a}$$

a, b, and c may be assigned as follows:

$$0 = a(x^2) + b(x) + c$$
$$0 = x^2 + 1.75 \times 10^{-5}x - 1.75 \times 10^{-6}$$
$$a = 1,\ b = 1.75 \times 10^{-5}, \text{ and } c = -1.75 \times 10^{-6}$$

$$x = \frac{-1.75 \times 10^{-5} \pm \sqrt{(1.75 \times 10^{-5})^2 - 4(1 \times -1.75 \times 10^{-6})}}{2}$$

It follows that:
either $x = 1.31 \times 10^{-3}$ or $x = -1.33 \times 10^{-3}$
x must be positive, since this represents a concentration of ions and it therefore follows that:

$$x = 1.31 \times 10^{-3}$$
$$x \equiv [H^+]$$
$$[H^+] \sim 1.31 \times 10^{-3} \text{ M}$$

Step 2: Take the $-\log_{10}[H^+]$ to give the pH for 0.1 M CH_3COOH.

$$-\log_{10} 1.31 \times 10^{-3} \text{ M} = 2.88$$

pH of 0.1 M $CH_3COOH = 2.88$

EXAMPLE 3.5 Calculate the pH of 0.1 M NH_3 aqueous solution.

Method

1. Calculate the OH^- concentration for the NH_3 solution using the dissociation constant, K_b, and a quadratic solution of the dissociation expression.
2. Calculate the H^+ concentration using the ionic product of water.
3. Calculate the pH of the NH_3 solution.

Step 1: Assign the stoichiometry to the dissociation of ammonia:

$$NH_3 + H_2O \rightleftharpoons NH_4{}^+{}_{(aq)} + OH^-{}_{(aq)}$$

so

$$(\alpha - x) + x \rightleftharpoons x + x$$

Now

$$K_b = 1.76 \times 10^{-5} = \frac{[NH_4^+][OH^-]}{[NH_3]}$$

Substitution leads to:

$$K_b = 1.76 \times 10^{-5} = \frac{x \times x}{(\alpha - x)}$$

so

$$K_b = 1.76 \times 10^{-5} = \frac{x^2}{(0.1 - x)}$$

Rearranging into the form of a quadratic gives:

$$1.76 \times 10^{-6} - 1.76 \times 10^{-5}x = x^2$$

so:

$$0 = x^2 + 1.76 \times 10^{-5}x - 1.76 \times 10^{-6}$$

This expression now needs solving as a quadratic equation by:

$$x = \frac{-b \pm \sqrt{b^2 - 4ac}}{2a}$$

and if one assigns a, b, and c:
$a = 1$, $b = 1.76 \times 10^{-5}$, and $c = -1.76 \times 10^{-6}$.
Substitution into the quadratic expression gives:

$$x = \frac{-1.76 \times 10^{-6} \pm \sqrt{(1.76 \times 10^{-5})^2 - 4(1 \times (-1.76 \times 10^{-6}))}}{2 \times 1}$$

$$= \pm 1.33 \times 10^{-3}$$

x corresponds to the concentration of OH^- ions and must therefore be positive.

$$\mathbf{[OH^-] = 1.33 \times 10^{-3}\ M}$$

Step 2: Calculation of $[H^+]$ from the ionic product of water:

$$K_w = 1 \times 10^{-14} = [H^+] \times [OH^-]$$

Since

$$[OH^-] = 1.33 \times 10^{-3}\ M$$

then:

$$[H^+] = \frac{1 \times 10^{-14}}{1.33 \times 10^{-3}} = 7.52 \times 10^{-12}\ M$$

Step 3: Take the $-\log10\ [H^+]$ to give the pH

$$-\log_{10} 7.52 \times 10^{-12} = 11.1$$

A 0.1 M aqueous NH_3 solution therefore has a pH 11.1

3.3 Buffers

The addition of an acid or base will normally cause a pH shift of a solution. Buffers are special solutions that resist changes in pH upon the addition of an acid or base.

Buffers are of great biological significance; a pH change, for example, of only 0.5 within the blood will normally cause death. In a similar manner, the pH of many industrial processes must normally be kept within very narrow limits.

A buffer is normally either prepared as:

1. a solution of a weak acid with one of its salts such as ethanoic (acetic) acid and sodium acetate, or
2. a solution of a weak base with one of its salts (e.g. aqueous ammonia and ammonium chloride).

We shall now consider how a buffer operates. Let us imagine a weak acid H-A in equilibrium with one of its salts M-A. H-A will only be slightly dissociated whereas MA will be fully dissociated, Eqns (3.12) and (3.13):

$$\text{H-A} \rightleftharpoons \text{H}^+ + \text{A}^- \tag{3.12}$$

$$\text{M-A} \rightarrow \text{M}^+ + \text{A}^- \tag{3.13}$$

The solution contains a relatively high concentration of H-A, which is an acid and A^- which can be considered as being a base.

Now if an acid is added to the solution, the excess H^+ ions will react with A^-, to minimize the effect on the pH. Provided there is a sufficiently large reservoir of A^-, the pH will only be slightly altered. H^+ ions will react with hydroxyl ions. Further HA will now dissociate to restore the $[H^+]$ and therefore the pH. Provided there is a sufficiently large reservoir of H-A the buffer will be able to resist pH changes upon the addition of further acid.

The $[H^+]$ and therefore the pH of the buffer is primarily governed by the dissociation of H-A, Eqn (3.14), which governs the equilibrium constant K_a.

$$K_a = \frac{[\text{H}^+][\text{A}^-]}{[\text{H-A}]} \tag{3.14}$$

It should be noted that if the buffer is diluted, then more H-A will dissociate to maintain the relative concentrations of H^+ and A^- and this has the consequence of also helping to maintain the pH of a buffer upon dilution.

A very similar argument may be used to describe the action of a buffer by a solution of a weak base B-OH in the presence of one of its salts, B-X.

The weak base, B-OH, only partially dissociates within solution, Eqn (3.15). In contrast, its salt B-X will be almost totally dissociated in solution, Eqn (3.16)

$$B + H_2O \rightleftharpoons BH^+_{(aq)} + OH^-_{(aq)} \quad (3.15)$$

$$B\text{-}X_{(aq)} \rightarrow B^+_{(aq)} + X^-_{(aq)} \quad (3.16)$$

The $[H^+]$ and therefore the pH are governed by the ionic product of water and so also the $[OH^-]$. Upon the addition of an acid, H^+ ions react with OH^- of the buffer; B-OH will dissociate further and the pH of the solution will be stabilized. If a base is added to the solution, the B^+ ions of the salt will react with the OH^- ions of the base, and once again the pH of the solution will be stabilized. Again, the buffer will be able to resist changes in pH provided there are sufficiently large reservoirs of B-OH and B-X.

EXAMPLE 3.6 A buffer is prepared, which contains 0.05 M sodium acetate, CH_3COONa, and 0.01 M acetic acid, CH_3COOH. Calculate the pH of the buffer. The k_a for $CH_3COOH = 1.7 \times 10^{-5}$.

Method

1. Calculate the concentration of undissociated acid.
2. Calculate the concentration of acetate ions.
3. Calculate the $[H^+]$ via the dissociation constant of CH_3COOH and hence the pH.

Step 1: Since the acid is weak, the concentration of the undissociated acid can be taken to be approximately that of the total acid concentration which in this case equals 0.01 M.
Step 2: Conversely, the salt can be considered to be fully dissociated in solution—in this case the $[CH_3COO^-]$ may therefore be taken to be approximately 0.05 M.
Step 3: K_a for the acid is given by:

$$K_a = \frac{[H^+][CH_3COO^-]}{[CH_3COOH]}$$

It follows that upon rearranging:

$$[H^+] = \frac{K_a[CH_3COOH]}{[CH_3COO^-]}$$

If one substitutes, the numerical values for K_a, the $[CH_3COOH]$, and the $[CH_3COO^-]$, then:

$$[H^+] = \frac{1.7 \times 10^{-5} \times 0.01}{0.05}$$
$$= 3.4 \times 10^{-6}\ \text{mol dm}^{-3}$$

so the $pH = -\log_{10} 3.4 \times 10^{-6}$
pH of the buffer = 5.47

We are now in a position to see how well a buffer can resist a change in pH, upon addition of even a strong acid or base. Let us see what happens when 10 cm^3 of 0.1 M NaOH is added to 1 dm^3 of the buffer used in Example 3.7.

EXAMPLE 3.7 Ten cubic centimetres of an aqueous 0.1 M NaOH solution is added to 1 dm^3 of a pH 5.47 sodium acetate (0.05 M)/acetic acid (0.01 M) buffer. Calculate the new pH.

Method

1. Interpret how NaOH will react with the buffer, giving the reaction stoichiometry.
2. Calculate the new concentration of the acid CH_3COOH.
3. Calculate the new concentration of the sodium acetate ion $[CH_3COO^-]$.
4. Calculate the $[H^+]$ and hence the pH by means of the dissociation expression for the acid.

Step 1: NaOH will react with CH_3COOH to form CH_3COONa, that is:

$$CH_3COOH + NaOH \rightarrow CH_3COO^- + Na^+ + H_2O$$

The concentration of CH_3COOH therefore falls while that of CH_3COO^- rises.

Step 2: 10 cm^3 of 0.1 M NaOH $\equiv$ 0.001 moles of OH^-.
It should be remembered that the total volume of buffer has increased from 1000 to 1010 cm^3.

There were originally 0.01 moles of CH_3COOH in the buffer. There will now be 0.01 − 0.001 moles of CH_3COOH in 1010 cm^3. There will, therefore, be 9×10^{-3} moles of CH_3COOH in 1010 cm^3.

The $[CH_3COOH]$ will therefore be

$$\frac{9 \times 10^{-3}}{1010} \times 1000 \text{ mol dm}^{-3}.$$

$$[CH_3COOH] = 8.91 \times 10^{-3} \text{ mol dm}^{-3}$$

Step 3: Calculate the $[CH_3COO^-]$:
There were originally 0.05 moles of CH_3COO^- in 1000 cm^3. On addition of the NaOH there will now be an additional 0.001 moles of CH_3COO^-. The volume of the buffer has also been increased from 1000 to 1010 cm^3. The new $[CH_3COO^-]$ may therefore be calculated as:

$$[CH_3COO^-] = \frac{0.05 + 0.001}{1010} \times 1000 \text{ mol dm}^{-3}.$$

$$= 0.0505 \text{ mol dm}^{-3}.$$

Step 4: Calculate the $[H^+]$ and therefore the pH.

$$K_a = \frac{[H^+][CH_3COO^-]}{[CH_3COOH]}$$

It therefore follows:

$$[H^+] = \frac{[K_a][CH_3COOH]}{[CH_3COO^-]}$$
$$= \frac{1.7 \times 10^{-5} \times 8.91 \times 10^{-3}}{0.0505}$$
$$= 2.99 \times 10^{-6}$$

Therefore, **pH = 5.52**

The new pH equals 5.52. The pH of the buffer before the addition was 5.47, which represents a shift of only 0.05 pH units.

By contrast, the same addition of alkali to neutral water (pH 7) will cause the pH to rise to approximately pH 3—which clearly represents a very much larger change in pH value.

Example 3.7 illustrates the action of an acid/salt buffer well. We will now look at a similar example for a base and its salt in Example 3.8.

EXAMPLE 3.8 Ten cubic centimetres of an aqueous 0.2 M HCl solution is added to 1 dm^3 of a 0.05 M NH_4OH/0.05 M NH_4Cl buffer solution. Calculate the pH of the buffer: (i) prior to; and (ii) following the addition of HCl. The K_b of NH_4OH is 1.88×10^{-5}.

Method

1. Calculate the $[NH_4Cl]$.
2. From the dissociation expression calculate the $[OH^-]$.
3. Calculate the $[H^+]$ from the ionic product of water.
4. Following the addition of the HCl—describe the reaction stoichiometry of how the HCl will react with the NH_4OH—and hence the new concentration of OH^- within the solution.
5. Calculate the new $[H^+]$ and hence the pH from the ionic product of water.

Step 1: NH_4Cl is a salt and can be considered to be fully dissociated, in which case the $[NH_4Cl]$ can be taken to be ~0.05 M.

Step 2: $NH_4OH_{(aq)} \rightleftharpoons NH_4{}^+{}_{(aq)} + OH^-{}_{(aq)}$

$$pK_b \text{ of } NH_4OH \equiv K_b = \frac{[NH_4^+][OH^-]}{[NH_4OH]}$$

Since the $[OH^-] = [NH_4^+]$ then:

$$[OH^-]^2 = K_b \times [NH_4OH]_{(aq)}$$

$$= 1.82 \times 10^{-5} \times 0.05$$

or

$$[OH^-] = \sqrt{9.1 \times 10^{-7}}$$

$$= 9.54 \times 10^{-4}\ M$$

Step 3: $K_w = 1 \times 10^{-14}\ mol^2\ dm^{-6} = 9.54 \times 10^{-4} \times [H^+]$

$$[H^+] = \frac{1 \times 10^{-14}}{9.54 \times 10^{-4}} = 1.048 \times 10^{-11}\ mol\ dm^{-3}$$

$$pH = -\log_{10} 1.048 \times 10^{-11}$$

or

$$\mathbf{pH = 10.98}$$

Step 4: $HCl_{(aq)} + NH_4OH_{(aq)} \rightleftharpoons NH_4Cl_{(aq)} + H_2O_{(l)}$
Now if 10 cm^3 of 0.2 M HCl is added, this is equivalent to:

$$\frac{10}{1000} \times 0.2\ mol\ HCl = 2 \times 10^{-3}\ moles\ of\ HCl$$

The acid/base reaction stoichiometry is 1:1

Initially there was 0.05 moles of NH_4OH present.

There will now be $0.05 - 2 \times 10^{-3}$ moles of NH_4OH left = **0.048 moles of NH_4OH**

The volume has now increased from 1000 to 1010 cm^3, so

$$[NH_4OH] = \frac{0.048}{1010} \times 1000\ mol\ dm^{-3}$$
$$= 4.75 \times 10^{-2}\ mol\ dm^{-3}$$

Now, the K_b of $NH_4OH = \dfrac{[NH_4^+][OH^-]}{[NH_4OH]} = 1.82 \times 10^{-5}\ mol^2\ dm^{-6}$

so

$$[OH^-][NH_4^+] = 1.82 \times 10^{-5} \times 4.85 \times 10^{-2}$$

$$[OH^-]^2 = 8.65 \times 10^{-5}\ mol\ dm^{-3}$$

or

$$[OH^-] = 9.30 \times 10^{-4}\ mol\ dm^{-3}$$

Step 5: $K_w = 1 \times 10^{-14}\ \text{mol}^2\ \text{dm}^{-6}$
So

$$1 \times 10^{-14} = [\text{H}^+] \times 9.3 \times 10^{-4}$$

$$[\text{H}^+] = \frac{1 \times 10^{-14}}{9.3 \times 10^{-4}} = 1.08 \times 10^{-11}\ \text{mol dm}^{-3}$$

or

$$\text{pH} = \log_{10} 1.07 \times 10^{-11}$$

so

$$\textbf{pH} = \textbf{10.97}$$

The pH of the solution has therefore only altered by 0.01 of a pH unit.

3.4 Acid and base interactions

Many chemical analyses depend on pH—and this, in turn, is often determined by how acids or bases interact. An acid may usually be considered to be strong or weak—and again a base may either be strong or weak. It therefore follows that:

1. a strong acid may interact with a strong base;
2. a strong acid may interact with a weak base;
3. a strong base may interact with a weak acid;
4. a weak acid may interact with a weak base.

The pH of a mixture is determined by the concentration of the *ionized* acidic or basic component.

The pH/concentration profile may therefore be complicated when an acid is added to a base or vice versa. We shall consider each of the four examples in turn.

3.4.1 Strong acid interacting with strong base

In this situation, both the acid and base may be considered to be fully dissociated in the solution. The acid and base will react to form a salt and water. If the acid and base are added in *exactly* equal molar quantities and both contain the same number of ionizable acidic or basic groupings, they will neutralize each other and the solution will have a neutral pH of 7. In all other situations there will be a net excess of acid or base and this will determine the pH of the final solution.

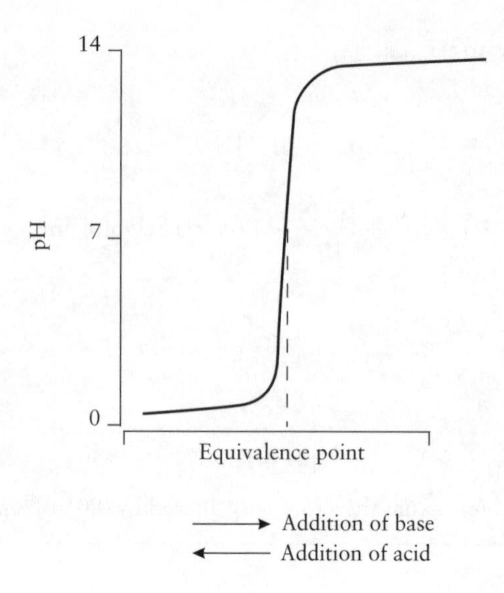

Figure 3.2 Interaction of strong acid and strong base.

If there is an excess of strong acid, then the pH will typically settle in the range of 2.5–1.0. Conversely, if there is an excess of strong base then the pH will typically settle in range of 12.0–13.0. Figure 3.2 shows how the pH changes as a strong acid is slowly added to a strong base—or, of course, conversely if a strong base is added to a strong acid. Remember that the pH is determined by whichever species remains in excess following reaction. The pH initially hardly changes until a point close to neutralization; following neutralization, the pH changes very rapidly as either the acid or base is being added in excess. The point at which the acid and base are in equal molar quantities is known as the ***equivalence point*** (neutralization), and the rapid change in pH around this region forms the basis of the widely used acid/base titrations. These will be considered in Section 3.6.

3.4.2 Strong acid interacting with weak base

The pH will again be determined by whichever species is in excess. The strong acid will again give rise to pH values of between approximately 1.0 and 1.5. The weak base, however, will cause the pH of the solution typically to be approximately 8.0–9.0. The equivalence point may again be identified by a rapid change in pH; Fig. 3.3.

3.4.3 Strong base interacting with weak acid

In this situation, the strong base will raise the pH to values of between 12 and 13, while the weak base will result in pH changes typically of

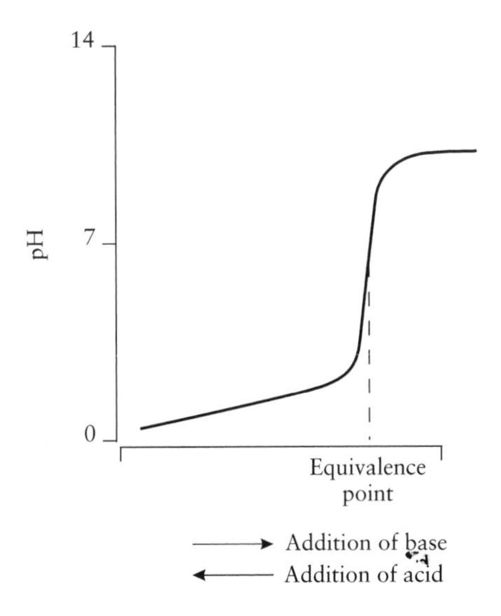

Figure 3.3 Interaction of strong acid and weak base.

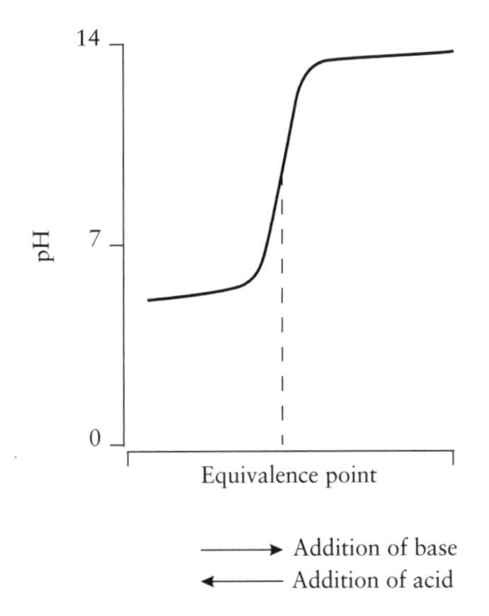

Figure 3.4 Interaction of strong base with weak acid.

between 4 and 5, when either one or the other species is in excess. The equivalence point is once again characterized by a rapid change in pH; Fig. 3.4.

3.4.4 Weak acid interacting with weak base

The change in pH is not so dramatic in this situation, since the weak acid may only lower the pH of the solution to values of between 4 and 5, while the base may only cause the pH to rise to around 8–9; Fig. 3.5. The equivalence point, even if it is characterized by a narrower change in pH value, is still identified by a *rapid change* in pH.

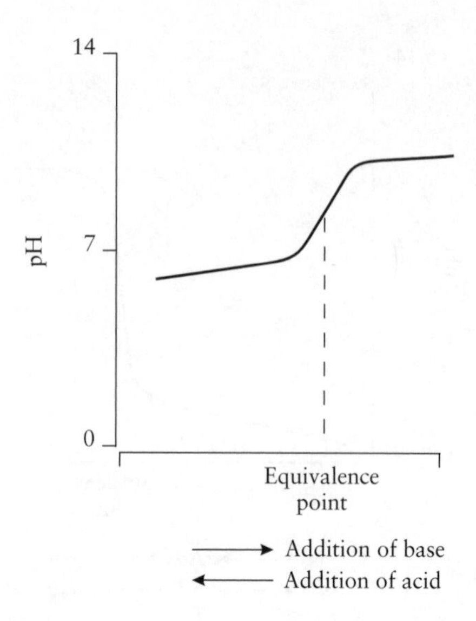

Figure 3.5 Interaction of weak acid with weak base.

3.5 The stoichiometry of titrations

All titrations are based upon knowing how different molar quantities of reagents react together. Reagents will normally react with each other in fixed molar quantities or ratios, and this is known as the ***stoichiometry*** of the reaction.

Hydrochloric acid, $HCl_{(aq)}$, (a monobasic acid), will react with sodium hydroxide NaOH (a monoacidic base) with a stoichiometry of 1 : 1 to form NaCl and H_2O, Eqn (3.17):

$$HCl_{(aq)} + NaOH_{(aq)} \rightarrow NaCl_{(aq)} + H_2O_{(l)} \tag{3.17}$$

By contrast, H_2SO_4 (a dibasic acid) will react with NaOH with a 1 : 2 stoichiometry, Eqn (3.18):

$$H_2SO_{4(aq)} + 2NaOH_{(aq)} \rightarrow Na_2SO_{4(aq)} + 2H_2O_{(l)} \tag{3.18}$$

If one knows the stoichiometry of how one reagent should react with another and a method for determining or following the molar concentrations of any one of the reactants or products, then one has the foundation for a ***titrimetric analysis***.

Titrimetric analyses are extremely useful—and may be followed by a variety of means. The most popular method still involves the use of an indicator (i.e. a chemical added to the solution), to follow the pH change associated with many equivalence points—especially when the titration is based on an acid/base reaction (see Section 3.4). Other methods for following titrations include electrochemical or photometric methods.

3.6 Acid–base titrations and indicators

Acid–base titrations are amongst the most widely used form of titration and find many areas of application. We shall as an example, first describe a simple titration for determining the alkalinity (HCO_3^-) of environmental (e.g. river or drainage) water samples.

Before we do, it is important to consider how indicators are used to determine equivalence points. It should be remembered that although there are other methods available for determining equivalence points, indicators are still very widely used and in most cases, offer the simplest of approaches. The indicator should be highly coloured (i.e. possess a large molar absorptivity, ε, in the visible region), so that only a drop or two is required in order to provide a vivid coloration to the reaction mixture.

Most indicators are organic molecules with one or more ionizable functional groups acting as chromophores. As a good example, we can consider methyl orange; Fig. 3.6.

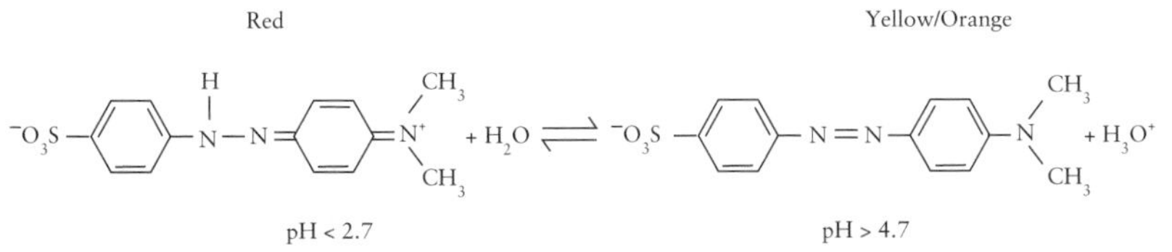

Figure 3.6 Colour change of methyl orange with pH.

The ideal indicator should show its colour change over a very narrow pH responsive range. Since many pH sensitive indicators undergo proton addition/loss reactions, this is quite achievable since the pH directly influences whether or not protonation or deprotonation occurs. An indicator should also ideally respond reversibly. Since the protonation or deprotonation is unique to each indicator—different indicators undergo colour changes over separate regions of the pH scale. This is extremely useful, since as we have already seen (Section 3.4) different equivalence points span different pH ranges. An indicator that will undergo a colour change over the pH range that spans the equivalence point should be selected. In practice, there are often two or three indicators that can be used, since the change in pH will, in many cases, span several pH units.

3.6.1 The determination of the alkalinity (HCO_3^-) content of environmental water samples

The HCO_3^- content of environmental water samples may be determined quite easily using a simple acid/base titration and forms a good illustrative example of a practical acid/base titrations.

Carbon dioxide is naturally formed from the growth and decay of biomass, and also by the combustion of fuels and this latter contribution is causing the global atmospheric content of CO_2 to increase year on year. Water droplets in clouds and rain return dissolved carbon dioxide to the ground as weak acidic solutions with a pH of around 5.4, Eqn (3.19):

$$CO_{2(g)} + H_2O_{(l)} \rightleftharpoons [CO_2.H_2O]_{(aq)} \rightleftharpoons H^+ + HCO_3^-{}_{(aq)} \quad (3.19)$$

Carbonic acid is a weak acid (and will, in limestone containing areas, facilitate the dissolution of calcium carbonate), Eqn (3.20):

$$H^+ + HCO_3^-{}_{(aq)} + CaCO_{3(s)} \rightleftharpoons Ca^{2+}{}_{(aq)} + 2HCO_3^-{}_{(aq)} \quad (3.20)$$

The HCO_3^- acts as a base and imparts an alkalinity (temporary hardness) to the water and this may be titrated against a strong acid such as HCl, Eqn (3.21):

$$H^+ + HCO_3^- \rightarrow H_2CO_3 \quad (3.21)$$

Between 2 and 3 drops of screened methyl orange indicator should allow identification of the equivalence points; the titration mixture should change from green through a grey coloration and finally to a magenta (purple-red) colour, at a pH of around 3–4.5.

It is commonplace in the water industry to report alkalinity as mg dm^{-3} $CaCO_3$ in which case we must consider the 1 : 2 stoichiometry of Eqn (3.20), that is, that 2 moles of HCO_3^- are produced for every mole of $CaCO_3$.

3.7 The determination of the hardness of tap water—two examples of compleximetric titrations

'Water hardness' is caused by the presence of metal cations, which form insoluble salts with long aliphatic carboxylates (soaps); other effects include the scaling of pipes and kettles when water is heated. The major cations in water contributing to water hardness are Ca^{2+} and Mg^{2+}. The contribution to water hardness is often sub-divided into either 'temporary' (due to Ca^{2+}) or 'permanent' hardness (due to Mg^{2+}) and for this reason it is often useful to identify the ion content of a particular water source. Temporary hardness, for example, may be removed by boiling but leads to the formation of kettle scale ($CaCO_3$). Magnesium salts by contrast (permanent hardness) may not be precipitated by boiling. Calcium and magnesium complexes also have differing stabilities with respect to pH and this may be exploited for an ethylenediaminetetraacetic acid (EDTA) titrimetric analysis for the Ca^{2+} and Mg^{2+} content of water samples.

3.7.1 Total (Ca^{2+} and Mg^{2+}) hardness

A titration for total hardness will give us the total Ca^{2+} and Mg^{2+} content in a sample. At a pH of ~10 (buffered with ammonia/ammonium chloride), Ca^{2+} and Mg^{2+} ions will form strong complexes with EDTA—both with a 1 : 1 stoichiometry. Upon addition of EDTA, Ca^{2+} or Mg^{2+} will only remain in free solution if there is insufficient EDTA for complete complexation. In a similar manner, the ions will also only complex with a weaker ligand than EDTA when insufficient EDTA is added to the solution. An indicator such as Erichrome Black acts as a weak ligand and will undergo a colour change from a wine-red to blue colour on the loss of a complexing ion.

If some solid Erichrome Black is placed into a water solution containing Ca^{2+} and/or Mg^{2+} ions, the complex will take on a wine-red coloration. Upon titration with EDTA, the Ca^{2+} and Mg^{2+} will complex to form a metal–EDTA complex. Once all of the free metal ions are consumed, the Ca^{2+} and Mg^{2+}/Erichrome Black complexed ions will become dissociated and will re-complex with the EDTA—and at this point the indicator will change from red to blue, corresponding to the equivalence point for the titration.

3.7.2 Temporary (Ca^{2+}) hardness

To determine the temporary (Ca^{2+}) water hardness, the pH of the water sample should this time be adjusted to a pH of ~12 or above (with, e.g. the addition of a few cm^3 dilute NaOH). The sample may then be titrated in the same way as before but with an indicator such as HSN. (HSN is itself an abbreviation for HHSNNA, which is used to denote the compound 2-hydroxy-1-(2-hydroxy-4-sulpho-1-naphthazo)-3-naphthoic acid.) The equivalence point in this case corresponds to the onset of a clear blue coloration.

3.7.3 Magnesium content

The Mg^{2+} ion content may be calculated following the determination of the total hardness and the Ca^{2+} content, for particular water samples. Since the total hardness represents the sum of the Ca^{2+} and Mg^{2+} contents, the Mg^{2+} content may be calculated by subtracting the value for the temporary hardness (Ca^{2+}) from the total hardness (Ca^{2+} + Mg^{2+}) concentrations.

It is helpful to think of a simple bar chart to help visualize and explain this reasoning; Fig. 3.7. The use of two or more pieces of analytical data to infer another is very common practice and we shall consider a few more examples in the following sections to illustrate this point.

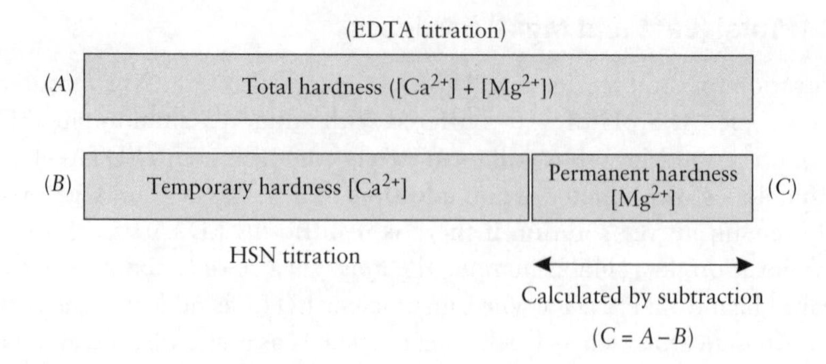

Figure 3.7 Determination of $Ca^{2+} + Mg^{2+}$ by titration.

3.8 The determination of chloride within environmental water samples: an example of a silver nitrate titration

The chloride content (salinity) of river water is of great significance to wildlife. Chloride concentrations have significant implications for corrosion as well as the taste of potable (drinking) water supplies. River water may become contaminated with excessive chloride due to the combustion of some coals that contain high chloride contents. Combustion of these fuels releases hydrogen chloride to the atmosphere and this returns to the earth as HCl (hydrochloric acid) in the rain, which ultimately finds its way to the rivers. Sewage treatment, industrial and farm effluents also contain chloride salts. During winter, chloride salts are used in many countries to de-ice roads and are washed away by rain, which finally collects in the rivers and lakes.

Fortunately, a simple $AgNO_3$ based titration may be used to determine the chloride levels within water samples.

Silver ions will react with chloride to form silver chloride, Eqn (3.22):

$$Ag^+_{(aq)} + Cl^-_{(aq)} \rightarrow AgCl_{(s)} \tag{3.22}$$

Silver chromate may be used as an indicator since silver ions will also react, though less readily, with chromate (CrO_4^{2-}), to form silver chromate, which is red, Eqn (3.23):

$$2Ag^+_{(aq)} + CrO_4{}^{2-}{}_{(aq)} \rightarrow Ag_2CrO_{4(s)} \tag{3.23}$$

The silver ions will, in fact, only react with chromate once all of the chloride within the solution has been consumed. This is convenient since potassium chromate may be used as an indicator once all of the chloride has been consumed. The original chloride concentration may in this way be calculated. This approach will permit chloride determinations down to parts per million concentrations.

3.9 The determination of dissolved oxygen: an example of a sodium thiosulphate titration

Thiosulphate iodide/iodate are widely used and we will describe one titration example here, originally developed by Winkler for the determination of dissolved dioxygen (O_2) in water. This analysis is again very simple to perform and is still widely used today. It is an invaluable test as the oxygen content of water can be easily upset by environmental pollution and this is clearly crucial to the fauna and flora that live in our rivers and lakes.

3.9.1 Sample preparation

Molecular dioxygen dissolved in water resides in a dynamic equilibrium with the air. For this reason, it is very important that samples are prepared on site before they are returned to the laboratory for analysis. The preparation involves 'fixing' the oxygen content to permit a simple laboratory determination.

Sample bottles of known volume (e.g. 100 cm^3) should be completely filled to the brim. Wearing gloves, 1 cm^3 of alkali-azide reagent should be micro-pipetted below the surface. One cubic centimetre of a $MnSO_4$ solution should next be added in a similar way. The bottle lids should finally be securely sealed and the contents gently mixed.

The alkali-azide reagent and $MnSO_4$ produce a precipitate of manganese(II) hydroxide, Eqn (3.24):

$$Mn^{2+}_{(aq)} + 2OH^-_{(aq)} \rightarrow Mn(OH)_{2(s)} \quad (3.24)$$

The dissolved molecular dioxygen then reacts with the manganese hydroxide to form manganese(III)oxohydroxide, Eqn (3.25):

$$4Mn(OH)_2 + O_2 \rightarrow 4MnO(OH) + 2H_2O \quad (3.25)$$

3.9.2 Titrimetric analysis

The samples may now be safely transferred to a conical flask to be titrated. An excess of potassium iodide is added together with a few cm^3 of phosphoric acid to produce I_3^- ions, Eqns (3.26) and (3.27), respectively:

$$4MnO(OH)_{(s)} + 12H^+_{(aq)} \rightarrow 4Mn^{3+}_{(aq)} + 8H_2O \quad (3.26)$$

$$4Mn^{3+}_{(aq)} + 6I^-_{(aq)} \rightarrow 4Mn^{2+}_{(aq)} + 2I^-_{3(aq)} \quad (3.27)$$

The iodine produced (I_3^- in the presence of excess iodide) may then be titrated against a dilute (0.01 M) sodium thiosulphate solution, Eqn (3.28), using a few drops of a starch suspension as an indicator.

The presence of iodine is shown by the onset of a deep royal blue coloration that is lost once all of the iodine has been consumed and this marks the end point. The final solution should take on a pale straw colour. If the concentration of the I_3^- is determined, this can then be related back to the O_2 concentration.

$$I_3^- + 2S_2O_3^{2-} \rightarrow 3I^- + S_4O_6^{2-} \qquad (3.28)$$

3.10 Titration of mixtures containing strong and weak bases (OH^- and HCO_3^-)

Titrimetric analyses are routinely used for a variety of theoretical and experimental purposes. Problems may occur, however, when a simple analytical approach cannot be performed due to the presence of one or more additional reagents, which compete or interfere with the reaction chemistry.

A good example, might, in this context, be a solution containing both hydroxyl and carbonate ions. Carbonate, as we have already seen in Section 3.7, is often present in natural water supplies due to dissolution of atmospheric carbon dioxide. If the water is contaminated with a strong base, then the analysis of samples becomes more complicated and titration with a strong base will not prove satisfactory. Initially, the high pH of the solution means that only OH^- ions will be protonated to water. However, since the pK_a of bicarbonate is about 10.3, significant quantities of carbonate become protonated before the last of the hydroxide can be consumed. To exacerbate the problem, the bicarbononate may be further protonated to carbonic acid, which has a pK_a of approximately 6.3, which renders most conventional indicators useless since their colour changes typically span a couple of pH units.

We can circumvent these problems, however, by mixing a couple of carefully chosen indicators to produce a colour change over a far narrower pH range. The mixture must contain indicators with closely lying pK values and colour changes that complement each other. In this way, a third colour change may be produced that corresponds strictly to pH changes intermediate to the two indicator pK values.

The mixed indicator most commonly used to facilitate titrimetric analyses of solutions containing both hydroxide and carbonate is composed of: ***six parts Thymol Blue to one part Cresol Red.*** Thymol Blue has a pK_a of ~8.9, while Cresol Red has a pK_a of ~8.2. The mixed indicator is violet at pH 8.4, blue at pH 8.3, and a rose pink at pH 8.2. It is therefore ideally suited to identify the end point for a carbonate protonation with acid that occurs at a pH of around 8.3. In other words, the mixed indicator

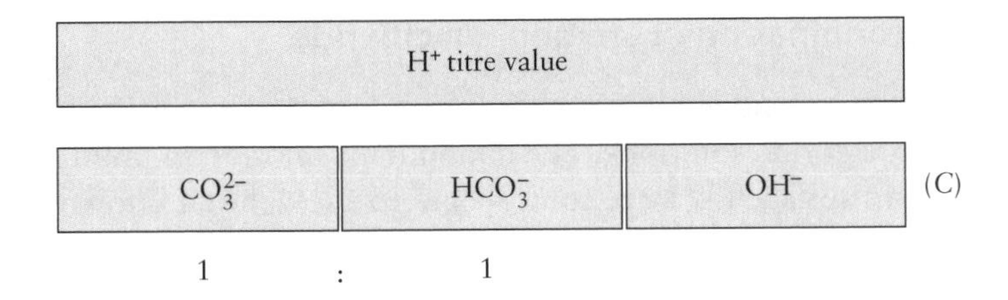

Figure 3.8 Determination of CO_3^{2-}, HCO_3^-, and OH^- by titration.

shows a rose-red coloration immediately after the carbonate equivalence point has been reached—*but* before the bicarbonate is protonated for a second time.

If we compare the titrations using mixed indicators with those obtained using methyl orange, which has a pK of ~3.7, the concentration of the bicarbonate may then be calculated by consideration of Eqns (3.29)–(3.31).

The reactions involved are shown below:

$$OH^- + H_3O^+ \rightarrow 2H_2O \quad (3.29)$$

$$CO_3^{2-} + H_3O^+ \rightarrow HCO_3^- \quad \text{(equivalence point shown by mixed indicator)} \quad (3.30)$$

$$HCO_3^- + H_3O^+ \rightarrow H_2CO_3 \quad \text{(equivalence point shown by methyl orange)} \quad (3.31)$$

The titration using methyl orange as an indicator allows calculation of the total number of moles of base (i.e. OH^-, CO_3^{2-}, and then, HCO_3^-). We should remember that the HCO_3^- is formed by the protonation of the CO_3^{2-} and will therefore be exactly equivalent in molar quantities. The titration with the mixed indicator solution will give the molar equivalent of the carbonate and bicarbonate together; the molar quantity of carbonate will correspond to half this value. The OH^- concentration in this case will correspond to the total concentration of base minus the CO_3^{2-} and the HCO_3^- content—this is depicted diagramatically in Fig. 3.8.

3.11 Back titrations

Some titrations cannot be performed directly. There are a number of reasons for this and can include the lack of a suitable indicator or, for example, the formation of a precipitate. In some cases the problem may be overcome by adding an excess of reagent and then *analysing* the amount of this that remains unreacted. This is the principle of a ***back titration*** since the analyte concentration is calculated by working backwards—that is, by determining the molar quantity of one of the initial reagents and not the analyte itself directly.

3.11.1 Determination of ammonium chloride

We shall first consider an example that allows the quantification of ammonium chloride solutions. The ammonium chloride solution should be placed in a sufficiently large conical flask to allow the addition of other reagents (that must then be boiled) before titration. An accurately known molar quantity of sodium hydroxide must now be added to be in excess, to react with the ammonium chloride, Eqn (3.32):

$$NH_4Cl_{(aq)} + NaOH_{(aq)} \rightarrow NH_{3(g)} + H_2O_{(l)} + NaCl_{(aq)} \qquad (3.32)$$

The solution should then be boiled *gently* (with anti-bumping granules) within a fume cupboard to ensure complete reaction, and care should also be taken to periodically add a little de-ionized water to maintain the fluid level. NH_3 gas will be evolved during the heating as the newly formed and volatile ammonia evaporates away. The reaction can be seen to reach completion when red moist litmus paper no longer changes colour when held above the flask. The flask will be ready for titrimetric analysis having cooled sufficiently to allow handling.

Two or three drops of screened methyl orange may then be added to permit the quantification of the excess and unreacted NaOH within the flask with standardized HCl, Eqn (3.33):

$$NaOH_{(aq)} + HCl_{(aq)} \rightarrow NaCl_{(aq)} + H_2O_{(l)} \qquad (3.33)$$

Since the exact molar quantity of NaOH originally added is known and the molar quantity following reaction has just been determined, we may easily find the molar quantity that must have reacted with the NH_4Cl. Since this reaction displays a simple 1 : 1 stoichiometry, the quantity and therefore the molarity of the NH_4Cl solution may be determined, as shown diagrammatically in Fig. 3.9.

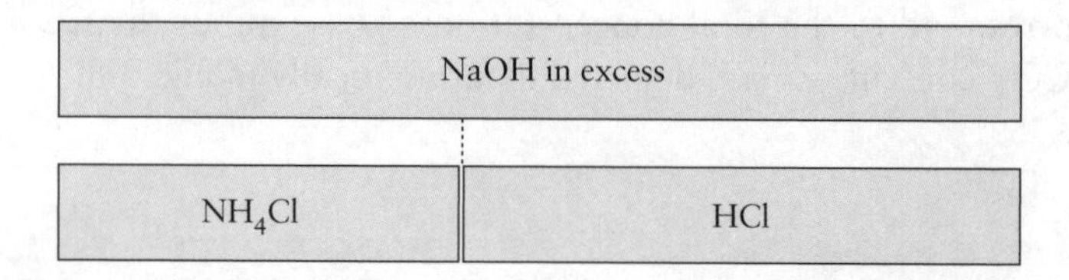

Figure 3.9 Determination of NH_4Cl by titration.

EXAMPLE 3.9 An unknown solution of ammonium chloride requires standardization. It is known that the solution cannot contain an NH_4Cl concentration greater than ~0.15 M. Fifty cubic centimetres of 0.101 M NaOH is added to neutralize 25 cm^3 of the unknown NH_4Cl solution. The remaining unreacted NaOH solution after reaction was then titrated against a standardized 0.104 M HCl solution. A number of titrations gave a mean titre value of 20.42 cm^3.

Determine the concentration of the NH_4Cl solution.

Refer to Fig. 3.9 throughout as this should help aid understanding the calculation.

Method

1. Determine the number of moles of HCl used within the titration.
2. Determine the number of moles of NaOH that must have reacted with the HCl.
3. Determine the number of moles of NaOH must have been added to neutralize the NH_4Cl.
4. From steps (b) and (c) determine how many number of moles of NH_4Cl there must have been present originally.
5. Calculate the molarity of the NaOH solution.

Step 1: 20.42 cm^3 of 0.104 M HCl were used as a mean titrant.
Calculate the number of moles of HCl used $\equiv \frac{20.42}{1000} \times 0.104$ mol HCl.

Number of moles of HCl used = $\mathbf{2.124 \times 10^{-3}}$.

(Note—it is perfectly acceptable to carry forward additional figures in your calculation provided the final answer is quoted to the correct number of significant figures.)

Step 2: Determine the number of moles of NaOH used within the reaction.
Sodium hydroxide and HCl react together with the stoichiometry shown below:

$$NaOH_{(aq)} + HCl_{(aq)} \rightarrow NaCl_{(aq)} + H_2O_{(l)}$$

that is,

$$1 + 1 \rightarrow 1 + 1$$

If 2.124×10^{-3} mol HCl were used there must also have been 2.124×10^{-3} mol NaOH consumed in the reaction.

Step 3: Initially there were 50 cm^3 of 0.101 M NaOH added into the reaction flask to neutralize the NH_4Cl.

Calculate the number of moles of NaOH in this aliquot.

There are $(50/1000) \times 0.101$ mol NaOH in the original aliquot
that is, there were 5.05×10^{-3} mol NaOH added to neutralize the NH_4Cl.

Step 4: Calculate the number of moles of NH_4Cl within the original sample. If there were originally 5.05×10^{-3} mol NH_4Cl added and after reaction there were 2.124×10^{-3} mol left of NH_4Cl with the NaOH, then there must have been:

$$5.05 \times 10^{-3} - 2.124 \times 10^{-3} \text{ mol } NH_4Cl \text{ within the original sample.}$$

that is, there were **2.926×10^{-3} mol of NH_4Cl within the original sample.**

Step 5: The original sample of NH_4Cl was of a volume of 25 cm^3.
There must therefore have been 2.926×10^{-3} mol NH_4Cl within 25 cm^3.
2.926 mol NH_4Cl within 25 cm^3 is equivalent to:

$$\frac{2.93 \times 10^{-3}}{25} \times 1000 \text{ mol dm}^{-3} = 0.1172 \text{ mol dm}^{-3}$$

or 0.117 mol dm^{-3} to four significant figures.

3.11.2 An iodimetric titration to determine the vitamin C (ascorbic acid) content of fruit juice

Potassium iodate(V), KIO_3, can be used as a primary standard to generate a known quantity of molecular iodine, I_2, to oxidize ascorbic acid within a sample of fruit juice or fruit-based drink. The unused iodine may then be titrated with a standardized sodium thiosulphate solution using starch as an indicator. This iodimetric titration therefore represents another very useful and commonly used form of a back titration.

The first step is to accurately produce a known molar quantity of molecular iodine. This is achieved by taking an accurate known molar quantity of potassium iodate(V) in solution, which should be reacted with an excess of potassium iodide (KI) in acidic conditions, Eqn (3.34):

$$IO_3^- + 5I^- + 6H^+ \rightarrow 3I_2 + 3H_2O \qquad (3.34)$$

Note the reaction chemistry shows that one IO_3^- ion yields three I_2 molecules—giving a working stoichiometry of 3 : 1.

In practice, the following quantities should produce a sensible working solution to permit practical vitamin C determinations within most fruit-based drinks: 1–1.2 g of KIO_3 should be accurately weighed and dissolved in a 200 or 250 cm^3 volumetric flask. KIO_3 is not very soluble in cold water, so it is sometimes easier to first dissolve the solid material in a little warm water within a beaker. The washings may then be further diluted and then carefully transferred to the volumetric flask to be made up to the mark. Do remember that volumetric flasks must not be heated since this will irreparably change and so ruin their calibrated volumes. This solution may then be added (pipetted) in known quantities to each fruit juice sample for titration within conical flasks.

The potassium iodide, KI (about 0.5 g should suffice) is added in excess to each conical flask; the exact quantity is unimportant since the reagent is being added in excess. A suitable acid could in this instance be bench top dilute H_2SO_4 (~1 M). About 20–5 cm^3 of the acid should again be added directly to the titration conical flask. It is perfectly acceptable to add the acid using a measuring cylinder since the exact quantity again does not have to be accurate; the purpose of the acid is simply to provide

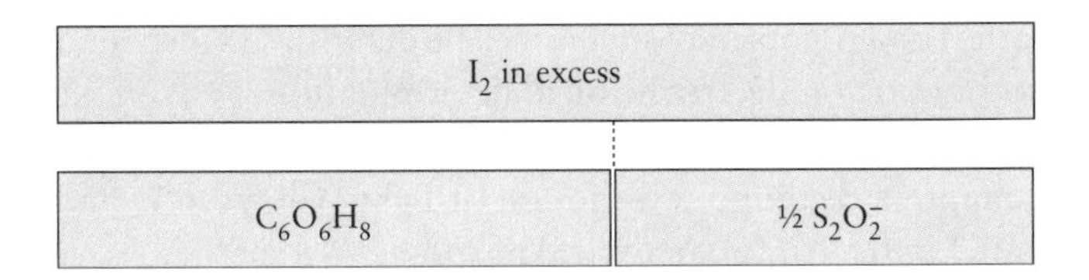

Figure 3.10 Determination of vitamin C via iodine sodium thiosulphate titration.

an excess of protons for the reaction of Eqn (3.32). The titration flask should be left to stand for around 10 min.

Vitamin C acts as a reducing agent and is oxidized by iodine, Eqn (3.35):

$$C_6O_6H_8 + I_2 \rightarrow C_6O_6H_6 + 2H^+ + 2I^- \qquad (3.35)$$

In this instance, the vitamin C and the molecular iodine react with a 1 : 1 stiochiometry, and so each molecule of vitamin C removes one molecule of iodine from the titrimetric mixture. Since iodine was initially added in excess it is therefore easy to determine the molar quantity of molecular iodine left within the solution.

The molecular iodine reacts with sodium thiosulphate with a 1 : 2 stoichiometry, Eqn (3.36):

$$I_2 + 2S_2O_2^- \rightarrow 2I^- + S_4O_6^{2-} \qquad (3.36)$$

The quantity consumed by the vitamin C may then be determined if a few drops of starch indicator are used to show us when all of the free iodine has been consumed and at this point the solution will turn from a deep blue to a pale straw colour. If we know how much iodine was present in the flask and how much was left in the solution following the reaction with the vitamin C, then we can deduce how much vitamin C was present within the fruit juice sample; this approach is shown diagrammatically in Fig. 3.10.

Great care should be taken when performing calculations of this type, because different stoichiometries are taken into account. The calculations at each step are very simple—and yet it is very easy to make simple mistakes.

3.12 Photometric and electrochemical titrations

We have until now described titrations that are based on the colour change of an indicator. The colour change is estimated by eye—and an equivalence point recorded. This is often perfectly sufficient, however, there are cases where the instrumental determination of the equivalence or an end point may be preferable, for example, to help remove human errors—for example, due to colour blindness. Many analytical laboratories are equipped with auto-titrators to speed up the throughput of samples. We also should not forget that for some titrations suitable indicators do not exist and in these cases titration end points may sometimes be determined instrumentally by either photometric or electrochemical determination.

We shall not consider titrations of these types in detail since there are separate chapters dealing with the use of light (Chapter 5) and electrochemistry (Chapter 10) for analytical purposes. We shall, however, briefly consider the basic principles and simple applications next.

3.12.1 Photometric titrations

Photometric titrations rely on changes in absorbance in the visible or UV spectrum that may be followed by UV/visible spectrophotometry.

Indicator-based titrations as described in Sections 3.6 to 3.11 may often be followed photometrically to facilitate automation.

Photometric determinations may also follow the consumption of a UV or visible absorbing reagent, or the generation of a UV or visible product, or simultaneous consumption and production of a reagent and product that both absorb in the UV or visible range.

The entire UV–visible range may be used since we are now not dependent on the limitations of the human eye, although it should be remembered that quartz should be used instead of ordinary glass if UV wavelengths are to be used (see Chapter 4).

A photometric titration curve is a plot of absorbance as a function of volume of the titrant; this plot must be corrected for volume changes since the volume will, of course, be changing as the titrant is added. The curve will typically consist of two linear (straight line) regions with different slopes. The end point of the reaction is taken as the intersection if the two linear portions of the graph are extrapolated back. The simplest of these plots, Fig. 3.11, corresponds to the onset of a colour (absorbance) change as the titrant is added. The absorption increases till it reaches a maximum. A graph of this type can correspond to the formation of an absorbing coloured complex as the titrant is added and the absorption increases till all of the analyte has been complexed. Figure 3.12 corresponds to the onset of absorption upon the formation of an absorbing product once the

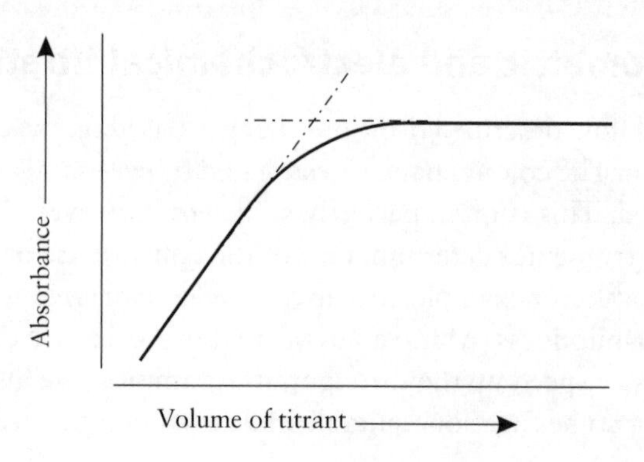

Figure 3.11 Onset of colour as titrate is added.

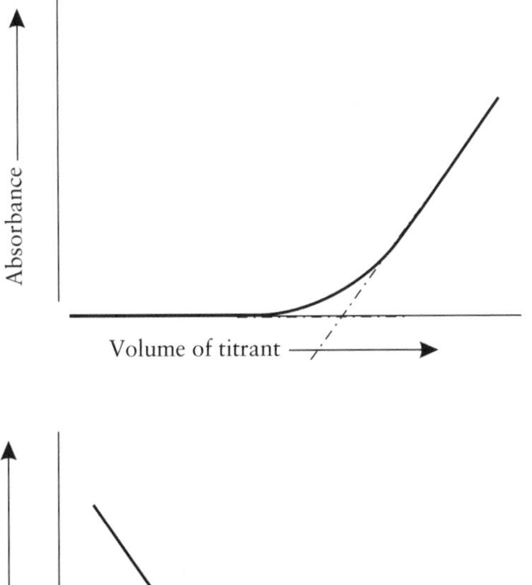

Figure 3.12 Formation of absorbing product.

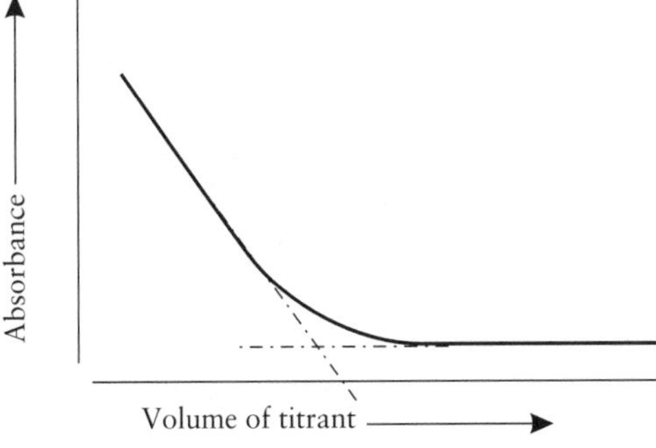

Figure 3.13 Absorption falling as titrant is added.

equivalence point has been reached, (an example here can include silver chromate formation in a silver nitrate based titration—Section 3.5). Figure 3.13, by contrast, shows an absorption titration curve that falls to a minimum as the titrant is added. In this case, for example, a coloured complex, may be changing in colour (or becoming colourless) as the titrant is added and reacts.

Photometric titrations are normally performed within photometers or spectrophotometers that have been constructed so that the titration vessel is held in the light path. It is not normally necessary to measure the *absolute* absorbance since relative values are all that is needed to follow the titration and identify the equivalence point. The species being monitored should, however, obey the Beer–Lambert law (Chapter 5) so that the aborbance measured is proportional to concentration.

Photometric determinations have been applied to many types of titration. Most acid–base titrations utilizing indicators may be followed photometrically. Photometric determinations have also proved extremely useful to EDTA and other compleximetric based titrations where colour changes are often associated with the formation or dissociation of transition metal ion based complexes.

Photometric determinations always dictate that the Beer–Lambert law be obeyed for the indicator system used.

3.12.2 Electrochemical titrations

Titrations may be in some cases be followed by amperometric or potentiometric electrochemical means. For further details of either of these approaches please, refer to Chapter 10.

Amperometric titrations

For a discussion of electrode and electrochemical cells, refer to Chapter 10.

Amperometric titrations typically involve the monitoring of the response of a working electrode as a function of the volume of titrant added with counter and reference electrode normally being included to complete the circuit. Micro-electrodes are often used as the working electrodes for amperometric titrations since their responses are independent of fluctuations in convectional (stirring) mass transport. In some arrangements, two polarizable micro-electrodes are employed without the use of a separate reference electrode and in these cases, the current that flows between them is simply measured as the titrant is added. Amperometry involves the continual oxidative or reductive removal of one of the reactants or products. The current response is moreover typically dependent on the rate of mass transport of solutes to and from the sensing (working) electrode. Micro-electrodes are moreover inherently associated with smaller rates of analyte or product consumption (see Chapter 10).

Amperometric titrations are often more accurate and reliable than volumetric titrations since they help minimize human error and imperfections in judgement.

Potentiometric titrations

Potentiometric titrations are based on the monitoring of a potential at a suitable sensing electrode (with respect to a reference electrode) as a function of the volume of a titrant. The sensing electrode is typically an ion selective or ion sensing electrode (ISE), see Chapter 10. The pH electrode is the most common form of ISE and is extremely useful for following any standard acid–base titration. The use of ISEs does incur some disadvantages especially in terms of maintenance requirements of these electrode systems although automated ISE-based instrumentation may overcome some of these drawbacks. ISE outputs are normally quoted in terms of potential differences with respect to a standard calomel or Ag/AgCl reference electrodes. pH electrodes (H^+ ion selective electrodes—see Chapter 10) normally have instrumentation to provide a reading directly in pH units.

Most potentiometric titration curves are plotted in the form of either changes in millivolts (with respect to a reference electrode) versus volume of titrant added, or pH versus titrant volume.

Potentiometric titrations provide data that are often inherently more reliable than that otherwise obtainable from standard indicator-based visually determined titrations. Again the technique minimises some experimental variations such as the amount of indicator present or estimations in colour change. Potentiometric titrations often offer additional advantages over amperometric approaches in that the technique is non-destructive; in other words, none of the reactants or products are consumed. A steady state equilibrium potential must be recorded and sometimes a little time must be allowed while the reagents are allowed to react and a steady potential obtained, although this can usually be minimized with (adequate) automated stirring of the titration flask.

3.13 Water determination by a Karl–Fischer titration

A number of organic solvents take up water over a period of time. ***The Karl–Fischer titration*** allows the determination of water via the use of a special reagent mixture known as the ***Karl–Fischer reagent***. The Karl–Fischer reagent contains iodine, sulphur dioxide, pyridine, and methanol as the solvent. The mixed reagent reacts with water according to Eqns (3.37) and (3.38):

$$C_5H_5N \cdot I_2 + C_5H_5N \cdot SO_2 + C_5H_5N + H_2O \rightarrow 2C_5H_5N \cdot HI + C_5H_5N \cdot SO_3 \quad (3.37)$$

$$C_5H_5N \cdot SO_3 + CH_3OH \rightarrow C_5H_5N(H)SO_4CH_3 \quad (3.38)$$

Equation (3.37) describes the oxidation of sulphur dioxide by iodine in the presence of water and pyridine to produce sulphur trioxide and hydrogen iodide. I_2, SO_2, HI, and SO_3 all exist as complexes in the presence of excess pyridine. It should be noted that *only* Eqn (3.37) involves the consumption of water—whereas Eqn (3.38) does not. Methanol is added to the reaction mixture to remove the $C_5H_5N \cdot SO_3$ formed within the reaction of Eqn (3.37) since this may itself consume water, which can of course affect the final end point of the reaction of Eqn (3.39).

$$C_5H_5N \cdot SO_3 + H_2O \rightarrow C_5H_5N \cdot SO_4H \quad (3.39)$$

The reaction may, however, be completely quenched by the addition of excess methanol, as shown within Eqn (3.38).

The stoichiometry involves 1 mol of iodine, 1 mol of sulphur, and 3 mol of pyridine for the consumption of 1 mol of water. In practice, a two-fold excess of sulphur dioxide and three- to four-fold excesses of pyridine are provided in the reagent mixture. The mixture partially decomposes with time and this means that the potency of the Karl–Fischer reagent can—(and indeed does) vary with time. For this reason, the reagent must typically be standardised by calibration with a known mass of water prior to the analysis of an unknown sample. It is normal practice to weigh a micro-syringe with a few milligrams of water (equivalent to a few microlitres) and place the water within the titration flask. The empty syringe may be re-weighed to determine the exact weight of the water transferred. A dummy titration is then carried out to calibrate the system.

The end point is noted by the appearance of the first excess pyridine/iodine complex, which has a deep brown coloration. The reaction mixture should remain colourless prior to the end point.

There are many commercial machines now available that are fully automated and rely on an amperometric determination of an end point via the twin micro-electrode approach, as described in Section 3.12.

The Karl–Fischer approach is still widely used to determine the water content within organic solvents, even though the technique does have some inherent problems that must be taken into account. The sample should ideally be prepared by dissolution in methanol. While samples that are only partially or even completely insoluble in methanol can be analysed this may lead to only partial recovery of the water content. The Karl–Fischer reagent must clearly be free from exposure to moisture itself, and care should be taken with the handling of pyridine due to its toxicity. Problems can also be encountered from chemical interferences present within some organic and inorganic reagents. Carbonyl containing compounds, for example, are capable of reacting with methanol to produce water, Eqn (3.40), which could clearly cause the distortion of the analyses:

$$RCHO + 2CH_3OH \rightarrow RCHOCH_3OCH_3 + H_2O \qquad (3.40)$$

In a similar manner, many metal oxides are capable of reacting with hydrogen iodide to form water (Eqn 3.41):

$$MO + 2HI \rightarrow MI_2 + H_2O \qquad (3.41)$$

where M = metal.

Reactions of the type seen in Eqns (3.40) and (3.41) often give rise to fading end points. Oxidizing or reducing agents may also interfere with Karl–Fischer titrations by re-oxidizing the iodide produced or by reducing the iodine within the regent mixture. There are now a number of commercial Karl–Fischer reagents with slightly altered chemistries to both enhance the stability of the preparation as well as overcoming some of the potential problems associated with possible interferents. The exact nature of chemical compositions of these preparations are, however, jealously guarded by the companies involved.

3.14 Titrations—best practice

The quality of the data that is obtainable from a titration is directly related to the care and time invested into performing the titration.

- If you are right-handed, hold and swirl the titration flask with your right hand and operate the burette tap by holding the tap around the body of the burrette, Fig. 3.14. If you are left-handed, hold the flask with your left hand and operate the tap with your right.
- Use a retort stand with a white base to aid seeing the colour change and therefore the equivalence point of the titration; alternatively place a white tile on the base of the retort stand. If you are unsure whether a colour change is starting to occur or not, hold the flask up to the light, Fig. 3.15.
- Only add sufficient indicator to produce a colour which can be easily identified. Further indicator may actually hide a subtle colour change.

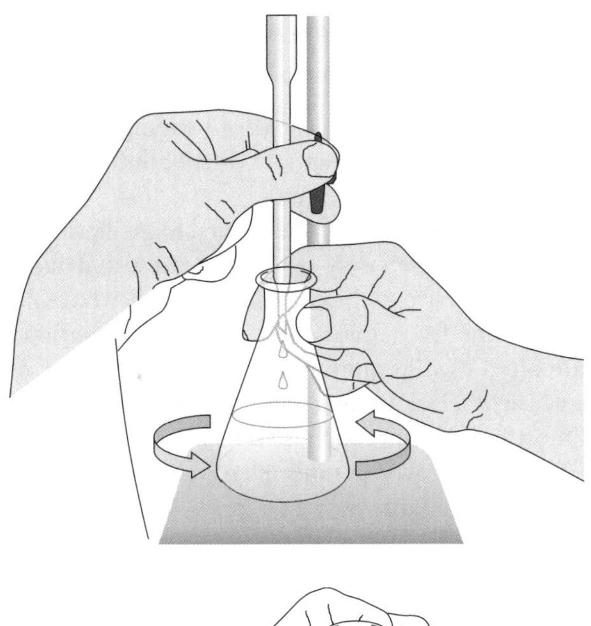

Figure 3.14 Correct way to operate burette.

Figure 3.15 Holding flask up to light to judge colour.

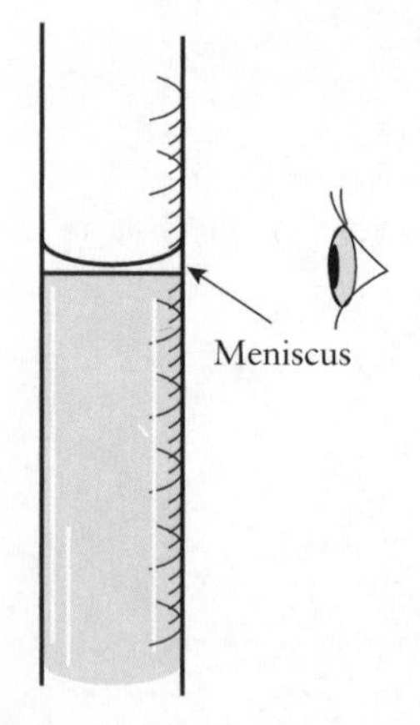

Figure 3.16 Viewing burette reading.

- The titrant should always be added ***dropwise and slowly*** to allow sufficient mixing of the reactants. It is extremely easy to consistently overshoot the end point by adding the reactants too hastily.
- Always perform titrations till you have at least three concordant results. It is perfectly acceptable to carry out an initial titration to determine approximately where the equivalence point lies.
- It is good practice to retain the first titration flask to compare with the colour of subsquent flasks, especially when the colour changes are subtle.
- ***Read the titre value from the burette by looking directly at the side of the burette at eye level with the bottom of the meniscus, Fig. 3.16.*** It is possible to make an estimation of where the bottom of the meniscus lies in-between the graduations of the burette. Many burettes, for example, have graduations every 0.1 cm^3. It is possible and indeed good practice to take a reading to within 0.01 cm^3.

3.15 Performing titration calculations

Many students have problems or even dread attempting titration calculations. This is a pity since the calculations are not difficult and only appear so in the way they are often presented.

There are many 'magic equations' designed to facilitate titration calculations, which, in fact, often only serve to confuse students; even worse, when a student does arrive at the correct answer they have no idea where the answer has come from apart from the magic equation. The magic equations are often used in quite inappropriate situations, such as back titrations because of a lack of understanding of what they represent. The easiest and most reliable approach is to tackle the calculation from first principles; experience shows that this invariably proves to be the simplest approach anyway.

Exercises and problems

3.1. Explain what is meant by a strong acid and a strong base.

3.2. Calculate the pH of an aqueous 0.15 M solution of HCl.

3.3. Calculate the pH of a 0.2 M H_2SO_4 solution.

3.4. Explain what is meant by a weak acid or a weak base.

3.5. Calculate the pH of a 0.15 M NaOH solution.

3.6. 20 cm^3 of a 0.153 M HCl solution is titrated with a 0.125 M solution of NaOH; calculate the expected equivalence point.

3.7. The K_a for acetic acid is 1.7×10^{-5}; calculate the pH of a 0.2 M CH_3COOH solution.

3.8. The K_b for NH_3 is 1.75×10^{-5}; calculate the pH of a 0.15 M aqueous NH_3 solution.

3.9. Explain what is meant by a buffer.

3.10. Calculate the pH of a 0.07 M sodium acetate CH_3COONa and 0.02 M acetic acid CH_3COOH buffer. The K_a for CH_3COOH is 1.7×10^{-5}.

3.11. Fifteen cubic centimetres of a 0.2 M HCl solution is added to 100 cm^3 of a 0.01 M NH_4OH/0.05 M NH_4Cl buffer solution. Calculate the pH of the buffer before and after the addition of the HCl (take the pK_b of the NH_4OH to be 1.88×10^{-5}).

3.12. The pH of a calcium containing solution is adjusted by the addition of a few drops of 2 M KOH and titrated with 0.01 M EDTA. If a 10 cm^3 aliquot of the Ca^{2+} containing solution requires 15.4 cm^3 of the EDTA to reach the equivalence point. What is the concentration of Ca^{2+} within the unknown sample?

3.13. A total of 10.5 cm^3 of a 0.0012 M $AgNO_3$ solution is required for the titration of a 20 cm^3 aliquot of a Cl^- containing water sample. Calculate the concentration of chloride within the water sample.

3.14. A 100 cm^3 of an oxygen containing water sample is firstly treated with excess alkali-azide reagent and $MnSO_4$. Excess KI is then added together with a few cm^3 of a phosphoric acid solution to release I_3^-. This is titrated with a 0.01 M sodium thiosulpahate and requires 5.23 cm^3 for the equivalence point to be reached. Calculate the concentration of O_2 within the water sample.

3.15. A basic solution is known to contain both OH^- and HCO_3^-. This solution is titrated with a 0.01 M HCl solution and two equivalence points are found using: (i) a mixed indicator of Thymol Blue and Cresol Red: and (ii) methyl orange. When a 20 cm^3 aliquot is titrated with the 0.01 M HCl solution, an equivalence point of 11.20 cm^3 is found using the mixed indicator. A total of 22.32 cm^3 is required for equivalence using methyl orange. Determine the concentration of OH^- and CO_3^{2-} within the solution.

3.16. The concentration of an ammonium chloride concentration is to be determined. 20 cm^3 of 0.2 M NaOH is added to a 10 cm^3 aliquot of the ammonium chloride solution. Following reaction and NH_3 evolution, the solution is titrated with 0.1 M HCl. A total 5.2 cm^3 of the HCl solution is required for equivalence. Calculate the concentration of the ammonium chloride solution within the unknown solution.

3.17. The vitamin C content of a fruit cordial is to be analysed. A total of 1.2 g of KIO_3 is added to a 250 cm^3 flask. KI is added in excess and the solution acidified by the addition of dilute acid. Twenty cubic centimetre aliquots of vitamin C require 25.3 cm^3 of a 0.05 M sodium thiosulphate solution for equivalence. Calculate the concentration of vitamin C within the cordial.

3.18. Silver nitrate based reactions to determine chloride concentrations may be followed photometrically since silver ion will react with chromate to form (red) silver chromate. The silver ions will, however, only react with chromate once all of the chloride has been consumed. The formation of silver chromate may therefore be used as an indicator. The data for a photometric silver nitrate titration are shown below. Plot the absorbance data for this titration and estimate the equivalence point for this titration.

Volume of silver nitrate (cm^{-3})	Absorbance
0	0.12
1	0.13
2	0.13
3	0.13
4	1.16
5	0.19
6	0.24
7	0.52
8	1.35
9	1.46
10	1.57
11	1.68
12	1.79
13	1.89
14	1.91

3.19. If the concentration of chromate used in Problem 3.18 is 0.12 M and a 20 cm^{-3} aliquot of the unknown chloride solution is being calibrated, calculate the chloride concentration.

Summary

1. The pH of a solution is = $-\log_{10}[H^+]$.

2. Strong acids can be assumed to fully dissociate when calculating pH values.

3. Strong bases can be assumed to fully dissociate when calculating pH values.

4. The dissociation constant for water is given by

$$K_c = \frac{[H^+][OH^-]}{[H_2O]}$$

The ionic product of water is given by

$$K_w = [H^+][OH^-]$$

and at 25°C $K_w = 10^{-14}$ mol^2 dm^{-6}.

5. The dissociation constant for an acid is

$$K_a = \frac{[H^+][OH^-]}{[HA]}$$

6. The dissociation constant K_b for a base is

$$K_b = \frac{[OH^-][BH^+]}{[B]}$$

7. Buffers resist pH changes. Buffers are formed by preparing a solution of a weak acid and one of its salts—or a solution of a weak base with one of its salts.

8. The equivalence point between acids and base titrations may be determined by following pH changes as an acid is added to a base—this may be achieved by careful choice of indicators. Strong acids and strong base interactions give the strongest shift in pH.

9. Compleximetric titrations involving ligands such as EDTA may be used to determine metal ion concentrations such as Ca^{2+} or Mg^{2+} (e.g. for water hardness).

10. Back titrations involve adding an excess of reagent and then analysing the amount of this that remains unreacted with an analyte. Two examples of back titrations are: (i) determination of ammonium chloride by using an excess of sodium hydroxide and titrating this with hydrochloric acid; and (ii) determination of vitamin C concentrations via the use of potassium(V) KIO_3 as a primary standard to generate excess I_2, which will react with vitamin C. The remaining I_2 may then be determined via titration with sodium thiosulphate and this in turn can be related to the concentration of vitamin C.

11. Some titrations may be followed by photometric or electrochemical approaches.

12. Chloride concentrations may be determined by titration with silver nitrate using potassium chromate as indicator. The titration can be followed colorimetrically to form silver chromate (which can be quantified colormetrically).

13. Dissolved O_2 concentrations may be determined via treatment with a manganese reagent followed by titration with sodium thiosulphate.

14. Mixtures of strong and weak bases (e.g. OH^- and HCO_3^-) may sometimes be determined via titration using two or more indicators to determine separate equivalence points.

15. The water content within some organic solvents may be determined by Karl–Fischer type iodine/pyridine titrations.

Further reading

Alcock, J. W. and Gilette, M. L. (1997). *Monitoring acid-base titrations with a pH meter: modular laboratory program in chemistry*. Chemical Education Resources.

De Levie, R. (1999). *Aqueous acid-based equilibria and titrations*. Oxford Chemistry Primers. Oxford University Press.

Hulanicki, A. (1987). *Reactions of acids and bases in analytical chemistry*. Ellis Horwood Series in Analytical Chemistry, Ellis Horwood, Chichester.

Oxlade, C. (2002). *Chemicals in action: acids and bases*. Heinemann Library.

Analyses based on solubility effects, precipitation, and the determination of mass

4

Skills and concepts

This chapter will help you to understand:

- The concept of using precipitation reactions for gravimetric analyses.
- What are meant by the terms: surface potential charge, primary adsorption layer, counterion layer, and zeta potential.
- How to collect and dry precipitates.
- How to use a dimethylgloxime (DMG) gravimetric approach to calculate the nickel content of an aqueous sample.
- What is meant by the solubility product and how this may be used to predict the formation of a precipitate.
- What is meant by the relative supersaturation of a solution and how to calculate it.
- What is meant by a chelating agent and how these are used in gravimetric analyses to form co-ordination compounds.
- The concept of thermogravimetric analyses and how the slope of mass/temperature thermogram profiles may be used to plot first derivative thermograms.

4.1 Introduction to gravimetric analyses

Gravimetric (or mass based) analyses are based upon weighing products and/or reactants before and after some chemical reaction. In many cases, a product of known composition is precipitated as an insoluble product from a reaction mixture. This precipitate is then normally filtered from the solution, dried, and weighed. If the composition of the precipitate is known we have the basis for a chemical analysis since weighing allows us to determine the mass of the product.

EXAMPLE 4.1 Calculate the percentage of copper in anhydrous copper sulphate (CuSO4).

Method

Relative molecular weight of $CuSO_4$ = 63.55 + 32.07 + (4 × 16) = 159.62
The relative atomic mass of Cu is 63.55
Therefore, % Cu = (63.55/159.62) × 100 = 39.81%

4.2 Complexation and precipitation

The majority of precipitation-based gravimetric analyses are performed with the reactants in aqueous solution. The product must be virtually totally insoluble if it is to be totally recovered—and this is essential if a quantitative analysis is to be attempted.

The solubility of a compound and therefore its dissolution and precipitation are primarily governed by the enthalpy and entropy associated with the compound in its immediate surroundings upon dissolution.

The most important factors determining the enthalpy of dissolution are the bond energies within the solid phase and the enthalpy of solvation for ions entering solution. Enthalpy will increase due to a loss of crystal structure as its component ions enter solution. Entropy will also be increased due to ion–solvent interactions upon solvation of the solid.

The net balance of these effects can be quite finely balanced. The sign of the Gibbs free energy change is normally the most important effect that determines whether or not a compound is soluble and this may vary between closely related compounds. Silver chloride and barium sulphate, for example, are not soluble in water, while sodium chloride and barium chloride are. As a general rule, however, compounds which are ionizable in water or at least highly polar will be aqueous soluble, whereas compounds that are non-polar and neutral will not be. It follows that if a non-polar and non-ionizable compound is formed from reactants that are soluble in water it will precipitate and this principle forms the basis of many gravimetric determinations. In many cases, an unknown quantity of an aqueous analyte is allowed to react with an excess of a second reagent. The precipitate which is then formed is filtered, dried and weighed. If the composition of the precipitate (product) formed is known then the molar quantity of the analyte may be determined.

The chemistry of the precipitation reaction should in all cases be highly selective and proceed through to completion. It is essential that there be no competing reaction chemistries that might form unwanted products and this applies to the analyte forming soluble products or indeed additional reagents being consumed in unwanted side reactions. The quantitative aspect of the analysis assumes that all of the analyte reacts to form the same products, which may be collected with close to 100% efficiency.

It is clear that if the product is to be collected with near total efficiency, then it must be virtually totally insoluble. The solubility of the product is described by the ***solubility product***, which, in this case, should be as low as possible.

If we consider the equation which describes the dissolution or precipitation of a complex A–B, Eqn (4.1):

$$xA - yB \rightarrow xA_{(aq)} + yB_{(aq)} \tag{4.1}$$

then the solubility product K_{sp} may be defined as:

$$K_{sp} = [A]^x \, [B]^y \tag{4.2}$$

Let us consider the complexation of nickel with dimethylgloxime (DMG), as an example of a reaction that gives a nearly totally insoluble product that may be exploited for a gravimetric determination. The solubility product of nickel dimethylglyoxime is 10^{-17} and so in practice DMG is almost totally insoluble. This means that almost all of the product precipitates, allowing its collection with close to 100% efficiency if experimental care is taken. Further details of this gravimetric approach are given in Section 4.4.2.

Precipitation reaction chemistries may be dependent on factors such as pH and temperature. The stability of the precipitate is another factor that should not be forgotten and it is normally essential that the precipitate is able to withstand the heating necessary to remove the residual water.

EXAMPLE 4.2 The K_{sp} of silver chloride (AgCl) at 25°C is 1.0×10^{-10}. Calculate the $[Ag^+]$ in a saturated AgCl solution.

Method

Calculate the molar concentration of Ag^+ from the K_{sp} expression.

$$K_{sp}\,AgCl = 1.0 \times 10^{-10} = [Ag^+][Cl^-]$$

Note the concentration of Ag^+ and Cl^- must be the same, so:

$$[Ag^+] = \sqrt{1.0 \times 10^{-10}}\ M$$

$$= 1 \times 10^{-5}\ M.$$

EXAMPLE 4.3 If the K_{sp} of AgCl at 25°C is 1.0×10^{-10}, calculate the molar solubility of AgCl.

Method

Calculate the concentration of AgCl from the K_{sp} expression.
The solubility of AgCl = $[Ag^+_{(aq)}] = [Cl^-_{(aq)}]$
From Example 4.2, $[Ag^+] = 1 \times 10^{-5}$ M
Therefore,
the solubility of AgCl = 1×10^{-5} M.

EXAMPLE 4.4 Calculate the concentration of Ag^+ that must be added to initiate precipitation of AgCl in a 1×10^{-4} M NaCl solution.

Method

Calculate the required $[Ag^+]$ from the K_{sp} expression.

$$[Ag^+] \times (1 \times 10^{-4}\,\text{M}) = 1 \times 10^{-10}\,\text{M Ag}^+$$

$$[Ag^+] \text{ required to initiate precipitation} = \frac{1 \times 10^{-10}}{1 \times 10^{-4}}\,\text{M}$$

$$= 1 \times 10^{-6}\,\text{M}$$

4.3 Formation of the precipitate and collection via filtration

4.3.1 Precipitates and colloidal suspensions

Precipitation of the product normally occurs following the addition of the reagent(s) to the analyte. Precipitates with larger particle sizes are more easily collected by filtration and inherently moreover contain fewer contaminants than precipitates with finer particles.

Particle size depends not only on the chemical nature of the precipitate, but also upon the conditions during the precipitation. The size of the particle may not vary consistently; colloidal particles, for example, range in size from 10^{-6} to 10^{-4} mm in diameter, and show no tendency to settle from the solvent. For this reason, they do not lend themselves to collection by filtration. At the other end of the spectrum, true precipitation occurs when particles form with sizes in the order of fractions of a millimeter. If colloidal particles are present within a solvent, the mixture is known as a ***colloidal suspension***, in which case the particles are kept in suspension by Brownian motion. Particles of larger sizes when agitated in the solvent form crystalline suspensions which spontaneously settle out allowing them to be easily collected by filtration.

The size of the precipitation particles may be influenced by the temperature, the concentration of the reactants and the rate at which the reactants are mixed. The variables may be accounted for by assuming that the particle size is related to a single cumulative parameter known as the ***relative supersaturation***. The relative supersaturation is defined by Eqn (4.3):

$$\text{Relative supersaturation} = \frac{Q - s}{s} \tag{4.3}$$

where Q is the concentration of the solute at any given time and s is the solubility of the precipitate at equilibrium.

It follows that during precipitation, the system is momentarily supersaturated and this situation is relieved by precipitation of the solid product. If the relative supersaturation is small then precipitates of larger particulate sizes are more likely to be formed.

4.3.1 Mechanisms of precipitation and experimental control of particle size

Solid particles crystallize and cease to be solvated via nucleation and particle growth mechanisms. The particle size is largely governed by the extent to which one or other of these processes predominates. The mechanism may, in turn, be related back to, and explained in terms of, the relative supersaturation within the solution as the precipitate forms.

Precipitation is normally initiated by ***nucleation***. The first stage of nucleation involves the spontaneous association of a few ions, atoms or molecules to form a stable second phase; this normally happens on some irregularly shaped surface such as a suspended dust particle within the solvent. Precipitation may now continue via further spontaneous nucleation reactions or via the growth of the existing nuclei. This latter process is known as ***particle growth***. If nucleation predominates, it follows that the average particle size will be small. Conversely if particle growth predominates, then the average particle size is likely to be larger. Increased levels of supersaturation are thought to increase the rate of nucleation and it therefore follows that increased levels of supersaturation typically cause smaller particles to be formed. By the same argument lower levels of supersaturation tend to favour precipitation via particle growth and this, in turn, leads to the formation of larger particles within the precipitate.

We can (at least in some situations) influence the size of particle precipitated. Increased temperatures help increase the solubility, s, and thus favour crystalline depositions. Dilute solutions and the addition of the precipitating reagent both minimize the concentration of the solute, Q, and therefore favour the formation of larger particles. pH may be also used to control the rate of precipitation and therefore the size of the crystals formed. If, for example, calcium oxalate is precipitated in mildly acidic conditions, large crystals are obtained, which may be easily filtered. The reason for this is that calcium oxalate is partially soluble in environments of low pH. The precipitation must clearly be brought to completion for a quantitative analysis and this can be performed by slowly adjusting the pH in a basic direction until all of the calcium oxalate has been forced out of solution.

We have assumed until now that all of the precipitates are finally formed as crystalline solids. If the solubility of this product is, however, very small indeed then the relative supersaturation will always be large during the precipitation process and it is probable that the solid will form as a non-crystalline colloidal suspension. Although colloids will not settle out and may not be filtered, it is fortunately often possible to coagulate or agglomerate the solid into a non-crystalline mass, which can then be separated by filtration.

Colloid particles are kept in a constant state of motion by Brownian motion effects. The particles do not associate with each other and therefore refuse to settle out of the solution due to a double layer association of charge around the particle and how this interacts with the polar H_2O solvent molecules. The particle will contain both negative and positive charges due to the different ions which form its structure. The particle is capable of attracting both anions or cations due to the differing charge associations that can occur. Ions will preferentially absorb to the colloidal particles if they are the same as ions that are found within the colloidal particles. If the solution contains an excess of the anions that contribute to the particles, then these particles will possess an excess of anions absorbed on its surface and so carry a net negative charge. The potential that exists at the surface of the particle is known as the ***surface potential charge***.

In a similar manner, if the particle is suspended in a solution containing a net excess of the cations that go to form the particle structure, the particle will absorb a net excess of cations and thus carry a net positive charge. Ions normally directly associated with the particle do so in the ***primary adsorption layer*** and ions of the opposite polarity now attempt to counteract this localized association of charge by forming the so-called ***counterion layer***. The net charge association depends on the balance of anions and cations within the solution.

If an electric field is applied across the solution, water molecules within the primary adsorption and counterion layers become orientated and move with the colloidal particle. The particle will move under the influence of electrical migration either to one electrode or the other. As the particle together with its associated ions move through the solution a potential resides between it and the bulk solution, which is known as the ***zeta potential***. If the *zeta potential* is of the same polarity as that of the surface potential and sufficiently large, repulsion effects will prevent the association, and therefore the coagulation of the colloidal particles.

Coagulation may generally be brought about by reducing the zeta potential of the particles. This can be achieved by either increasing the electrolyte concentration of the solution or temporarily heating the solution whilst stirring.

The addition of a suitable ionic compound to increase the electrolyte concentration effectively reduces the volume in which the ions that can neutralize the zeta potential. The effect is to force the colloidal particles together, which, in turn, favours coagulation and precipitation.

If the colloidal suspension is heated, then the number of surface adsorbed ions decreases and so the zeta potential also decreases. The inter-particle repulsion decreases and this in turn decreases the inter-particle spacing, which again favours precipitation.

4.3.2 Collection of the precipitate

The precipitate once formed is normally collected using a ***Büchner flask*** together with a ***sintered crucible***: Fig. 4.1. The solution from which the solid was precipitated, (which is now known as the ***mother liquor***) is slowly poured through the crucible to collect the solid material or ***filtrate***.

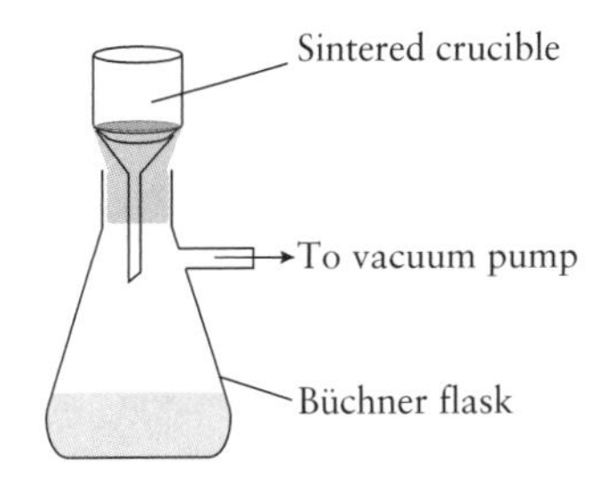

Figure 4.1 Collection of precipitate.

The crucible must first be prepared by baking it in an oven at a temperature in excess of 100°C to remove any adsorbed water. The crucible and precipitate must then be allowed to cool in a desiccator, at room temperature. The weight should then be recorded on a four figure balance. *It is very important that the crucible should have truly reached the temperature of its surroundings, since thermal convection currents above a warm crucible will cause fluctuations in the observed weight.*

Sintered glass crucibles are normally supplied in three grades—fine, medium, and coarse. The side arm of the Büchner flask is attached to a vacuum line and this may be supplied either from a central vacuum pump or alternatively by a local water vacuum pump attached to a laboratory tap. The mother liquor should not be discarded but rather retained. It is in most cases virtually impossible to collect all of the filtrate by one filtration, since some of the solid material will almost always adhere to the walls of the beaker or vessel in which the precipitation was performed. *Remember it is absolutely essential to collect as much of the solid product as possible if a quantitative analysis is to be even approximately attempted!*

The mother liquor may then be used to wash out the precipitation vessel, and this may then be re-filtered. It is preferable to use this mother liquor instead of using fresh distilled water for two reasons.

First, the mother liquor may still contain some solid material of smaller particulate size which escaped capture during the first filtration process. The sintered glass crucible will now be partially clogged by the filtrate already captured—and a second filtration often serves to capture some of the very small quantity of the remaining solid material.

Second, it should also be realized that the solid precipitate will in fact possess a finite solubility, however, small this may be. The mother liquor must be saturated, and its use prevents the re-solubilization of any of the precipitate either within the precipitation flask or indeed within the sintered glass crucible.

As we have just mentioned, it is sometimes very hard to remove the last of the solid material that adheres to the walls of the glass vessel. A glass rod with a rubber sleeve at one end may be used to dislodge the material, which, if kept stirred, may be poured into the crucible. A glass rod of this type is sometimes known as a ***policeman*** due to its resemblance to a truncheon. The policeman may be finally washed with the mother liquor to transfer the last of the material that may be clinging to the rod. This process may have to be repeated four or even five times in order to transfer the last remnants of the solid material. The quality of the result will be dependent on the time taken in performing the analysis.

The crucible is made of a ceramic material capable of withstanding baking in an oven. The crucible is normally placed in an oven set at around 90–95°C to drive off the water; higher temperatures may be used if the filtrate is thermally stable. If the material decomposes below this temperature, then a lower temperature must obviously be used—in which case the crucible must be baked for a longer period of time. The crucible must then be removed and allowed to cool to room temperature within a desiccator (to prevent water absorption from the air). The crucible may now be re-weighed on a four figure balance taking great care that the crucible has firstly *fully* cooled to room temperature. If the weight of the crucible drops the process must be repeated until consecutive concordant results are obtained. When successive weighings reach a concordant value, the mass of the precipitate may be calculated.

The analysis should be repeated at least four times to obtain a mean analyte content for the sample. Since it is almost impossible (and indeed unnecessary) to start with three samples of exactly equal mass, we should not expect to obtain concordant weights for different crucibles after precipitation and filtration. The crucibles themselves will possess different masses prior to the analysis and, for this reason, it is impossible to ascertain whether or not the results appear to be concordant with each other until calculations are performed. This contrasts with, for example, many volumetric analyses where discrepancies between different titrations are often immediately obvious as each titre is read.

4.4 Some practical gravimetric analyses

4.4.1 The gravimetric determination of the chloride ion content in an aqueous sample

The chloride content of a soluble salt may be precipitated as silver chloride, AgCl, by reaction with silver nitrate, $AgNO_3$, Eqn (4.4):

$$Ag^+_{(aq)} + Cl^-_{(aq)} \rightarrow AgCl_{(s)} \tag{4.4}$$

The precipitate may be collected in a sintered glass crucible and its weight determined after drying to constant weight at a temperature of around 100°C. The reaction mixture should be kept acidic by the addition of a small quantity of nitric acid to prevent the formation of salts with weak organic acids such as CO_3^{2-}.

Silver chloride initially forms as a colloid but may be coagulated by heating (see Section 4.3). Traces of nitric acid in the precipitate decompose to form volatile products upon heating to dryness and are lost to the atmosphere by evaporation or sublimation.

The precipitate may acquire a mild violet colouration due to the elemental formation of silver, which occurs following the natural photodecomposition of silver chloride (Eqn (4.5)). This process may lead to an underestimation of the silver content of an aqueous sample.

$$2AgCl_{(s)} \rightarrow 2Ag_{(s)} + Cl_{2(g)} \tag{4.5}$$

If photodecomposition of the AgCl occurs before the solid is filtered and collected, the free chlorine in solution may cause further AgCl to be formed, Eqn (4.6), which in turn may erroneously raise the results obtained.

$$3Cl_{2(aq)} + 3H_2O_{(l)} + 5Ag^+_{(aq)} \rightarrow 5AgCl_{(s)} + ClO^-_{3(aq)} + 6H^+_{(aq)} \tag{4.6}$$

For this reason, it is prudent to minimize exposure of the reaction mixture as far as possible from daylight since these species may precipitate along with AgCl, following reaction with $AgNO_3$. In the same context, tin (Sn) and antimony (Sb) may also form oxychloride precipitates.

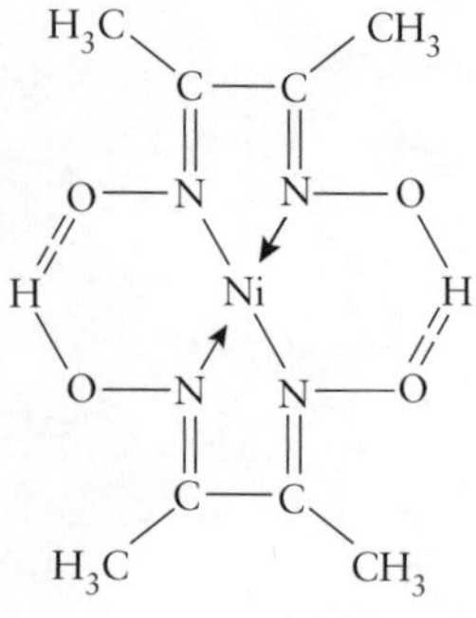

Figure 4.2 Structure of nickel dimethylglyoxime.

4.4.2 Determination of nickel using a DMG gravimetric approach

In mildly basic aqueous solutions, the organic compound DMG, will form a precipitate with nickel with near total specificity. The nickel content of even very complex samples may be determined with ease. Nickel dimethylgloxime is bright red and has the structure shown in Fig. 4.2.

DMG is an example of an organic precipitating agent known as a ***chelating agent*** and forms a non-ionic precipitate known as a ***coordination compound.*** Chelating agents such as DMG or 8-hydroxyquinoline possess at least two functional groups that are capable of forming covalent bonds by donation of a pair of electrons to a cation.

The precipitate that is formed with nickel is very bulky and this greatly facilitates its collection. The very intense colour also helps identify the presence of any remaining solid adhering to the glass walls of the precipitation vessel. The solid is, moreover, thermally stable and may be dried in an oven between 100°C and 110°C without any fear of decomposition.

Methodology

The term *Chelate* come from the Greek 'χηλή' (pronounced 'Chel-a') for 'a Crab's claw'.

A sample containing an unknown quantity of a nickel salt is analysed as follows. Add a little dilute hydrochloric acid (~1 M) to ensure that all of the salt is fully dissolved prior to the analysis. The sample should then be heated in a fume cupboard at a temperature of approximately 75°C (on a steam bath) and an aqueous solution of DMG may then be added in excess. Dilute aqueous ammonia (~0.5 M) should be added immediately to bring the pH up to a value of about 9. The pH may be monitored with a moistened strip of pH paper as the ammonia is added. The solution should be constantly stirred as the ammonia and the DMG are added to ensure thorough mixing throughout. It is difficult to say how much DMG should be added if the concentration of the nickel is completely unknown at the outset. The precipitate does, however, coagulate naturally very well and should readily collect at the bottom of the flask. The addition of further DMG should therefore either immediately cause the formation of further precipitate or alternatively have no effect whatsoever if all of the nickel has already been precipitated as nickel dimethylgloxime. The relative molecular mass of $Ni(C_4H_7O_2N_2)_2$ is 288.9344 whilst that of Ni alone is 58.71. It therefore follows that 20.319% of the mass of the precipiate is due to the nickel within the dry sample.

The nickel content of many different types of sample may be analysed. For example, the nickel content of a steel sample may be determined by dissolution of the metal in hot HCl of around 6 M concentration. In this case, a small quantity of nitric acid should also be added to remove any oxides of nitrogen that might have been introduced to the analyte sample. The sample should first be neutralized and adjusted to a pH of around 9 by the addition of dilute NH_3 as before. At this stage, DMG should be introduced to precipitate the nickel content of the sample.

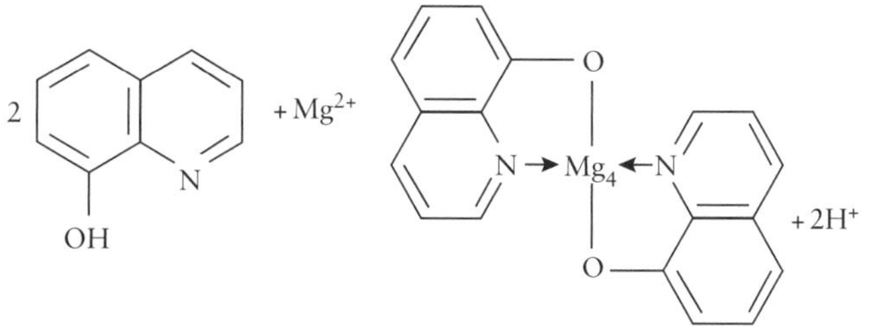

2 [8-hydroxyquinoline] + Mg^{2+} → [Mg^{2+} oxinate chelate] + $2H^+$

Figure 4.3 8-hydroxyquinoline (oxine) chelation with Mg^{2+}.

4.4.3 Gravimetric analyses utilizing 8-hydroxyquinoline as a precipitating agent

The gravimetric determination of Ni via its precipitation with DMG has an unrivalled specificity for gravimetric analyses. Many precipitating agents are capable of precipitating many different, *if similar*, analytes. The use of ***8-hydroxyquinoline***, which is also known as ***oxine*** (Fig. 4.3), may, however, offer a route to the specific precipitation of a number of different metal ions.

At first this statement might seem to be a contradiction in terms: while it is true that oxine may cause the precipitation of up to a couple of dozen cations to form metal oxinates, their solubilities are highly variable and are dependent on solution pH. The pH dependency of metal oxinates is due to the 8-hydroxyquinoline always deprotonating as a result of the chelation process.

4.4.4 Gravimetric determinations using sodium tetraphenylboron as a precipitating agent

Sodium tetraphenylboron $(C_6H_5)_4B^-Na^+$ may be used for the gravimetric determination of the potassium and ammonium content of samples. Sodium tetraphenylboron shows good specificity towards these two ions and forms salt-like precipitates, which may be readily collected by filtration and heating at 100–110°C without any fear of decomposition. Mercury(II), caesium, and rubidium ions may cause some partial interference and should be removed prior to the ammonium or potassium analyses.

4.5 Time requirements—sensitivities and specificities associated with gravimetric analyses

The sensitivity of a gravimetric analysis can be hard to rival with even the most sophisticated instrumental approaches. It is perfectly possible to determine weights to within a few micrograms and, with a suitably large mass, this may correspond to within a few parts per million of the mass of precipitate collected. Sensitivities in this range may often only be challenged by techniques such as atomic absorption spectroscopy. The sensitivity and accuracy of gravimetric analyses may often be limited by physical losses in the mass of the precipitates as it cools. Chemical interferences can cause precipitation of unwanted materials or solubility losses associated with products possessing more than a negligible solubility. These errors may usually be minimized by forcing the precipitation reaction to completion, careful experimental approaches for the collection of the filtrate and prudent choice of the precipitating agent (Section 4.3).

Gravimetric analyses are often considered to be extremely time consuming and laborious and it is certainly true that the total time taken to perform an analysis may be rather long. However, it should be realized that much of this time is associated with the drying and subsequent cooling of the sample. This time does not require the attention of the operator and, with careful planning, other tasks may be carried out at the same time. Gravimetric analyses should therefore not be labour intensive from the analyst's point of view. A student who is working through a series of practicals in analytical chemistry can easily perform another experiment (e.g. a titration analysis), while his or her crucibles are drying in the oven and/or cooling in a desiccator.

4.6 Thermal degradations and thermogravimetric analyses

We have already seen how heat may be used to dry a precipitate for a gravimetric analysis. In this situation, it is very important that the sample is weighed in thermal equilibrium with its surroundings. By contrast, ***thermal gravimetric analysis*** (TGA) uses temperature as an experimental variable. Thermal gravimetric studies are also very useful for determining the optimal temperature for drying samples within conventional gravimetric determinations.

Thermal gravimetric analyses involve monitoring the weight of a sample as the temperature is raised, (normally linearly), with time. The resulting mass–temperature profiles are known as ***thermograms***. The balance pan and sample are enclosed within a small closed oven to: (i) ensure that accurate, stable, and uniform temperatures may be achieved; and (ii) prevent thermal conduction air currents that will affect the balance. Weight variations in the order of micrograms can be easily followed. The balance should be calibrated each time it is used with a known mass; this deflection is normally recorded in a thermogram curve.

Let us consider a typical thermogram for silver chromate, $AgCrO_4$, as an example (Fig. 4.4). The initial drop in weight corresponds to the

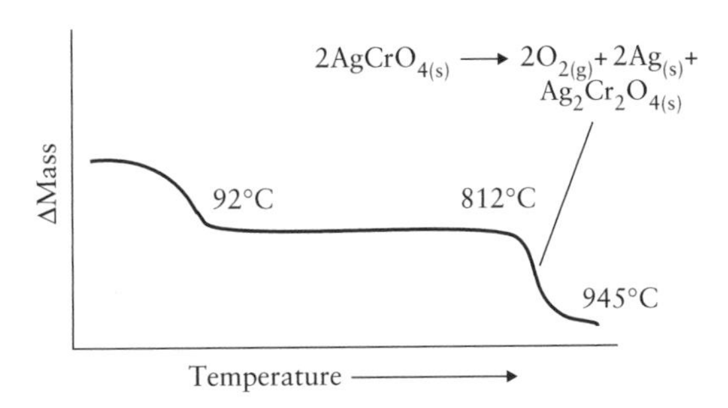

Figure 4.4 Thermogram for silver chromate, $AgCrO_4$.

evaporation of water. The compound will reach a stable weight once all of the water has been removed from the sample. The compound remains thermally stable as it is heated until a temperature of approximately 812°C is reached. The silver chromate starts to decompose thermally above this temperature with the loss of oxygen, to give molecular oxygen, metallic silver, and silver chromate, Eqn (4.7):

$$2AgCrO_{4(s)} \rightarrow 2O_{2(g)} + 2Ag_{(s)} + Ag_2Cr_2O_{4(s)} \quad (4.7)$$

Two important points should be noted: first, the thermogram demonstrates that the solid may be dried at any temperature from above 100°C to 800°C without any fear of thermal degradation. The second point relates to the temperature at which Ag_2CrO_4 ***does*** start to decompose, that is, above approximately 812°C. A thermogram and the temperature at which weight losses are seen may be treated as 'fingerprints' to help identify the presence of a compound within a sample. The weight loss meanwhile may be used to quantify the molar quantity of any particular analyte in a sample of variable composition.

A thermogram for mercurous chromate, Hg_2CrO_4 (Fig. 4.5), exhibits broadly similar behaviour to the thermogram for silver chromate. Mercurous chromate shows a loss of weight as water evaporates from the sample. Between the temperature of around 100 and around 250°C, mercurous chromate remains thermally stable. Above 256°C mercurous chloride begins to decompose thermally to give mercurous oxide and chromium trioxide, Eqn (4.8):

Great care should always be taken with thermogravimetric analyses involving mercury compounds due to their toxicity.

$$Hg_2CrO_{4(s)} \rightarrow Hg_2O_{(g)} + CrO_{3(s)} \quad (4.8)$$

The mercurous oxide is lost by sublimation leaving the solid chromium trioxide above temperatures of around 670 °C.

Another example could include a comparison of the thermal gravimetric analysis of silver and cupric nitrates; Fig. 4.6.

Silver nitrate, $AgNO_3$, Fig. 4.6(a) is thermally stable and does not exhibit any weight loss till a temperature of around 470°C is reached at

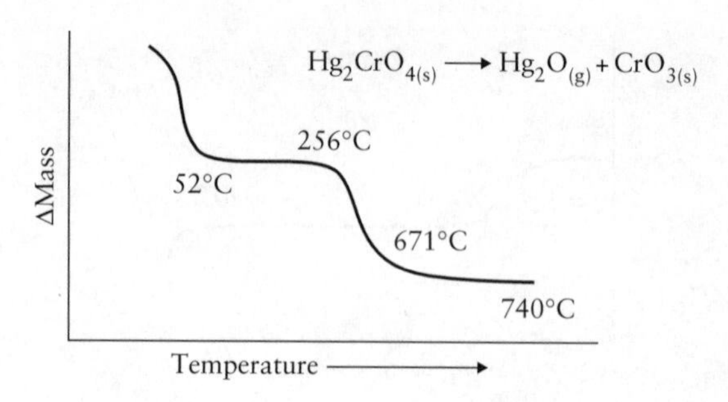

Figure 4.5 Thermogram for mercurous chromate, Hg_2CrO_4.

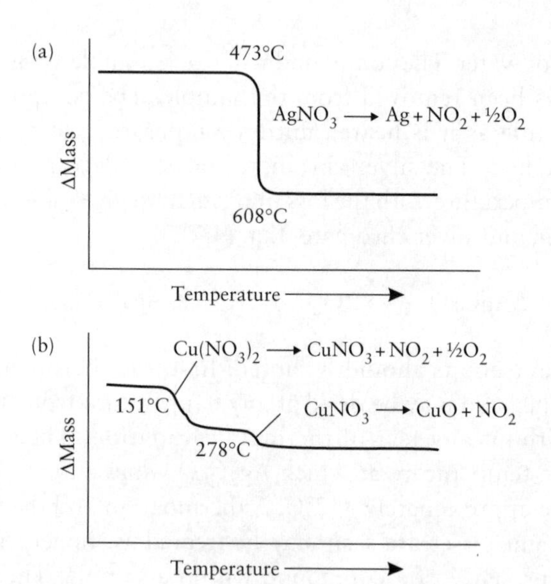

Figure 4.6 Thermograms for: (a) silver nitrate, $AgNO_3$; and (b) cupric nitrate, $Cu(NO_3)_2$.

which stage it begins to decompose to give NO_2, O_2, and metallic silver, Eqn (4.9).

$$2AgNO_{3(s)} \rightarrow 2NO_{2(g)} + O_{2(g)} + 2Ag_{(s)} \quad (4.9)$$

Cupric nitrate, $Cu(NO_3)_2$, by contrast decomposes thermally in two steps; Fig. 4.6(b).

It should be realized that as the temperatures are scanned, it is likely that the system rarely reaches *equilibrium when undergoing a weight loss, as the temperature is being scanned* at a constant rate. It follows that *the slope of the weight loss curve will (in most cases) depend on the rate of change of temperature*. The form and shape of the thermograms will therefore change from machine to machine or indeed even if the temperature scan rate is changed on the same instrument.

EXAMPLE 4.5 A 50 cm^3 sample of water is analysed for its calcium content by precipitating and collecting calcium oxalate (CaC_2O_4). The precipitate of CaC_2O_4 is slowly heated to 900°C. On heating, the oxalate initially loses water to form anhydrous calcium oxalate. As the temperature continues to be raised the calcium oxalate decomposes to form calcium carbonate ($CaCO_3$) and carbon monoxide (CO). As the temperature is raised still further the calcium carbonate decomposes to finally leave calcium oxide (CaO) and carbon dioxide (CO_2). (One mole of CaC_2O_4 gives one mole of CaO.)

If the mass of the crucible before collection of the precipitate was 25.7932 g, and then after precipitation, heating, and cooling the mass was 25.8216 g, calculate the mass of Ca in the water sample and the mass in 1 dm^{-3}.

Method

1. Calculate the mass of CaO.
2. Calculate the number of moles of CaO.
3. Calculate the mass of Ca in the 50 cm^3 water sample.
4. Calculate the mass of Ca in 1 dm^{-3}.

Step 1: Calculate the mass of CaO
Mass of CaO = 25.8216 − 25.7932 g = 0.0284 g

Step 2: Calculate the number of moles of CaO
Molecular weight of CaO = 40.08 + 16.00 = 56.08

Step 3: Calculate the mass of Ca in the 50 cm^3 water sample

$$5.064 \times 10^{-4} \text{ mol of Ca is equivalent to } 5.064 \times 10^{-4} \times 40.08 \text{ g} = 0.0203 \text{ g Ca}$$

Step 4: Calculate the mass of Ca in 1 dm^{-3}

$$\text{Mass of Ca in 1 dm}^{-3} = \frac{0.0203}{50} \times 1000 \text{ g} = 0.4059 \text{ g Ca dm}^{-3}$$

4.6.1 Derivative thermogravimetric analyses

The experimental procedure for performing ***derivative thermogravimetric analyses*** is exactly the same as for normal TGA determinations. The difference lies in an additional treatment of the data. Losses in weight give staircase-like thermograms, which, despite being easy to understand conceptually, may be a little difficult to fully interpret. This is particularly true if the weight losses are either small due to a limited sample volume, or indeed if the sample contains a number of differing compounds which decompose at similar or overlapping temperatures. A plot of the first derivative of the weight–temperature profile may be highly beneficial in such cases. The new thermogram is known as ***a first-derivative***

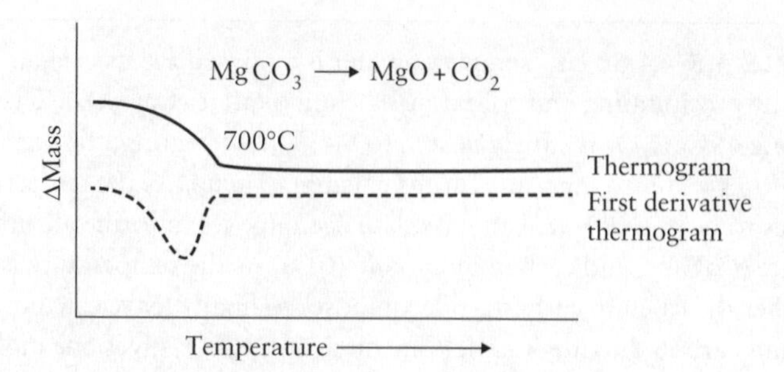

Figure 4.7 Comparison of thermogram and first derivative thermogram for magnesium carbonate, $MgCO_3$.

thermogram, an example of which is shown in Fig. 4.7 along with the standard thermogram from which it is derived. Weight losses are converted from a staircase like appearance to give a series of peaks. A simple consideration of the slope of the standard thermogram shows why this happens.

If no weight loss is observed, the thermogram will have a slope of zero (i.e. a horizontal line). The first derivative of this portion of the graph is also zero. If the *sample starts to undergo a weight loss* then the slope of the thermogram (i.e. the first derivative) will also be negative (a peak starts to form). As material is lost from the sample, the rate of weight loss decreases; the slope (first derivative) of the thermogram now passes through a minimum and begins to rise as the sample weight begins to stabilize once more. If the sample ceases to lose any more weight, the first derivative thermogram peak returns once more to a baseline value. The identification of a peak is very much easier if two or more thermal degradation processes occur over overlapping temperature ranges.

Exercises and problems

4.1. Calculate the percentage of chlorine in a pool and spa chlorinating agent that contains 95.3% $C_3Cl_3N_3$.

4.2. Calculate the percentage of iron in $FeSO_4$.

4.3. Calculate the percentage of silicon in SiO_2.

4.4. The K_{sp} of $PbSO_4$ at 25°C is 1.6×10^{-8}. Calculate the concentration of Pb^+ and SO_4^- in a saturated solution of $PbSO_4$.

4.5. If the K_{sp} of $PbSO_4$ at 25°C = 1.6×10^{-8}, calculate the molar solubility of $PbSO_4$.

4.6. The K_{sp} of $Fe(OH)_3$ is 4×10^{-38}. Calculate the $[OH^-]$ in a saturated solution of $Fe(OH)_3$.

4.7. Calculate the concentration of Ag^+ that must be added to initiate precipitation of AgCl in a 1×10^{-6} M NaCl solution.

4.8. A total of 284.45 g of a thermally stable precipitate is collected and left to dry overnight in an oven at 95°C. In the morning, constant weights of 222.45 g are recorded at three successive hourly intervals. Calculate the percentage of water in the sample.

4.9. A nickel determination for a nickel containing steel alloy is performed via a DMG gravimetric analysis. A total of 121.45 g of $Ni(HC_4H_6O_2N_2)_2$ is collected following drying careful of the precipitate. The original sample of the alloy weighed 125.15 g. Calculate the percentage of nickel within the iron.

4.10. A sodium salt, $Na(NH_4)HPO_4 \cdot 4H_2O$ loses four water molecules upon heating. As the temperature is increased further another water molecule is lost and when the temperature is increased further still a molecule of NH_3 is lost to finally yield $NaPO_3$. Write down the thermal degradation steps and sketch the shape of an expected thermogram with increasing temperature.

4.11. Sketch the expected shape of the first derivative thermogram of the thermogram of Problem 4.10.

4.12. A 50 cm^3 sample of water is analysed for its calcium content by precipitating and collecting calcium oxalate (CaC_2O_4). The precipitate of CaC_2O_4 is slowly heated to 900°C. On heating, the oxalate initially loses water to form anhydrous calcium oxalate. As the temperature continues to be raised the calcium oxalate decomposes to form calcium carbonate ($CaCO_3$) and carbon monoxide (CO). As the temperature is raised still further the calcium carbonate decomposes to finally leave calcium oxide (CaO) and carbon dioxide (CO_2). (One mole of CaC_2O_4 gives one mole of CaO.)

If the mass of the crucible before collection of precipitate was 25.7824 g, and then after precipitation, heating, and cooling the mass was 25.9625 g, calculate the mass of Ca in the water sample and the mass in dm^{-3}.

Summary

1. Gravimetric (or mass based) analyses are based on weighing products and/or reactants before and after some chemical reaction.

2. The solubility product, K_{sp}, for a compound A–B = $[A]^x [B]^y$.

3. The relative supersaturation is given by $(Q - s)/s$ where Q is the concentration of the solute at any given time and s is the solubility of the precipitate at equilibrium.

4. Precipitates may be collected via the use of Büchner funnels, sintered crucibles, and careful drying.

5. Widely used gravimetric determinations include: (i) the determination of chloride ion concentrations via precipitation with silver nitrate; and (ii) the determination of nickel contents within samples via the formation of precipitates of nickel dimethylgloxime (DMG).

6. Further gravimetric reactions include precipitation of a number of metal ions via the use of oxine (8-hydroxyquinoline) or sodium tetraphenylboron.

7. Thermal gravimetric analysis (TGA) may be used to follow the thermal decomposition of compounds such as $AgNO_3$, $AgCrO_4$, and $HgCrO_4$.

8. Derivative thermogravimetric analyses involve recording the slope (first derivative) of thermograms and in some cases this approach facilitates determinations that would otherwise prove problematic.

Further reading

Duval, C. (1963). *Inorganic thermogravimetric analysis.* Elsevier.

Erdey, L. (1965). *Gravimetric analysis.* International Series of Monographs in Analytical Chemistry, Vol. 7. Pergamon Press.

Hawkins, M. D. (1970). *Calculations in volumetric and gravimetric analysis.* Butterworth.

Rattenbury, E. M. (1966). *Introduction to titrimetric and gravimetric analysis.* Pergamon Press.

An introduction to the use of visible and ultraviolet light for analytical measurements

5

Skills and concepts

This chapter will help you to understand:

- The wave-like properties of electromagnetic radiation.
- The wavelength ranges across the electromagnetic spectrum (such as the ultraviolet (UV) and visible region of the spectrum).
- The quantized nature of UV and visible light and how this causes absorption effects.
- How to use the Beer–Lambert law for the calculation of analyte concentrations within unknown samples.
- The electronic transitional basis for molecular fluorescence effects and how fluorescence may be exploited for analytical purposes.
- Why absorption is a unit-less quantity.
- Why measured absorptions should only be trusted in the range of 0–2.
- The operation of tungsten filament lamps, hydrogen/deuterium lamps and the advantages each offer as light sources for UV–visible spectroscopy.
- The operation of: photo-tube, photo-multiplier, silicon photo-diode, and photo-voltaic cell based detectors and the relative advantages and disadvantages of each for use in UV–visible spectroscopy.
- The operation of single- and double-beam UV–visible spectrophotometers and their associated advantages and disadvantages.
- The wavelength ranges over which plastic, glass, and quartz cuvettes may be used.
- How to perform a number of practical colour-based complexation reactions for practical analyses.
- What is meant by an organic chromophore.
- What is meant by hypsochromic and bathochromic shifts and how these may be used to facilitate UV–visible determinations.

- How to relate intensities of fluorescent radiation to fluorescent analyte concentrations.
- How compounds act as fluorescent quenching agents.
- What is meant by optical activity and how this can be measured via polarimetry to determine the concentration of optically active compounds.
- How to relate the specific rotation of a compound and the measured optical rotation to the concentration of an optically active compound within an unknown sample.

5.1 An introduction to the use of visible and UV light and the electromagnetic spectrum

The interaction of light with different compounds offers many possibilities for performing both qualitative and quantitative measurements. Many chemical reactions generate vivid colours, which, as well as being fascinating, often provide sufficient information to perform an analysis. Colour changes may, however, be subtle or indeed difficult to distinguish due to the limitations our eyes have, and this is especially true if the results are to be compared from one day to another. In these and many other similar situations, an instrument may be used to study the way in which light interacts with a sample.

Visible light forms part of the electromagnetic spectrum (Fig. 5.1) with γ-rays at one end of the spectrum having wavelengths of the order of about 10^{-14} m, and radio-waves at the other end having wavelengths of 3×10^3 m or greater. Our eyes detect or 'see' radiation in a fairly narrow part of the spectrum that spans wavelength from about 400 to 750 nm. Radiation falling within this region of the electromagnetic spectrum is therefore classified as the 'visible' part of the spectrum. Many instruments designed to measure and quantify the interaction of visible radiation with matter have the capability, however, for operating at wavelengths that extend into the UV range of the electromagnetic spectrum, which covers wavelengths from around 400 down to 180 nm or so. This region is known as the 'ultraviolet' region since it is beyond the range of our vision

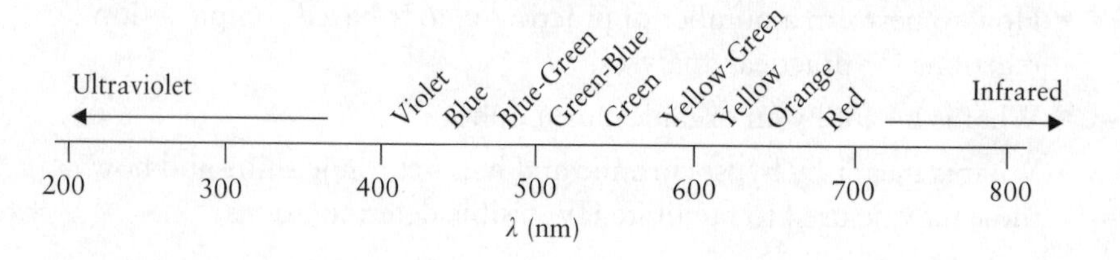

Figure 5.1 UV and visible region of the electromagnetic spectrum.

and extends from the violet region of the spectrum at the boundary of our vision.

The interaction of light with matter and the exploitation of colour is truly an everyday event that we often take for granted. We *perceive* each part of the visible electromagnetic spectrum as light of a different colour; these colours include red, orange, yellow, green, blue, indigo, and violet. The interaction of light with matter gives us the perception that objects are 'coloured'. Colour plays a very important role in many aspects of our lives and, for example, we use coloured traffic lights to help with traffic safety, watch colour television sets, decorate our houses, and choose clothes based on their colours—to cite but a very few cases.

We often use light in everyday life to determine the *quality* of a number of goods such as food. We might, by experience, expect vegetables such as runner beans, to display a pleasant and reasonably homogeneous green coloration, which we know corresponds to a certain level of freshness. By the same argument, we are likely to avoid those which show signs of decay and rotting which we detect or 'see' as having unpleasant dark brown patches. We also know that drinking water should be colourless if held up to the light and that the water might not be good to drink if we can see suspended or dissolved materials that absorb and scatter light thereby discolouring the water. Both of these tests are qualitative in nature.

We also perform some rather rudimentary semi-quantitative tests by eye. If we add milk to a cup of tea or coffee, we judge the correct amount to add based on the colour of the drink; we are using colour to estimate concentration. The judgement is obviously rather crude, but the principle is clear. The use of instrumentation to perform far more accurate estimates of the intensity of the colour of a sample forms the basis of a ***colorimetric measurement.***

Before we consider the details of any spectroscopic analysis, we should consider the reason why some compounds appear coloured. Our eyes detect the light that impacts on a retina. If green light reaches the eye, the eye perceives the colour green. If the eye perceives the colour red, then it is responding to red light entering the iris of the eye and striking the retina. The light may come from a light emitting source, be reflected from the surface of some object, or reach the eye having been transmitted through a transparent object that absorbs all other wavelengths than those that correspond to 'red' wavelengths.

When light impacts upon a material it may be reflected, be absorbed, or simply pass through unaffected in which case it is said to be *transmitted*. Many light sources are said to be 'white'—that is, they emit radiation across the entire visible region of the electromagnetic spectrum. It is quite common for some wavelengths of light to be absorbed by a material, while others pass through unaffected. Our eyes detect and perceive the colour of the light which *is not* absorbed and, by default, is therefore transmitted. If light is reflected from the surface of a material, the same argument applies. Some of the light striking the object will be absorbed. Our eyes, however, detect and therefore perceive the light that is reflected—that is, the component that is not absorbed.

If we perceive a fruit cordial drink to be orange in colour this is because the drink absorbs all of the visible electromagnetic spectrum except for the orange light (~600–650 nm), which is transmitted unaltered through the drink. Similarly, if we see an object such as a solid ball as orange, this is because most of the visible light striking the surface is absorbed, except for the orange light that is reflected. It should be remembered that a totally reflective surface is a mirror and conversely a totally absorbing surface appears matt black.

To conclude, the basis of absorption, transmission, and/or reflection of light is due to interactions of radiation with the molecules that go to make up a material and we consider these in greater detail throughout this chapter.

5.2 The quantization of light and electronic energy levels

In order to understand how light interacts with matter, we must first consider the atomic, and in particular, the electronic structure of molecules. Each atom consists of a positively charged nucleus that is surrounded by a series of electrons. Electrons travel around the nucleus in areas of space known as orbitals and possess differing energies. The choice of orbitals in which the electrons reside depend on their energetic states. The energetic states of orbitals and therefore of the electrons that reside in them correspond to a series of separate and well-defined energy levels and are said to be ***quantized***. It therefore follows that the promotion of an electron from one energy level to another must correspond to a quantized energy change.

The promotion of a valence electron from one orbital to another involves the absorption of radiation, and this is normally in the UV–visible range of the electromagnetic spectrum. Electrons are said to be promoted from a ***ground state*** to an ***excited state***.

In some circumstances, light may be primarily considered to behave as a wave while in others appears to possess ***particle-like behaviour***. In one model, light may be considered to consist of a series of discrete 'packets' of energy or ***photons***. The energetic state of each photon is quantized and is proportional to the frequency of the radiation, ν. The energy of the photon may be quantified from the product of the Planck constant, h, and the frequency, ν, Eqn (5.1):

$$E = h\nu \tag{5.1}$$

Since the frequency of the radiation is given by

$$\nu = \frac{c}{\lambda} \tag{5.2}$$

it follows that the energy of a photon may also be given by

$$E = \frac{hc}{\lambda} \tag{5.3}$$

If visible or UV light is to cause the promotion of a valence electron from the ground to an excited state, the energy of a photon must correspond exactly to the energy gap associated with the electronic transition.

Photons that are insufficiently energetic or indeed too energetic will not cause the electronic transition.

5.3 Absorption: when photons give up their energy

Photons of UV and visible light may sometimes impart their energy to materials by interaction with individual atoms or molecules. Energy is imparted to the atoms or molecules, causing the excitation of valence electrons. Molecules with excited electronic states represent an unstable state and will relax by allowing their electrons to fall to the ground state as soon as possible (ca. 10^{-16} s). The energy that is lost in a transition of this type is normally dissipated as heat. It follows that objects that are irradiated with light will often adsorb radiation and so be heated and we all know, for example, that objects placed in sunlight are warmed. There are circumstances, however, when some of the energy may be dissipated by heat while some is lost by the emission of a photon (light) of a longer wavelength. In this situation, the material emits light of a wavelength having been irradiated by light of another wavelength. This effect is known as ***fluorescence*** and we shall look at this phenomenon in Section 5.13.

5.4 Absorption: how much radiation is absorbed?

Let us consider a simple transparent cell through which an incident beam of light of intensity, I_0 is passed; Fig. 5.2. Some of the light is absorbed and therefore a transmitted beam of light of a lower intensity, I, emerges from the cell. Light is a form of energy and therefore the intensity of a light beam is a measurement of power (energy per unit of time) and therefore

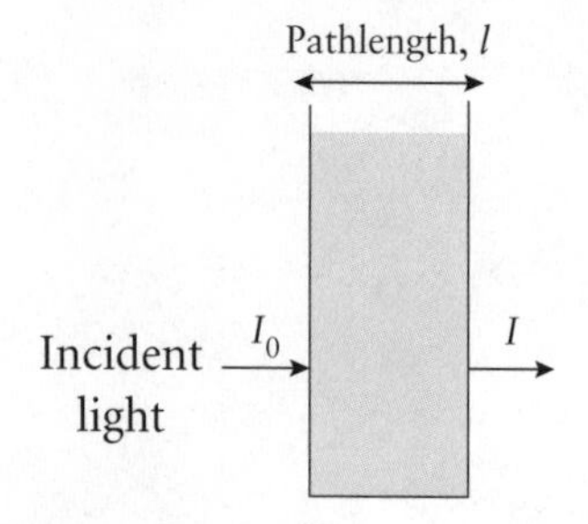

Figure 5.2 Absorption of light by an analyte sample within a cuvette.

has the units of $J\,s^{-1}$ or watts (W). The absorbance, A, is defined as the $\log_{10}$ of the ratio of the incident to transmitted light beams, Eqn (5.4):

$$A = \log_{10}\left(\frac{I_0}{I}\right) \tag{5.4}$$

Since I_0 and I are both measured in units of power (W), their units cancel each other out and so ***absorption is a unit-less quantity***. The implications of absorption being defined from a logarithmic expression are worth noting as we shall see shortly.

If I and I_0 are equal, then it follows that the intensity of the emergent beam is equal to the incident beam and that I_0/I equals 1. The $\log_{10}$ of 1 is 0, which corresponds to an absorbance of zero. Now, an aborbance of 1 means that (I_0/I) equals 10 (i.e. 10/1) and that 90% of the light is being absorbed. Similarly, an absorbance of 2 means that I_0/I equals 100; in this case 99% of the incident light is absorbed and only 1% is transmitted. It is therefore normally only feasible with most instrumentation to *measure* absorbance on a scale of 0 though to 2, since absorbances of greater than 2 correspond to an immeasurably small fraction of the incident radiation being transmitted.

Absorbance is measured by comparing the intensity of an incident and transmitted light beam through a sample. The intensity of the light beams may be detected by devices such as a photo-multiplier tubes. The ratio of the incident and transmitted light beam intensities may then be used to calculate the absorbance of the sample.

EXAMPLE 5.1 A coloured solution is placed within a UV–visible spectrophotometer. At 465 nm the sample shows an absorbance of 0.79. Calculate the percentage of light that is being absorbed.

Method

Use the relationship absorbance $= \log_{10}\frac{I_0}{I}$ to calculate the percentage of incident light absorbed.

$$A = 0.79 = \log_{10}\frac{I_0}{I}$$

Therefore,

$$6.166 = \frac{I_0}{I}$$

If $I_0 = 1$ then $6.166 = \frac{1}{I}$

Therefore,

$$I = \frac{1}{6.166} = 0.1622$$

16.22% of the incident light is being transmitted.

It follows that (100 − 16.22) = 83.78% of the incident light is being absorbed.

EXAMPLE 5.2 A solution within a UV–visible spectrophotometer shows an absorbance of 0.67 at a wavelength of 560 nm. What is the percentage of radiation that is transmitted?

Method

From the relationship absorbance $(A) = \log_{10}\frac{I_0}{I}$ calculate the percentage of incident light absorbed.

$$A = 0.67 = \log_{10}\frac{I_0}{I}$$

Therefore, $4.677 = \frac{I_0}{I}$

If $I_0 = 1$ then $4.677 = \frac{1}{I}$

Therefore,

$$I = \frac{1}{4.677} = 0.2138$$

21.38% of the incident light is being transmitted.

5.5 The Beer–Lambert (or absorption) law

The collision of a photon of suitable energy with the appropriate molecule results in absorption of light. It follows that if a greater number of molecules are placed in the path of the light beam, then there will be a greater chance of a collision occurring and therefore of absorption. Our everyday experience knows this to be true. Think of a glass of blackcurrant cordial that we look through; the drink appears purple in colour. We know that certain wavelengths are absorbed while others are transmitted and pass through easily and it is this that gives the drink its characteristic colour. Now if we hold another similar glass of the drink behind the first, the colour appears twice as intense. We know that less light is able to pass through the two glasses than the one alone. It is clear that the probability of a collision between a photon and a light absorbing molecule has increased. By the same argument, we know that if we dilute the drink with water, its colour intensity will decrease and this is because we have decreased the concentration of the molecular absorbing species. In this situation, we have decreased the probability of a photon colliding with a light absorbing molecule.

Let us imagine a simple transparent cell or ***cuvette*** (Fig. 5.2) containing a solution that absorbs radiation of a particular wavelength.

From the previous discussion it follows that the absorption will be proportional to the concentration of the absorbing molecular species and also to the path length through which the light has to travel.

The absorption or ***Beer–Lambert*** law relates the absorption of most molecular species to the concentration, c_n, the path length, l, and the molar absorptivity, ε, Eqn (5.5):

$$A = \varepsilon c_n l \tag{5.5}$$

ε is sometimes known as the ***extinction coefficient.*** The Beer–Lambert law is sometimes expressed in terms of transmittance, T, where $T = 1/A$. Unfortunately, there are differing ways of defining the units of ε and l. ε is most commonly described in terms of $dm^3\ mol^{-1}\ cm^{-1}$ in which case, l, the path length must be quoted in terms of cm. Some reference books, however, will quote ε values in terms of $dm^3\ mol^{-1}\ m^{-1}$ in which case the path length, l, must also be quoted in terms of, *m*. A path length of 1 cm is normally chosen to simplify the calculation of the absorbance or molar absorptivity. The Beer–Lambert law is one of the most widely used relationships within analytical chemistry and is at the heart of the majority of UV and visible quantitative analyses.

See the Box on page 136 for a description of a practical method that applies the Beer-Lambert Law.

The Beer–Lambert law holds for the majority of compounds over a wide range of experimental conditions. It should be noted, however, that the absorption of light is highly wavelength specific. We have already seen that radiation of a specific wavelength must be supplied if an electronic excitation is to occur. It should be remembered that this corresponds to the absorption of light.

Molar absorptivities are therefore quoted for individual compounds at a specific wavelength and pathlength.

The majority of compounds will only absorb radiation at specific wavelengths and it this that gives rise to their colour. If the compound is coloured then there will be at least one and even a number of differing wavelengths which show maximal absorbances.

A plot of *absorbance* versus *wavelength* is known as a ***UV–visible spectrum*** and may be measured by a ***UV–visible spectrophotometer***. Most UV–visible spectrophotometrs will allow the scanning of a wavelength range. The UV–visible spectrum of a potassium permanganate ($KMnO_4$) solution is shown in Fig. 5.3. $KMnO_4$ absorbs strongly across a wide range of wavelengths but has little absorption in the red and blue regions of the visible spectrum. Light corresponding to these colours passes through the solution with relative ease and is thus transmitted. The mixing of blue and red transmitted light is seen as purple, which is how our eyes perceive the solution to be coloured. The wavelength of maximum absorption is known as the λ_{max} and is usually used as the wavelength which is quoted within data values for the molar absorptivity, ε (Section 5.4).

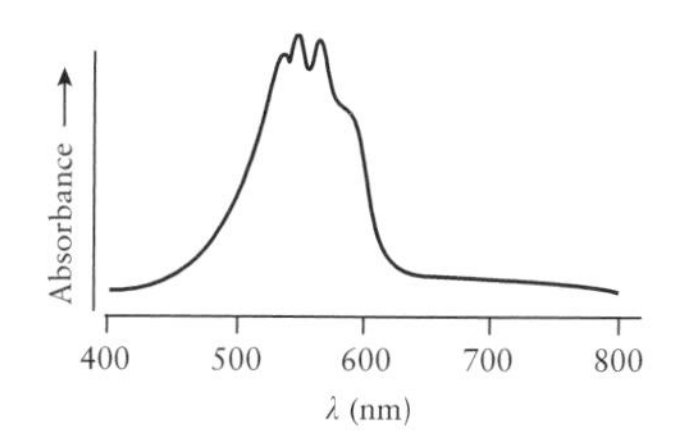

Figure 5.3 Absorbance spectrum for $KMnO_4$ in water.

Since the absorption of light is so wavelength specific, it is crucial that the wavelength of as well characterized and indeed as narrow a wavelength range as possible is used to irradiate the sample under study.

EXAMPLE 5.3 A 0.1 M solution within a 1 cm path length placed within a UV–visible spectrophotometer shows an absorbance of 0.95. Calculate the molar absorptivity for this compound.

Method

Calculate molar absorptivity from the relationship $A = \varepsilon cl$

$$A = \varepsilon(0.1 \times 1)$$

It follows that: $0.95 = 0.1\varepsilon$

Therefore $\varepsilon = 10 \times 0.95 = 9.5\ \mathrm{dm^3 mol^{-1} cm^{-1}}$

EXAMPLE 5.4 A compound with a molar absorptivity of $3578\ \mathrm{dm^3\ mol^{-1} cm^{-1}}$ (at 650 nm) exhibits an absorbance of 0.78 when placed within a 1 cm path length cuvette in a UV–visible spectrophotometer. Calculate the concentration of the compound.

Method

Calculate the concentration from $A = \varepsilon cl$

$$0.78 = 3578(c \times 1)$$

Therefore,

$$c = \frac{0.78}{3578}\ \mathrm{mol\ dm^{-3}}$$

Concentration of the compound $= 2.18 \times 10^{-4}\ \mathrm{mol\ dm^{-3}}$

5.6 The nature and use of UV and visible absorptions—chromophores

The UV–visible spectrum of a compound may often be used to identify its presence within a sample. Many compounds, and especially those which are highly coloured, absorb radiation over characteristic and often comparatively narrow wavelength ranges. Differing molecular groupings give rise to the absorption of light at characteristic wavelengths and are known as ***chromophores***. Chromophores are functional molecular groups that cause compounds to be coloured—that is, to absorb radiation at particular wavelengths.

Chromophores are moieties that possess electrons capable of being readily promoted by the absorption of UV or visible light. Many highly coloured compounds contain either a ***transition metal ion*** or a number of ***unsaturated*** carbon–carbon bonds. Very small changes in structure may often, however, give rise to extremely different absorption phenomena. It is changes of this type which often form the basis of many pH sensitive indicators such as phenolphthalein (Fig. 5.4). Within environments of ($pH > 8.1$), phenolphthalein resides in a de-protonated structure that absorbs light across a broad band of the visible spectrum; in this state the indicator appears red. If the pH falls below ~8.1, the structure becomes protonated by conversion of the two carbonyl (–C=O) groups to hydroxyl groups (–C–OH) and this corresponds to a colourless state. The hydroxyl moieties are more electron withdrawing towards the aromatic carbon ring than the electron-rich $C-O^-$ moiety. The subtle change in the structure of the chromophores causes the absorption band to shift to shorter wavelengths.

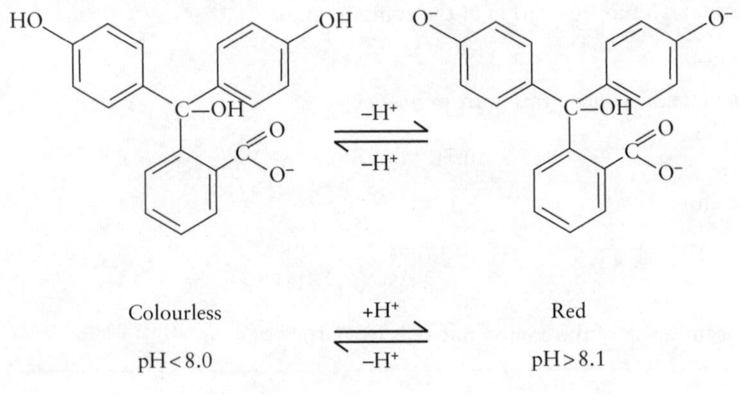

Moieties are functional groups within molecules that impart certain features—in the case of chromophores these cause absorption in the UV–visible range and so impart colour.

Figure 5.4 Structural and colour-change properties of phenolphthalein with pH..

5.7 Light sources and monochromators

5.7.1 Light sources

The irradiation of a sample for UV or visible spectroscopy requires a light source with a constant output intensity. The light source should also be sufficiently intense so as to allow sufficient transmitted radiation to be detected when the absorption falls within a range of 0–2. The sample should at any one time be irradiated with as ***monochromatic*** a source of radiation as possible (i.e. as narrow a wavelength range as possible). A monochromator is the most commonly used device to select a wavelength of light for the irradiation of a sample.

The radiant power of many light sources increases essentially exponentially with the applied voltage and, consequently, even very small fluctuations in the applied electrical supply voltage may cause significant variations in the intensity of the incident radiation and for this reason it is common practice to employ voltage regulators in the light source power line.

Tungsten filament lamps

The majority of UV and visible spectrophotometers employ tungsten filament lamps, Fig. 5.5(a), to supply radiation in the wavelength range of around 320–2500 nm, which covers most of the visible part of the spectrum. The energy output characteristics with respect to wavelength are shown in Fig. 5.5(b). It can be seen that the power output of lamps such as these dramatically deceases as the UV region is approached. The normal

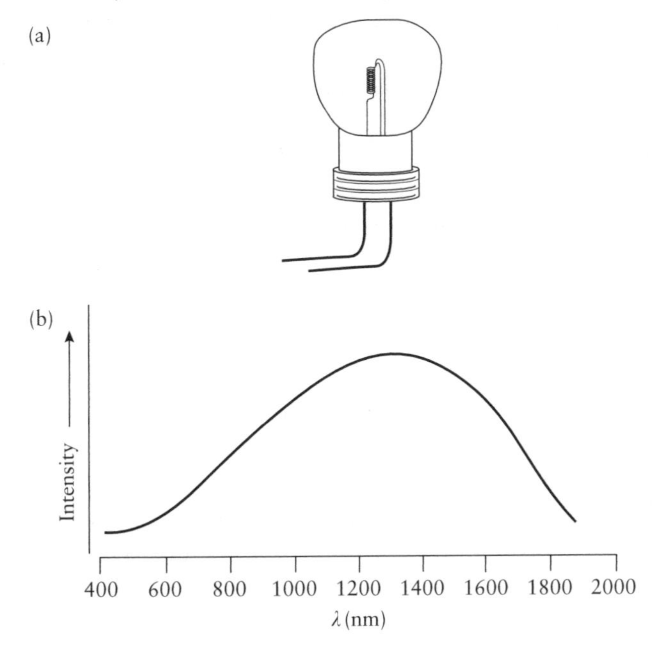

Figure 5.5 (a) Tungsten filament lamp and (b) its emission spectrum output.

operating temperature of tungsten filament lamps may be in the range of ~2900–3000 K.

Tungsten–halogen lamps contain small quantities of iodine, and the lamp is enclosed within quartz (as opposed to glass) housings. The halogen gas allows the temperature of the lamp to be raised to about 3500 K and this permits the intensity of the output radiation to be increased—as well extending the output down to ~190 nm, which extends well into the UV region. The quartz envelope in turn allows the transmission of the UV radiation (glass, in contrast, would absorb and therefore block UV light). Despite the higher operating temperatures, tungsten–halogen lamps typically have operating lifetimes that are double those of standard tungsten filament lamps. The lifetime of both of these lamps is limited due to the sublimation of the tungsten, W, from the filament. If trace amounts of iodine are introduced then sublimed tungsten molecules react with iodine to give WI_2 molecules which diffuse back to the hot filament, where they decompose to re-deposit metallic tungsten once more. Tungsten–halogen lamps are more expensive to manufacture, although their greater longevity and performance normally more than justify their cost.

It is extremely important that tissue paper or gloves are used to handle tungsten filament or tungsten–halogen lamps, since even tiny traces of grease from skin may cause tiny fractures of the glass or quartz casings at the extreme operating temperatures reached, and this shortens the lifetime of the lamp.

Hydrogen and deuterium lamps

Many instruments employ a tungsten filament lamp for the visible and fringe UV–visible ranges in conjunction with a hydrogen or deuterium lamp to produce high-intensity UV radiation.

The electrical excitation of hydrogen or deuterium at low pressure produces a continuous spectrum of UV radiation. Hydrogen (or deuterium) gas may be excited by electrical energy to produce two hydrogen atoms with the release of a photon of energy, Eqn (5.6):

$$H_2 + E_e \rightarrow H_2^* \rightarrow H + H + h\nu \tag{5.6}$$

The total excitation energy input, E_e, must be distributed between the two hydrogen atoms and the photon. The energy distribution between the two atoms is random. If two hydrogen atoms of low energy are produced, it follows that the photon will be highly energetic. Conversely, if the hydrogen (or deuterium) atoms are highly energetic, the photon of light emitted will be of a correspondingly lower energy value. The result is that deuterium or hydrogen lamps both give uniquely uniform outputs in intensity of radiation over a wavelength range of 160–375 nm.

At longer wavelengths, the lamps may produce emission lines that are superimposed on the otherwise stable output. Many spectrophotometers employ a tungsten or tungsten–halogen lamp for supplying radiation of wavelengths longer than 360 nm and a hydrogen or deuterium lamp for wavelengths below this value.

5.7.2 Monochromators

All of the light sources used within UV–visible spectrophotometers simultaneously produce a spectrum of radiation across a wide range of wavelengths. Photo-multiplier tubes and other radiation detectors are indiscriminatary in detecting radiation of different wavelengths. It therefore follows that if a spectrum is to be measured then some method must be employed to select as narrow a wavelength range as possible to irradiate the sample with. It is important to note that it is impossible to truly select one wavelength alone and that is why we say we will select as narrow a wavelength range as possible. The wavelength of the radiation source will tend to follow a Gaussian distribution around a mean value of wavelength known as the ***nominal wavelength***. The ***effective bandwidth*** is defined as the wavelength range which corresponds to the half peak height width of the wavelength distribution profile; Fig. 5.6.

There are many different designs for the construction of monochromators: low specification instruments sometimes use ***optical filters*** to select wavelengths with a bandwidth of around 30 nm or so. It is far more common, however, to employ a ***monochromator*** for the wavelength selection.

A collimator is a device for producing a parallel beam of radiation

Monochromators use a series of lenses, mirrors, slits and windows together with either prisms and/or diffraction gratings to isolate a narrow band of wavelengths. There are many designs for monochromators. However, we shall only consider in detail two of the most popular designs that are based on ***refracting prisms*** and ***diffraction gratings***.

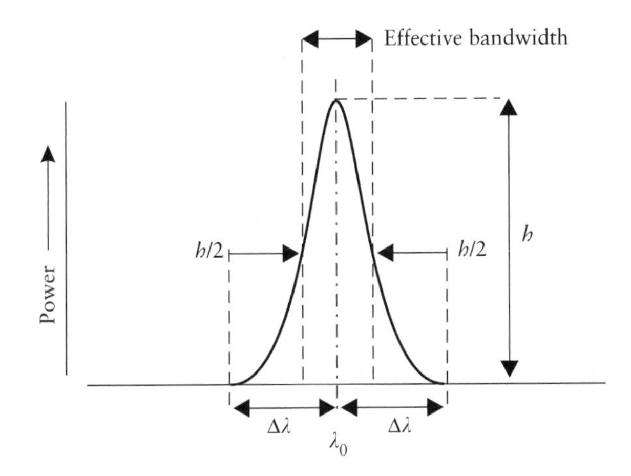

Figure 5.6 Nominal wavelength λ_0 and the effective bandwidth.

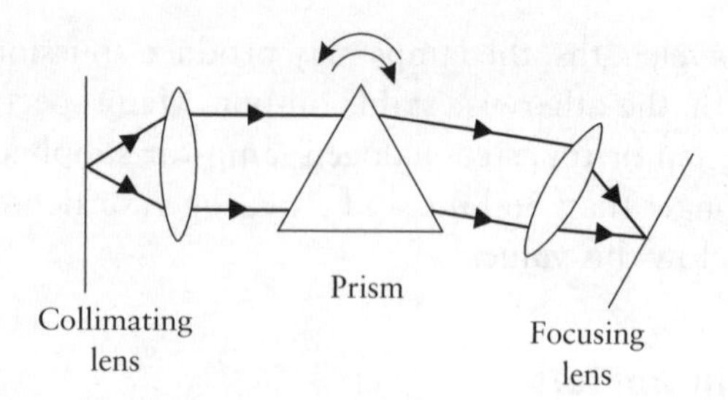

Figure 5.7 Simplified schematic of a prism monochromator.

Monochromators based on refracting prisms

Monochromators of all types tend to be encased in discrete housings within the spectrophotometer to prevent the ingress of dust and other contaminants. In refracting prism monochromators, the 'white' light enters the monochromator via an ***entrance slit*** before being collimated by a ***collimating lens***; Fig. 5.7. The light then passes through a ***refraction prism*** that disperses the light into its component wavelengths. The light is then focused by another lens towards an exit slit, which is situated at the focal plane. The prism is rotated by means of a stage and a stepper motor to select radiation with different frequencies to pass through the exit slit.

Monochromators based on diffraction grating

While the light path through a refracting prism-based monochromator might initially appear to be rather different to that of a diffraction grating, the principle is fairly similar.

White light passes through an entrance slit and is focussed towards a diffraction grating via a concave mirror; Fig. 5.8. The diffraction grating disperses the light into its component wavelengths and reflects the light onto a second concave mirror. The grating is mounted on a stage which may be rotated via a stepper motor. Light may then be reflected and

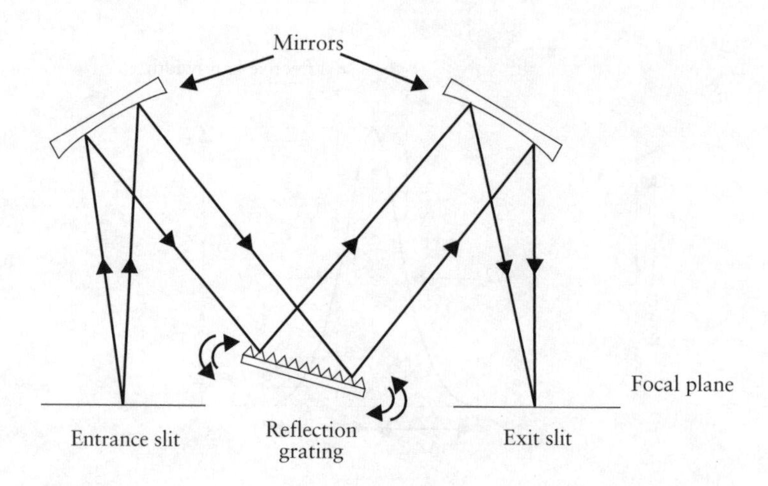

Figure 5.8 Simplified schematic of grating monochromator.

focussed by this concave mirror towards the exit slit. As the grating is rotated different wavelengths can be selected.

5.8 The detection of light—photon detectors

If the absorption of an analyte is to be determined, the intensity of the transmitted light must be monitored and there are several ways in which this can be achieved.

The intensity of electromagnetic radiation may be measured by: (a) the photo-emission of electrons; (b) the electronic excitation of valance electrons following the absorption of electromagnetic radiation; or (c) the measurement of the heat imparted to a material as a result of the absorption of electromagnetic radiation.

Within the UV–visible range the most popular approaches are based on photo-emission and/or electronic transition principles, with both of these techniques essentially counting photons and hence monitoring the intensity of the light. Four types of detector based on these approaches are commonly employed within UV–visible instrumentation namely; photo-tubes, photo-multiplyer tubes, silicon photo-diodes, and photo-voltaic cells, each of which are described in the following sections.

5.8.1 Photo-tubes

A photo-tube consists of an evacuated tube with a quartz window, behind which is placed a large cathode; Fig. 5.9. The cathode is coated with a layer of photo-emission material such as a metal oxide or alkali metal. A smaller wire anode is situated in front of a cathode and a polarizing potential of 90 V or greater is imposed between them. Photons enter the tube via the quartz window and strike the cathode. This gives rise to the photo-emission of electrons which travel towards the anode. The

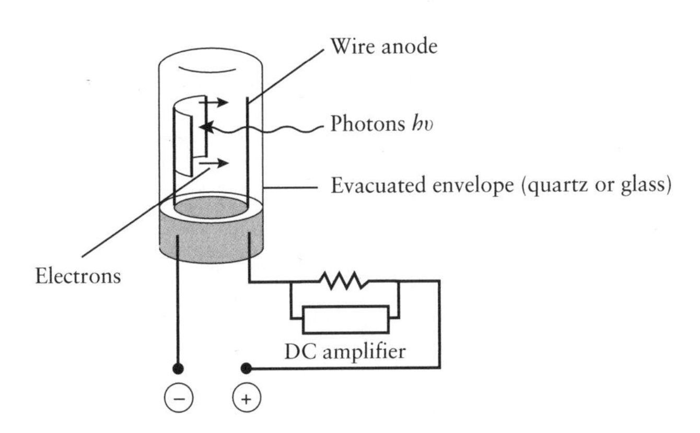

Figure 5.9 Photo-tube and monitoring circuit.

photo-current that passes between the electrons may be measured and so related to the intensity of the incoming light.

5.8.2 Photo-multiplier tubes

Photo-multiplier tubes operate in essentially the same way as photo-tubes, but produce a cascade of photo-emitted electrons via a series of accelerating and electron emitting electrodes. Photons again enter an evacuated tube by means of a quartz window; Fig. 5.10. The photons strike the cathode, which again causes the emission of electrons, which in this case are accelerated towards the first of a series of ***dynodes*** that are polarized at +90 V relative to the cathode. The electrons strike the dynode, which causes a series of further electrons to be emitted, which are then accelerated towards another dynode that is polarized at +90 V relative to the first dynode. The process continues till the electron cascade is finally collected at the collecting electrode. This process, known as the ***cascade effect***, causes some 10^6 to 10^7 electrons to be collected for every photon entering the tube.

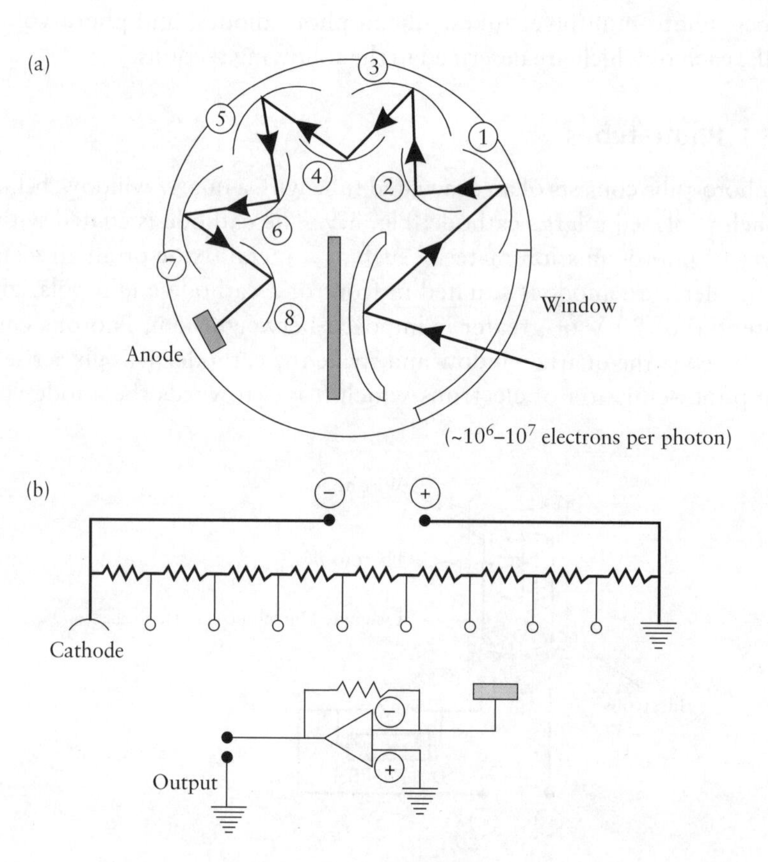

Figure 5.10 (a) Photo-multiplier and (b) associated circuitry.

5.8.3 Silicon photo-diodes

Photo-diodes comprise specially constructed silicon in which the conductivity may be modulated by UV or visible light illumination. Silicon, a Group IV element is a semiconductor, and as such, its conductivity is less than that of a metal but greater than that of an insulator. Each silicon atom is covalently linked within a covalent lattice super-structure to four of its neighbours. Thermal agitation at room temperature allows an occasional electron to leave a silicon atom to move within the lattice. The unoccupied position is known as a ***hole***, which effectively represents a positive charge. Conduction occurs by the movement of electrons and holes in opposite directions. The conductivity may be greatly increased by the addition of trace amounts of either Group III or Group V elements. The addition of Group III elements produces a so-called p-type semiconductor that is rich in holes. The addition of a Group V element produces an n-type semiconductor rich in electrons.

If a piece of n-type silicon is joined to a piece of p-type silicon, a so called **p–n *junction*** diode is formed. These p–n junctions conduct electricity if polarized in one direction, (called the ***forward bias***) but block the passage of current if they are ***polarized with a negative bias***. The forward biasing of a p–n junction involves the *n* region being negatively polarized and the p region being positively polarized. An excess of electrons is made available to the n-type semiconductor. Similarly, electrons are drawn away from the p-type semiconductor, which creates more holes in this region. Holes and electrons neutralize each other in the vicinity of the p–n junction and conduction is permitted as more electrons and holes are effectively made available. By contrast, if the p–n junction diode is reverse biased, then both the holes and electrons move away from the p–n junction region to form a so called ***depletion layer***, which now becomes non-conductive; in this arrangement the p–n diode impedes the passage of current.

Silicon photo-diodes are specially adapted p–n junction diodes with optically transparent widows to allow the illumination of the p–n junction region by UV or visible light. Photons that pass through the window and are absorbed in the vicinity of the p–n junction may, if sufficiently energetic, cause the excitation of electrons to form holes and free electrons; Fig. 5.11. The generation of electrons and holes in the depletion layer causes a significant increase in the conductivity of the diode, which is used to measure the intensity of the incident radiation.

5.8.4 Photo-voltaic cells

Photo-voltaic cells are the simplest but least sensitive type of cell used for the detection of visible light. They are insensitive to—and therefore cannot be used for the detection of—UV light. Photo-voltaic cells also

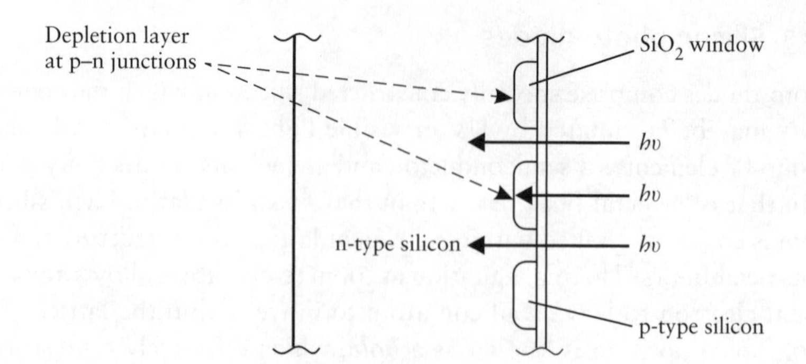

Figure 5.11 Schematic of silicon photo-diode.

suffer from fatigue, that is, their response fades with constant illumination. Despite these disadvantages, photo-voltaic cells are used in simpler instrumentation due to their simplicity and the lack of need for an external power supply.

Most photo-voltaic cells consist of a copper or iron electrode coated with a semiconducting material such as copper(I) oxide or selenium, which, in turn, is coated with a film of gold, silver, or lead sufficiently thin so as to be optically transparent. This metallic film is polarized with respect to the copper or iron electrode and acts both as an optical window and as a second electrode. Light that reaches the semiconductor causes the formation of electrons and holes which migrate away from each other and towards the two electrodes. If the electrodes are connected to a low resistance circuit, the current that flows may be related to the intensity of the incident light.

5.9 Spectrometers, spectrophotometers, and UV–visible cells

There is often some confusion as to the difference between a spectrometer and a spectrophotometer.

A ***spectrometer*** is a monochromator equipped with a fixed slit at the focal plane. A spectrometer equipped with a phototransducer is known as a ***spectrophotometer***. In turn, a spectrophometer that is capable of measuring the absorbance across a range of wavelengths by linearly varying the incident wavelength is known as a ***scanning spectrophotometer***.

5.9.1 Single-beam spectrophotometers

As the name suggests, ***single-beam spectrophotometers*** use a single light beam in order to irradiate the cell in an arrangement as depicted in

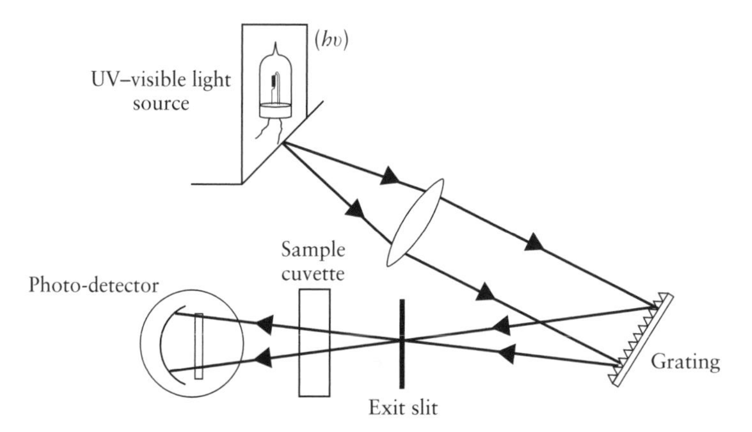

Figure 5.12 Schematic of a single-beam spectrophotometer.

Fig. 5.12. The light enters the sample and a phototransducer monitors the transmitted radiation as it emerges from the sample. There is a problem, however, with this simple arrangement that must be addressed. The cell and the solvent in which the sample is dissolved will both absorb radiation at every wavelength to a finite extent. We wish to record the absorbance spectrum of the analyte and not the spectrum superimposed upon a background. Instruments of this type necessitate the measuring of a ***baseline spectrum***. The baseline spectrum is normally recorded by placing a cell filled with the appropriate solvent (minus analyte) into the spectrophotometer. The baseline spectrum is then subtracted from all subsequent spectrums for samples containing the analyte; the spectrum that is produced now corresponds to the absorption of the analyte alone and is sometimes known as a ***normalized spectrum***. Most modern spectrophotometers, however, now store baselines within the memory of a computer and perform this function electronically.

5.9.2 Double-beam spectrophotometers

Double-beam spectrophotometers, as shown schematically in Fig. 5.13, employ two light beams of equal intensity together with two photo-multipliers to record and subtract the baselines from spectra. One light beam irradiates the cell containing the analyte. The other irradiates a cuvette containing the appropriate solvent only. In essence, two spectra are recorded simultaneously as the wavelength of light is scanned through the desired range. The baseline spectrum is then subtracted from the spectrum corresponding to the analyte sample and a normalized UV–visible spectrum is obtained.

Several assumptions must be made if we use this approach. First, the two light beams must be of exactly the same intensity. Second, the two cells must possess exactly the same absorptivity, and for this reason, should be cells of the same make and type. Third, the solvent must be

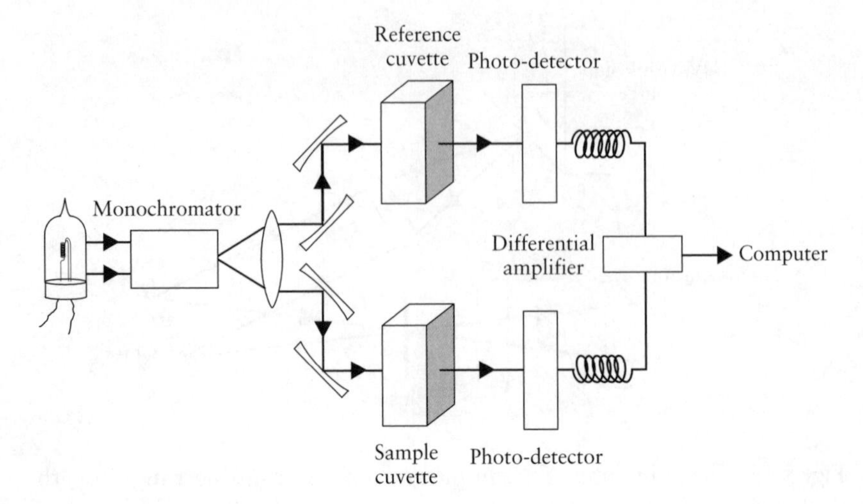

Figure 5.13 Schematic of a double-beam spectrophotometer.

exactly the same in each cuvette. Finally, it must be assumed that the photo-transducers exhibit exactly the same sensitivities.

5.9.3 The use of UV–visible cells and cuvettes

An optically transparent cell with a volume of a few cubic centimetres (known as a *cuvette*) is used to contain the analyte samples. The cuvette is normally designed with an internal cross-sectional area of 1×1 cm to give an optical path length of 1 cm, although cells of differing dimensions may be purchased for specialized applications. The cuvette normally has a height of several centimetres to allow easy insertion and removal from the spectrophotometer.

The cell is normally provided with two facing optically transparent faces and two frosted optically opaque faces. The opaque faces are designed to provide surfaces that may be handled without placing fingerprints on the faces through which the optical beam must pass.

Fingerprints, grease, or other trace contaminants may have a profound effect on the spectrum obtained, so it is imperative to keep the optically transparent faces of the cuvette as clean as possible. For this reason, cuvettes should always be polished with a piece of lens tissue every time they are handled.

For similar reasons, it is most important that if a spectrum is taken using a double-beam spectrophotometer, that matched optical cells are used. The optical properties of matched cells may be verified by running a baseline spectrum with the two cells in the path of each beam. If the cells are well matched and contain only solvent, a blank spectrum should be obtained, that is, a constant absorbance $A = 0$.

Cuvettes are typically made of one of one three main classes of material and these are:

1. optically transparent plastics;
2. optical grade glass;
3. fused silica or quartz cells.

Optically transparent plastics

For many purposes, an optically transparent plastic cuvette suffices for the measurement of a visible spectrum within the wavelength range of ~480–900 nm. Beyond these limits, the cuvette may begin to absorb radiation to a significant extent and should therefore not be used. The precise optical properties of cells may vary from one manufacturer to another, though the limits should be supplied with the cuvette. Plastic cuvettes are the least expensive type of cells that may be used for determinations in the visible region, although care should be taken since these cells can easily become scratched, rendering them unsuitable for further use.

Optically transparent glass cells

Many glasses have slightly wider optically transparent wavelength ranges of ~400–900 nm in comparison to optical plastics. Glass has, moreover, the additional advantage of being far more resistant to scratching than plastic.

Fused silica and quartz cells

Fused silica and quartz cells offer the optimal optical properties available and allow UV and visible spectra to be taken throughout the full range of modern instrumental capabilities (approximately 190–750 nm). Fused silica and quartz cells are, unfortunately, very much more expensive than their glass counterparts and their use is primarily reserved for situations where the UV part of the spectrum must be recorded (i.e. 190–400 nm).

5.10 Qualitative UV–visible measurements

5.10.1 Qualitative applications of UV–visible spectroscopy

Comparison of the UV–visible spectra of the two structures of phenolphthalein would easily allow us to deduce whether the molecule is in the protonated or de-protoanted form. It is clear therefore that UV–visible spectroscopy can be used both to *identify the presence* of a particular molecular species as well as for an analytical 'fingerprinting' technique.

Analyte samples, however, rarely contain a single absorbing species. The UV–visible spectrum of real samples is normally the summation of several molecular absorption spectra and the absorption at any one particular wavelength will be equal to the sum of the individual absorptions of each component within the solution. For simple solutions, it is often possible to identify the presence of individual solutes from wavelength maxima of different absorption peaks, although it should be noted that the identification of compounds should normally be confirmed by infrared (IR) spectroscopy, nuclear magnetic resonance (NMR), or melting point data.

5.10.2 The effect of different organic chromophores

Organic moieties that act as chromophores possess electrons that may be easily excited and thereby promoted to higher energy levels via the absorption of light in the UV or visible ranges. Organic chromophores often contain one or more double or triple bonds and/or an aromatic ring. Many chromophores absorb radiation over a wavelength range of 20 nm or greater, which may cause the overlapping of one or more absorption peaks. The exact wavelength range over which the absorption occurs is also dependent upon the electron withdrawing/donating properties of the rest of the molecule, so it is impossible to identify with certainty the presence of a specific moiety in the same way that one can with IR spectroscopy (see Chapter 12).

A few examples of some of the most common chromophores are given in Table 5.1.

Table 5.1 Examples of UV or visible absorptions for a number of organic functional moieties acting as chromophores

Chromophore	Functional groups	Typical λ_{max} values (nm)
Alkene	–CH=CH–	175–185
Alkyne	–C≡C–	175–195 and 220–30
Amines	$-NH_2$	195–200
Carbonyl	–CH=O	186 and 280
Nitro	$R-NO_2$ (Nitro-alkanes)	280
Nitroso	(Nitrosamines) N=O	300 and 665
Aromatic	(Benzene)	200

5.10.3 UV and visible light absorption by inorganic compounds

Many inorganic compounds absorb UV and/or visible radiation and possess absorption spectra with broad and frequently overlapping absorption spectra.

Compounds of the first two transition series are among the most highly coloured inorganic compounds. Absorption by these compounds involves transitions of electrons between the unfilled and filled d-orbitals, and thus the wavelength at which absorption occurs depends on the atomic number, the oxidation state of the metal, and the ligand to which it is bonded. The detailed treatment of the colour chemistry of inorganic compounds is beyond the scope of this book although the interested reader is referred to other works.

5.10.4 Charge transfer UV and visible light absorption processes

Many inorganic and organic compounds absorb UV or visible radiation due to charge transfer processes, and are thus known as ***charge transfer complexes***. ε_{max} values are frequently in the order of 10 000 $dm^3\,mol^{-1}\,cm^{-1}$ or more, which makes them both highly coloured and easy to quantify even at very low concentrations.

A charge transfer complex contains an electron donor group together with an acceptor group. Upon absorption of light, an electron is transferred from an orbital largely associated with the donor group to an orbital largely associated with the acceptor grouping. This behaviour contrasts with the absorption of an organic chromophore in which electrons are associated within *shared* molecular orbitals.

In many examples, a metal ion acts as the electron acceptor. One of the most familiar examples of this type of absorption is potassium permanganate $KMnO_4$ that appears purple in aqueous solution.

5.10.5 The choice and effect of using different solvents

The overwhelming majority of UV or visible analyses demand that the analyte be dissolved within a solvent. The solvent must of course first and most importantly solvate the analyte so that it is distributed homogeneously in the path of the incident radiation beam. Water will often be the solvent of choice, however, many organic compounds require that an aprotic solvent such as acetonitrile or dimethylformamide (DMF) be used.

An aprotic solvent is one that does not contain protons due to dissociation of the solvent molecules. Aprotic solvents are almost always organic solvents. Dimethylformamide (DMF) is a good example.

It should be remembered that the light must, pass through the solvent itself (i.e. be transmitted), although solvents are never perfectly optically transparent and in all cases exhibit their own absorptions. It is therefore crucially important to choose a solvent that allows the optimal transmission of the light throughout the wavelength region of interest. Water and many organic solvents appear to be colourless yet possess significant absorption spectra of their own in the UV range, which of course our eyes are insensitive to. In practice, it is as we approach the UV range that many of the most commonly encountered solvents begin to absorb significantly. It is therefore especially important that consideration be given to the choice of solvent when absorption spectra below around 250 nm are required.

A list of commonly used polar and non-polar solvents are shown in Tables 5.2 and 5.3, respectively.

It is essential that solvents of very high purity are always used (preferably high-performance liquid chromatography—HPLC grade) since many technical grade solvents such as ethanol and hexane contain contaminants of, for example, benzene that absorb at wavelengths below 280 nm.

However suitable the choice solvent (and indeed however good its purity), it should be remembered that the solvent will always exhibit a finite absorption, which must be accounted for in all cases. A blank spectrum of the solvent must either be run as a baseline that can be subtracted from all subsequent spectra, or alternatively, if a double beam spectrophotometer is being used, then twin cuvettes containing the solvent and analyte samples respectively may be used in tandem (see Section 5.9).

Table 5.2 Polar solvents

Solvent	Lowest wavelength beyond which solvent must not be used for analysis (nm)
Water	200
Ethanol	220
Diethyl ether	210
Acetonitrile	185

Table 5.3 Non-polar solvents

Solvent	Lowest wavelength beyond which solvent must not be used for analysis (nm)
Hexane	200
Cyclohexane	200
Benzene	280
Carbon tetrachloride	260
Dioxane	320

5.11 Colour-based complexation analyses

Many transition metal ions form highly coloured complexes. The colour can be exploited as the basis of simple and highly specific spectrophotometric determinations. For example, solutions of aqueous iron(II) react with *ortho*-phenanthroline (1,10-phenanthroline) to form an orange-red complex that may be easily quantitatively determined spectrophotometrically. The extinction coefficient for this complex is ~1.08×10^4 mol dm^{-3} cm^{-1}.

Aqueous *ortho*-phenanthroline acts as a weak base and dissociates in the presence of an acid to form phenanthrolinium ions, $PhenH^+$, Eqn (5.7). At pH values of 3.5 or less phenanthrolinium ions react quantitatively with Fe^{2+} to form the $Fe(Phen)_3{}^{2+}$ complex.

$$Fe^{2+} + 3PhenH^+ \rightarrow Fe(Phen)_3{}^{2+} + 3H^+ \qquad (5.7)$$

The iron content of aqueous solutions may be determined by the addition of an excess of a reducing agent such as hydroquinone or hydroxylamine. The reducing agent ensures that all of the iron resides in the +2 oxidation state and so is ready to complex with the $PhenH^+$ ion.

See the Box on page 136 for a description of the method for determining the iron content in an unknown sample

The $Fe(Phen)_3{}^{2+}$ complex exhibits a sharp absorption maximum (λ_{max}) at approximately 508 nm. The absorption of a series of standardized iron solutions should then be determined at concentrations corresponding to absorbances of around 0.1–1. A calibration graph may then be plotted, to allow the concentration of an unknown aqueous iron containing sample to be determined.

5.12 Bathochromic and hypsochromic shifts

If the λ_{max} of an analyte overlaps with the absorption spectra of any of the reagents and/or any other chemical species that may be present within real analyte samples, then a quantitative spectrophotometric determination will certainly be complicated. One way around the problem would be to further react the complex with another reagent to form a *new* complex, with a λ_{max} sufficiently far away from the absorption peaks of any interfering reagents and/or analytes.

A shift in the λ_{max} from a shorter to a longer wavelength is known as a ***bathochromic shift***.

A shift in the λ_{max} from a longer wavelength to a shorter wavelength is known as a ***hypsochromic shift***.

To understand how this is performed in practice, we need to consider a couple of examples:

Aqueous solutions containing tin may be spectrophotometrically determined by means of a bathochromic shift. Tin(IV) may be complexed with

Practical Methodology for an Iron Determination in an Unknown Sample

Standardized solutions of Fe^{2+}, hydroxylamine hydrochloride, *ortho*-phenanthroline and sodium acetate should first be prepared.

Fe^{2+} solution: Accurately weigh and dissolve approximately 0.07 g $Fe(NH_4)_2(SO_4)_2 \cdot 6H_2O$ in a 1 dm^3 volumetric flask. Add 2 cm^3 of concentrated H_2SO_4 and dilute to the mark with de-ionized water.

Hydroxylamine hydrochloride solution: Dissolve 10 g $H_2NOH{\cdot}HCl$ in 100 cm^3 of de-ionized water.

Ortho-phenanthroline solution: Dissolve 1.0 g of *ortho*-phenanthroline monohydrate in 1 dm^3 of water; *this solution must be prepared freshly on a daily basis.* *Sodium acetate*: Dissolve 166 g of $NaOAc \cdot 3H_2O$ in 1 dm^3 of de-ionized water.

Procedure

A series of secondary iron standards (four to five should suffice here) need to be made using the stock solution of Fe. This may be done by introducing 1 cm^3 of the hydroxylamine, 10 cm^3 of the sodium acetate, and 10 cm^3 of the *ortho*-phenanthroline to each of the flasks. Iron solutions (5, 10 15, 20 cm^3, . . .) should then be added to each of the flasks, which should then be made up to the mark with dissolved water. A blank should also be prepared without the iron but containing each of the other reagents, that is, the sodium acetate, *ortho*-phenanthroline, and hydroxylamine hydrochloride.

We have already said that the wavelength of maximum absorption (λ_{max}) occurs at 550 nm. A UV–visible spectrum should, however, be run between the wavelengths of around 460–560 nm ($\pm$50 nm of the expected λ_{max}) to determine as accurately as possible the λ_{max} (as determined by the machine you are to perform the analysis with). In this way you are: (a) ensuring that your experimental procedure is providing results which are broadly in line with the literature; and (b) ensuring that you are maximizing the analytical sensitivity of your own experimental apparatus.

The absorptions recorded for each of the samples may then be plotted as a concentration calibration curve that should be seen to follow the Beer–Lambert law (i.e. following a good straight line fit). The absorption of the unknown sample may then be read off against the corresponding concentration. If the absorption of the unknown sample falls beyond the range of the calibration graph, then the sample should be diluted by a known factor (e.g. dilution by 2- or 10-fold) to bring the absorption of the sample to within the experimentally determined range. The concentration of the unknown sample is then simply given by multiplication of the concentration as given by the calibration curve by the appropriate factor. Dilution of known samples to bring the experimentally determined absorption values to within the calibration range is important since deviations from the Beer–Lambert law are often observed at higher concentrations as we shall see in Chapter 6.

the dye catechol violet. The catechol violet–tin complex exhibits a strong absorption λ_{max} at 555 nm; unfortunately, the cathechol violet also absorbs radiation to a significant extent at this wavelength. Since the cathechol violet has to be added in excess in order to ensure complete complexation of the tin and the tin–cathechol violet complex cannot be easily separated from the excess cathechol violet, a quantitative spectrophotometric determination of the tin–cathechol violet complex is not possible.

The tin–cathechol violet complex, may, however, be further reacted with a further ligand, cetyltrimethylammonium bromide (CTAB). CTAB does not exhibit any absorption maxima in the 400–700 nm wavelength range. This then forms a tin–cathechol violet–CTAB complex, which shifts the absorption maximum to 662 nm. The addition of the CTAB ligand to the complex has also added a new absorption maximum, which

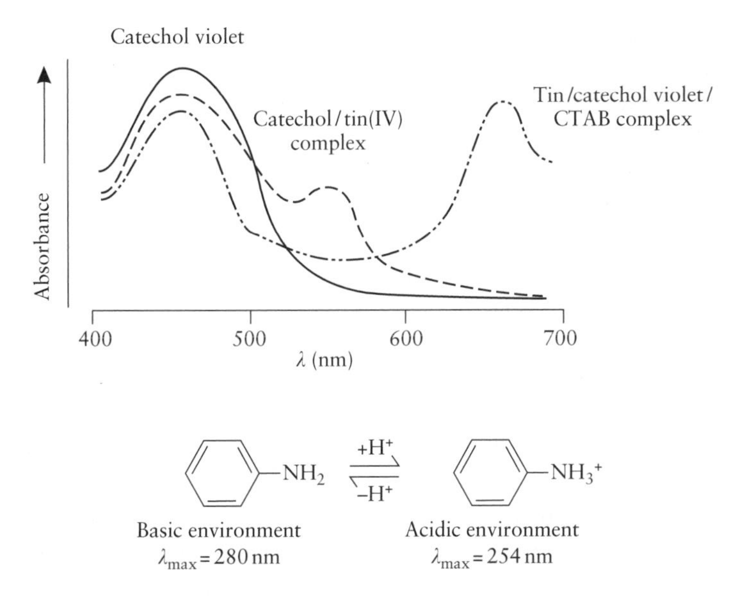

Figure 5.14 Bathochromic shift—showing shift in absorbance to longer wavelength.

Figure 5.15 Hypsochromic shift—aniline absorbance (λ_{max}) shifts to a shorter wavelenth as pH is lowered.

now allows the quantitative determination of tin in the presence of both the cathechol violet and CTAB reagents; Fig. 5.14.

An aqueous solution of aniline provides us with a good example of a hypsochromic shift. In basic solutions aniline resides in a neutral form and exhibits a λ_{max} at 280 nm. As the pH is lowered, the λ_{max} is shifted to 254 nm; Fig. 5.15.

5.13 Introduction to fluorescence and fluorometric determinations

So far we have considered how molecules absorb UV and visible radiation by the excitation of their valence electrons. The excited electrons are now energetically less stable than the ground state and must relax at some time. Upon relaxation, some or all of the energy gained by the capture of photons upon the absorption of light must be lost, and this normally occurs by dissipation of heat. In some cases, however, some of this excess energy is dissipated by the emission of a photon, that is, the emission of light.

Fluorescent reactions may be of great use to the analytical chemist since the emission of fluorescent radiation is highly specific to each individual compound. The intensity of the fluorescent radiation may, moreover, be directly related to the concentration of an analyte. In reality very few compounds fluoresce. A number of species may, however, be made to fluoresce by attaching or 'labelling' the compound with a molecule that *does* fluoresce. We shall consider this approach of 'fluorescent labelling' in Chapter 6.

The emission of photons (light) as a result of an electronic relaxation process is known as fluorescence.

Let us consider further how photons are emitted by fluorescence. The energy which is imparted to the photons that comprise the fluorescent radiation is derived from photons that were originally captured by the fluorescent molecule. As we have seen, some of the excess excitation energy that is lost by an electronic relaxation process is dissipated as heat, and the remaining energy is lost by the emission of a photon. It follows that the energy of the emitted photons will be of a lower energy than the absorbed photons and, since energy is proportional to frequency, the frequency of the emitted radiation will be lower than that of absorbed radiation. If the frequency of the emitted radiation is lower than that of the absorbed radiation, it follows that the fluorescent radiation must be of a longer wavelength than that of the radiation which was originally absorbed.

The intensity or power of the fluorescent radiation, P_f, will be proportional to the quantity or power of the light absorbed. The radiation power that is absorbed may be given by $(P_0 - P)$, where P_0 is the power of the incident radiation and P is the power of the radiation that is not absorbed—or, in other words, which is transmitted. Only some of the absorbed photons will give rise to the emission of fluorescent photons, since many electronic relaxations proceed solely as a result of molecular collisions and energy dissipation as heat. The ***quantum efficiency***, ϕ, is a proportionality constant that describes the proportion of absorbed incident photons that give rise to the emission of fluorescent photons.

If we are to measure the intensity of the emitted fluorescence, then a detector must be placed to monitor the photons emitted by the fluorescing species. The detector can only detect the radiation that enters it and therefore only monitors the light that is emitted in one direction from the analyte sample. Photons are, however, emitted in all directions (i.e. 360°) and it is therefore necessary to include another proportionality factor so that we can make a quantitative estimate of the total fluorescent output. This factor, k', is known as the ***geometric factor***. In practice, the fluorescent detector is normally placed at an angle of 90° (Fig. 5.16) to the incident radiation beam, to help avoid any interference with the incident or transmitted light sources, even though the fluorescent radiation will be of a different wavelength and will be monitored using a monochromator in conjunction with the detector.

We are now in a position to relate the fluorescent power output, P_f, to these other parameters, Eqn (5.8):

$$P_f = \phi k'(P_0 - P) \tag{5.8}$$

The incident light absorbed $(P_0 - P)$ is related to the concentration of the absorbing species, and if this is taken into account we can describe the

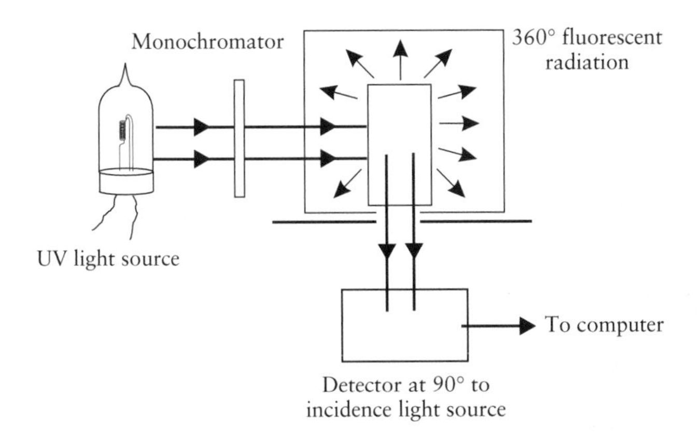

Figure 5.16 Schematic of a fluorescence spectrophotometer.

fluorescent power output according to

$$P_f = \phi k' P_0 (1 - e^{-\varepsilon cl}) \tag{5.9}$$

If one expands the exponential term to allow the substitution and expression of P_f in terms of $\log_{10}$ we can estimate that

$$P_f = \phi k' \times 2.303 \varepsilon cl \tag{5.10}$$

Eqn (5.10) is an approximation since it assumes that the absorbance is small. In practice, this equation is normally taken to be valid as long as εcl (i.e. the absorbance) is <0.05, and in this case the error incurred will be $<5\%$.

Practical Example 5.1

Quinine is a naturally fluorescent molecule that is found in many proprietary soft drinks such as sparkling bitter lemon or tonic water. The quinine content of a drink may be determined as a simple demonstration of a fluorometric analysis.

Method

Prepare the following solutions:

(a) A total of 2 dm^3 of a 0.05 M H_2SO_4 solution.

(b) A 1 ppm quinine sulphate standard. This may be prepared by weighing 0.1 g quinine sulphate (to within 0.5 mg), which should then be dissolved volumetrically in 1 dm^3 0.05 M H_2SO_4. Ten cm^3 of this solution should then be transferred to another 1 dm^3 flask and diluted to the mark with 0.05 M H_2SO_4. This latter solution will now contain 1 ppm quinine and should be prepared freshly and kept refrigerated in the dark between measurements since it is easily photo-oxidized.

Quinine fluoresces with an emission maximum at approximately 450 nm and so the emission wavelength of the fluorimeter should be set to this or a similar wavelength. A calibration curve should be determined by analysing a series of standardized samples, which are made by diluting 10 cm^3 and lower (8, 6, 4, and 2 cm^3) of the stock 1 ppm dm^{-3} solution to 1 dm^3 with 0.05 M H_2SO_4.

The unknown sample may then be analysed by diluting the sample with 0.05 M H_2SO_4 to bring the fluorometric reading within the linear region of the calibration curve.

EXAMPLE 5.5 Quinine fluoresces with an emission maximum of 450 nm. A series of quinine samples were prepared to known concentrations and the fluorescence recorded as in the table below:

Quinine concentration (ppm)	Fluorescence (arbitrary units)
0.0	0.10
0.2	0.21
0.4	0.32
0.6	0.40
0.8	0.51
1.0	0.62

If an unknown sample exhibits a fluorescence of 0.37, estimate the concentration of quinine within the sample.

Method

Calculate the gradient m by the least-squares approach and then use the $y = mx + c$ equation to calculate the fluorescence.

By the least-square method the equation for a line gives

$$m = 0.5114$$

$$\bar{y} = 0.36, \bar{x} = 0.5, \text{ and } c = 0.104$$

so substituting a value for a fluorescence of 0.37 into $y = mx + c$:

$$0.5114x + 0.104 = 0.37$$

$$0.5114x = 0.266$$

Therefore x (unknown) = 0.52 ppm quinine

If a fluorometric determination is to be performed, then we must normally determine: (a) the λ_{max} corresponding to the absorption peaks of the compound; and (b) the λ_{max} of the fluorescent emission spectrum. In practice, pinpointing the exact fluorometric emission λ_{max} is slightly less crucial since fluorometric emissions occur across a broader range of wavelengths than absorption spectra. This is to be expected since the emitted light may be thought of as the residual energy that is not lost as a result of molecular collision and subsequent thermal dissipation. Electrons may relax in a number of separate (albeit quantized) steps so the fluorescent emission spectrum peaks tend to span a few tens of nanometres. The majority of modern spectrophotometers permit a continuous variation of both the incident radiation wavelength and fluorometric detection wavelengths although some simpler instruments rely on the interchanging of interference filters (which are normally supplied in 10 or 20 nm increments) if the output fluorescence wavelength is to be altered.

5.13.1 Fluorometric quenching

Fluorometric quenching is a term used to describe processes that suppress molecular fluorescence. We have already seen that the fluorescence occurs when the relaxation of excited electrons gives rise to the emission of photons (light) at longer wavelengths than the incident light. This process is accompanied by the dissipation of heat due to molecular collisions. It follows that if more energy is lost as the result of molecular collisions, less energy is available for emission in the form of photons as light. This is the process underlying fluorescence quenching. The more molecular collisions that occur in the fluorescent mixture, the less fluorescence will occur. Any process that increases the frequency at which molecular collisions occur will therefore quench the fluorescence.

The majority of fluorescent reactions occur in solution, in which case molecular collisions between the fluorescent molecule and: (a) other fluorescent molecules; (b) the solvent molecules; and/or (c) any other solutes will all serve to quench the fluorescence to some extent.

Brownian motion will always be present and again any factor that increases Brownian motion and/or diffusion of solvent/solute molecules will increase the frequency of molecular collisions and so increase the quenching of the fluorescent reaction.

Ions and/or other solutes may be added to the mixture to act as quenching agents; larger molecules will be involved with a greater number of molecular collisions than smaller solutes and it therefore follows, for example, that K^+ will act as a more effective quenching agent than Na^+.

5.14 Polarimetry and optical rotations

Many inorganic and organic compounds possess the capability of *rotating the plane* of a source of polarized radiation. Materials that display this behaviour are said to be ***optically active***. Some of the most popular examples include, for example, quartz as well as the mono- and disaccharide sugars such as glucose. If an observer looks towards the light source and the light is rotated towards the right (i.e. clockwise), then the direction of rotation is said to be ***dextro*** or (+) ***rotatory***. Conversely, if the direction of rotation is towards the left (anticlockwise) then the direction of rotation is said to be ***laevo*** or (−) ***rotatory***.

The extent of rotation depends on the *number* of atoms or molecules in the path of the light source and hence the concentration of a solution (c) and the path length (l). The degree of rotation also depends on the wavelength of the radiation (λ) and the temperature (t).

Note: For historical reasons, the pathlength of the cell is quoted in dm rather than m, and the concentration in $g\,cm^{-3}$ rather than $mol\,dm^{-3}$.

The ***specific rotation***, $[\alpha]^t$ is defined as the extent of rotation in degrees of a plane of polarized light, α, at a specified wavelength through a 1 dm

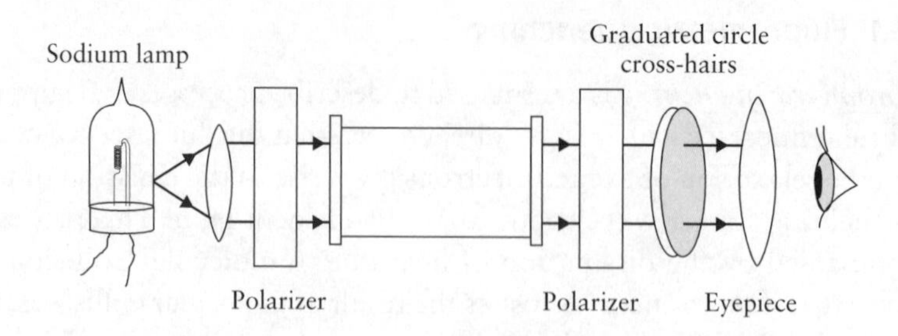

Figure 5.17 Schematic of a simple polarimeter.

(10 cm) pathlength of a solution of concentration, c (quoted in $\mathrm{g\,cm^{-3}}$) at temperature t, Eqn (5.11):

$$[\alpha]^t = \frac{\alpha}{lc} \tag{5.11}$$

A sodium vapour lamp is often chosen as the source of radiation, with an emission at 589.3 nm corresponding to the D line of a sodium emission.

For discussion of atomic emission lines, please refer to Chapter 7.

Optical rotation may be measured most commonly by an ***automated polarimeter***, a simplified schematic of which is shown in Fig. 5.17. The light must be: (a) monochromatic; (b) polarized; and (c) focused so that all of the light is travelling in the same direction. The light originates either from a white light source or, more commonly, may be generated from a monochromatic source, such as, for example, a sodium lamp. The light is then passed through a collimator and, in turn, onto a calcite prism to produce the plane-polarized focused light beam. An auxiliary calcite prism is sometimes used to intercept or divide the incident radiation source into two beams of equal intensity—*but* orientated a few degrees of rotation apart relative to each other.

The two light beams are passed through a glass tube of known length (normally 10 cm) and an 'analyser' is placed at the far end of the sample tube together with an eyepiece. The intention is to determine the angle through which a polarized light beam is rotated by the sample. There are a number of ways of achieving this. A polarizing filter ***will fail to transmit any light*** if this is placed at an angle of 90° to the plane of a polarized beam of light. The detector at the far end of the sample, however, may also be rotated; in this way the intensity of the light emerging from the sample may be monitored *at any angle*. The angle of rotation of the detector, which corresponds to a total blocking of the transmitted light, will therefore correspond to the angle through which the polarized light beam has been rotated. This method suits automated methods for determining the optical rotation of samples, since detectors readily detect signal minima or maxima, which in this case correspond to the output from a photomultiplier tube or other optical detector.

Many polarimeters rely on the manual determination of the optical angle of rotation and, if used correctly, provide results that compare favourably with many automated and therefore rather costly instruments. Manual polarimeters also use two light beams. In this arrangement a polarizing filter or prism is placed directly in front of, and in the path of, the sample at an angle of 90° to the plane of the incident light source. This prism splits the light beam into two separate beams that have a few degrees of rotation clockwise and anticlockwise, respectively, to the incident light. The two light beams are then passed through the sample and the transmitted light from both beams of light are then passed through a polarizing filter, which may be rotated. *It follows that there will be one angle at which the intensity of the radiation of the two light beams will be of very small but equal intensity.* The polarizing filter is rotated until the intensity of the two beams is equal. This angle corresponds to the ***angle of optical rotation***. At this angle the light beams possess angles of rotation equidistantly spaced either side of the detector; the midpoint between these angles corresponds to the angle through which the sample has rotated the incident light beam. Two light beams are used to compare intensity, since our eyes more easily match light intensities than determining the angle corresponding to the absolute extinction of the light as it emerges from the sample.

5.14.1 Practical determinations based on polarimetry

Many organic compounds contain asymmetric arrangements of atoms within a group that allow mirror images of the same molecule to exist; such compounds are said to possess ***chirality***. Chiral and therefore optically active compounds are extremely important to many biological systems. Biological substrates often have chiral centres and are acted upon by enzymes that only recognize the dextro- or laevo-rotatory forms. The dextro and leavo forms are known as rotamers. Many biological samples contain only one rotamer of a given compound and for this reason, many bio-analytes may be quantified by polarimetry.

Sucrose polarimetric determinations

The most widely used application for polarimetry is in the sugar industry for the determination of sucrose due to its huge commercial significance. Although Eqn (5.11) allows us to quantitatively determine solutions of sucrose in the absence of other optically active materials, the analysis becomes more difficult if other naturally occurring sugars are present. Practical determinations are normally to be undertaken within sugar solutions derived from plants (sugarbeet and cane) that contain a number of sugars other than sucrose.

Fortunately, sucrose (a disaccharide) may be hydrolysed in the presence of a dilute acid to yield glucose and fructose, Eqn (5.12). This hydrolytic splitting of the saccharide is known as an ***inversion*** reaction, and the resulting mixture of the two monosaccardies, glucose and fructose—as the ***invert sugar***.

$$\begin{array}{lll} C_{12}H_{22}O_{11} + H_2O \rightarrow & C_6H_{12}O_6 + & C_6H_{12}O_6 \\ \text{(Sucrose)} & \text{(Glucose)} & \text{(Fructose)} \\ [\alpha]_D^{20} = +66.5^\circ & +52.7^\circ & -92.4^\circ \end{array} \quad (5.12)$$

Glucose and fructose, however, have very different specific rotations from sucrose, Eqn (5.12). As the hydrolysis proceeds (i.e. the inversion reaction) the *specific rotation* for a solution initially containing only sucrose will change from +66.5 to −19.8°. The angle of rotation measured will of course be dependent on the concentration of sucrose and whether or not any other optically active compounds are present. The *change* in the angle of rotation will, however, depend on the concentration of sucrose in the sample. It therefore follows that if we note the change in the angle of rotation we can calculate the sucrose concentration.

Ten cm^3 of concentrated HCl are typically added to every 100 cm^3 of the sucrose sample and allowed to stand for at least 24 h at room temperature. The hydrolysis (or 'inversion') reaction can, however, be accelerated by heating the mixture at 70°C, and this should allow the reaction to reach completion within approximately 15 min.

Penicillin—penicillinase polarimetric determinations

The antibiotic *penicillin* is another example of an optically active biological compound that can be determined polarimetrically. The dextrorotamer form is the biologically active form of the molecule and may be metabolized by the enzyme *penicillinase*. A solution of penicillin and penicillinase will therefore show a decreasing angle of optical rotation as penicillin is metabolized by the enzyme. The enzyme has a very slow turnover rate and is effectively saturated in all but the most dilute solutions. It follows that until almost all the penicillin is consumed, the rate of consumption of the substrate will be dependent on the concentration of the enzyme and not the penicillin and so the reaction will proceed at a near constant rate until all of the penicillin is consumed; the time for the reaction to reach completion may, however, be used to determine the concentration of the penicillin within the sample.

Conversely, the same reaction chemistry may be used to determine the concentration of a penicillinase antibiotic solution. In this situation, the time taken for the consumption of a known molar quantity of penicillin may be monitored polarimetrically and, in this way, it is possible to calculate the concentration of penicillinase that is present in an unknown sample.

EXAMPLE 5.6 A solution of sucrose causes a 32.05° *dextro* rotation of plane polarized light within a polarimeter. Assume that $[\alpha]^t$ ($dm^3\,g^{-1}\,dm^{-1}$) at 20°C = 66.5°. Calculate the concentration of sucrose within the solution.

Method

From the expression $[\alpha]^t = \frac{\alpha}{lc}$ calculate α:

$$[\alpha]^t = 66.5 = \alpha/(1 \times c)$$

Rearrange and substitute in values: $c = \frac{\alpha}{66.5} = \frac{32.05}{66.5}$

Therefore, concentration of sucrose = $\mathbf{0.482\ g\,cm^{-3}}$. Since the RMM of sucrose = 342 this is equivalent to $\frac{0.482}{342} \times 1000$ mol dm^{-3} = **1.4 mol dm^{-3}**.

Note: it is normal to quote concentration as g cm^{-3} in polarimetry. If concentration is to be quoted in terms of molarity then a correction must be made. Note also the path length is quoted in dm. One decimeter is 10 cm.

Exercises and problems

5.1. Which electrons within molecules are normally involved in absorption of UV and visible radiation?

5.2. A student performs a set of UV–visible determinations on a set of unknown samples for an analyte, *a*, with a known λ_{max}. All of the student's results give absorbances of between 2 and 3. Why may these results not be used for the quantitative determination of *a*, and what must the student do next?

5.3. Explain why absorbance is a unit-less quantity.

5.4. Why will many UV–visible spectrophotometer possess more than one lamp?

5.5. Sodium chloride and potassium chloride both act as quenching agents for a fluorometric determination of quinine. Explain: (a) why this is the case; and (b) which salt will act as the most effective quenching agent.

5.6. A cuvette with a path length of 1 cm and solution containing 8.96 mg dm^{-3} of a dye with an RMM of 107.4 gives an absorbance of 0.8. Calculate the molar absorptivity of the dye.

5.7. A compound of molecular weight 245 is found to have an absorptivity of 298 $dm^3\,g^{-1}cm^{-1}$. Calculate its molar absorptivity, ε.

5.8. Three standard $Fe^{2+}{}_{(aq)}$ solutions are found to have absorbance values as shown below:

Conc. Fe^{2+} (mol dm^{-3})	Absorbance
0.010	0.21
0.025	0.53
0.052	1.00

(a) Show whether or not the absorbances recorded for these solutions obey the Beer–Lambert law.

(b) An iron salt containing solution is found to exhibit an absorption of 0.2 at 510 nm following the addition of 1,10-phenanthroline (together with excess hydroxylamine) to form a coloured complex. The molar absorptivity, ε, for the iron–phenanthroline complex is 1.08×10^4 $dm^3\,mol^{-1}\,cm^{-1}$. Determine the concentration of iron within this solution.

5.9. A solution of the drug tolbutamine is found to exhibit an absorbance of 0.85 in a 1 cm pathlength cuvette. The molecular weight of Tolbutamine is 270 and the molar absorptivity at 262 nm is 703 $dm^3\ mol^{-1}\ cm^{-1}$. What is the molar concentration of tolbutamine?

5.10. A 0.15 M solution within a 1 cm path length placed within a UV–visible spectrophotometer shows an absorbance of 0.62. Calculate the molar absorptivity for this compound.

5.11. A compound with a molar absorptivity of 32 667 $dm^3\ mol^{-1}cm^{-1}$ (at 740 nm) exhibits a concentration of 0.81 when placed within a 1 cm path length cuvette in a UV–visible spectrophotometer. Calculate the concentration of the compound.

5.12. You have a compound which you know has λ_{max} at 215 and 244 nm and you wish to determine its ε value in a range of solvents. It is soluble in acetone, acetonitrile, benzene, carbon tetrachloride, dioxan, methanol, toluene, and water. Which of these solvents can you use for this experiment?

5.13. You have three dilute solutions of equal molarity of pentane, 1,3-pentadiene, and 1,4-pentadiene. How would you be able to tell them apart using UV spectrometry?

5.14. Define what is meant by a hypsochromic shift a bathochromic shift.

5.15. A solution that absorbs radiation with a λ_{max} at 475 nm is placed within a UV–visible spectrophotometer. An absorbance of 0.82 is recorded. Calculate the percentage of the incident light that is being absorbed.

5.16. A solution within a UV–visible spectrophotometer shows an absorbance of 0.72 at a wavelength of 489 nm. Calculate the percentage of radiation that is transmitted.

5.17. The co-enzyme nicotinamide adenosine dinucleotide (NADH) fluoresces when irradiated with radiation of wavelength 340 nm. A fluorescence emission maximum is observed at 465 nm. An analyst measured the relative intensities of fluorescent radiation at 465 nm for varying concentrations of NADH as shown below:

Conc. NADH (μmol dm^{-3})	Relative intensity
0.2	8.92
0.4	18.00
0.6	27.43
0.8	35.85

Plot a best-fit calibration curve using the least-squares approach. Calculate the concentration for an unknown NADH sample exhibiting a fluorescence with a relative intensity of 20.

5.18. Quinine fluoresces with an emission maximum of 450 nm. A series of quinine samples were prepared to known concentrations and the fluorescence recorded as in the table below:

Quinine concentration (ppm)	Fluorescence (arbitrary units)
0.0	0.123
0.2	0.258
0.4	0.394
0.6	0.492
0.8	0.627
1.0	0.763

If an unknown sample exhibits a fluorescence of 0.63, estimate the concentration of quinine within the sample.

5.19. A solution of sucrose causes a 42.07° *dextro* rotation of plane polarized light within a polarimeter. Assume $[\alpha^t]$ ($dm^3\,g^{-1}\,dm^{-1}$) at 20°C = 66.5°. Calculate the concentration of sucrose within the solution.

5.20. A fructose solution of unknown concentration causes a *laevo* rotation of plane polarized light of 62.76°. What is the concentration of fructose? Assume $[\alpha^t]$ ($dm^3\,g^{-1}\,dm^{-1}$) at 20°C = −92.4°.

5.21. A 0.24 M glucose solution is analysed within a polarimeter. If α^t for glucose at 20°C is 66.5°, calculate the expected angle of rotation for this sample.

Summary

1. Visible light forms part of the electromagnetic spectrum and extends from wavelengths of approximately 400 to 750 nm.

2. Ultraviolet (UV) radiation (of wavelengths that may be exploited for analytical purposes) extends from wavelengths of approximately 180 to 400 nm.

3. The energetic states of orbitals and their electrons are quantized.

4. The promotion of a valence electron from one orbital to another involves absorption of radiation normally in the UV or visible range of wavelengths.

5. The energy of electromagnetic radiation $= h\nu = hc/\lambda$.

6. The Beer–Lambert law describes the absorption of radiation by compounds with a molar absorptivity of ε, and a concentration, c, through a pathlength, l, and states that $A = \varepsilon cl$.

7. Absorbance is a unit-less quantity. $A = \log_{10}(I_0/I)$ where I_0 is the intensity of the incident radiation and I the intensity of the transmitted radiation.

8. A number of different light sources may be used for UV–visible spectroscopy such as tungsten filament, hydrogen, and deuterium lamps.

9. Monochromators are used for wavelength selection and are based on diffraction prisms or reflection gratings.

10. A number of photon detectors are used with UV–visible spectrometers and include photo-tubes, photo-multiplier tubes, silicon photo-diodes, and photovoltaic cells.

11. The energy of absorbed photons is normally dispersed as heat when electrons relax to their ground states.

12. Fluorescence gives rise to the emission of a photon; the fluorescent radiation will be of a longer wavelength than the incident radiation.

13. Fluorescence quenching is the term used to describe processes that suppress molecular fluorescence.

14. UV–visible spectrometers are based either on a single- or double-beam formats. Single-beam spectrometers necessitate separate baselines to be determined with blank samples. Double-beam spectrometers, by contrast, allow two cells to be determined—one as a blank to establish the baseline and one for the sample itself.

15. Cuvettes may be made of quartz, glass, or plastic depending on the wavelength range over which spectra are to be run. Quartz cuvettes must always be used for wavelength ranges <300 nm.

16. A shift in λ_{max} from a shorter to a longer wavelength is known as a bathochromic shift.

17. A shift in the λ_{max} from a longer to a shorter wavelength is known as a hypsochromic shift.

18. Compounds that are capable of rotating the plane of polarized radiation are said to be optically active. Compounds that rotate the light in a clockwise (+) manner are said to be dextrorotatory. Those that rotate radiation in a anticlockwise (−) manner are said to be laevorotatory.

19. The specific rotation $[\alpha]^t$ is defined as the extent of rotation in degrees, that is shown by a solution of concentration 1 g cm^{-3} in a 1 dm (10 cm) cell. It may be calculated by:

$$[\alpha]^t = \frac{\alpha}{lc}$$

20. The study of optical rotation is known as polarimetry and may be used to quantify solutions of, for example, sugars such as sucrose or antibiotics such as penicillin.

Further reading

Duckett, S. and Gilbert, B. (2000). *Foundations of spectroscopy*. Oxford Chemistry Primers, Oxford University Press.

Lakowicz, J. R. (1999). *Principles of fluorescence spectroscopy*, Kluwer Academic.

Thomas, M. J. K. (1996). *Ultraviolet and visible spectroscopy*. Analytical Chemistry by Open Learning Series. Wiley.

Valeur, B. (2002). *Molecular fluorescence: an introduction – principles and applications*. Wiley.

Part III

The key analytical techniques

6 Further applications of UV–visible absorption and fluorescence phenomena including X-ray fluorescence, Raman, Mössbauer, and photoelectron spectroscopic techniques

Skills and concepts

This chapter will help you to understand:

- What is meant by the uncertainty principle.
- What are meant by the principal and secondary quantum numbers and how electronic shells are occupied as the atomic number of elements increase.
- How to interpret a Grotian diagram for an element.
- How deviations from the Beer–Lambert law arise.
- What is meant by, and to be able to differentiate between, fluorescent, phosphorescent, and incandescent effects.
- The analytical application of fluorescent labelling.
- What is meant by chemifluorescence and how this can be applied.
- What is meant by the Raman effect as well as how and why Stokes and anti-Stokes transitions lead to Raman shifts.
- How to identify which molecules are likely to exhibit Raman behaviour and why.
- The analytical applications of Raman spectroscopy.
- What is meant by the rotational quantum number and rotational constant.

- The origins of X-ray fluorescent effects and the analytical applications of X-ray fluorescence spectroscopy.
- The analytical applications of UV–visible and X-ray photoelectron spectroscopy.

6.1 Introduction

In Chapter 5 we saw how UV and visible light might be used to identify, and in some cases, quantify a wide range of analytes. We shall in this chapter first look in slightly greater detail at the theoretical basis underpinning UV and visible absorption and then consider a number of other regions of the electromagnetic spectrum that may be used for analytical purposes. We shall also examine a number of practical considerations including analyte samples that do not obey the Beer–Lambert law and mixtures that contain a number of species capable of absorbing radiation over similar wavelength regions.

6.2 Allowed and forbidden electronic transitions

In Chapter 5 we introduced the concept of how valence band electronic transitions were at the theoretical heart of absorption and fluorescent spectroscopy. Whilst we discussed how energy levels and thus the transitions between them were quantized, we did not mention that *only certain transitions are allowed*. Transitions between different energy levels are governed by selection rules and we shall now consider these and the factors that contribute to these rules in this following discussion.

We must first consider a brief recap on the electronic structure of atoms and molecules. The ***uncertainty principle*** tells us that at any one moment in time we cannot exactly simultaneously define where an electron will be and what momentum it will have. The ***orbital approximation***, however, allows us to map out specific areas around the nucleus in which the electrons spend almost all of their time; these regions are known as ***orbitals***. There are different types of orbitals that can be organized into the quantized energy levels of the electrons that reside within them and this allows us to predict the ***electronic configuration***—in other words, we can predict in which orbitals electrons will reside.

The number of electrons in a neutral atom will equal the number of protons within the nucleus. Electrons normally (unless otherwise excited) sequentially fill orbitals at the lowest energy levels available. The energy levels of electrons are denoted by the ***principal quantum number, n***, which are given the values $n = 1, 2, 3$, etc. Each principal quantum energy level may contain a maximum number of orbitals and each of these a number of electrons.

The orbitals and electrons that comprise a principal quantum energy level are collectively known as a ***shell***. Different shells corresponding to $n = 1$ or 2, for example, contain differing numbers of orbitals and hence electrons.

Shells are sometimes alternatively described by letters where: $n = 1 \equiv$ k shell; $n = 2 \equiv$l shell; $n = 3 \equiv$m shell; and $n = 4 \equiv$n shell.

The orbitals within shells are arranged into ***sub-shells*** and these are labelled by the ***secondary quantum number***. Each type of sub-shell possesses characteristically shaped orbitals and these are denoted s, p, d, or f.

It follows that sub-shells of an atom may therefore be denoted by both the ***principal and secondary quantum numbers*** in the form of 1s or 2p, etc. s sub-shells may contain only one orbital, and so two electrons. p sub-shells contain three orbitals and hence up to six electrons. d sub-shells contain five orbitals and up to 10 electrons, and finally f sub-shells contain seven orbitals and a maximum of 14 electrons when completely filled.

The ***Pauli exclusion principle*** states that an orbital may only contain two electrons and an orbital that contains two electrons is therefore said to be filled. Electrons possess a direction of spin and this is labelled as being either (↑) or (↓). The spin of an electron may impart angular momentum to an orbital. A pair of electrons within one orbital will always have their ***spins paired*** (i.e. their spins will be in opposite directions) and this means that a filled orbital will have a net zero angular momentum; electrons with paired spins within a filled orbital are denoted ↑↓.

If we wish to say how many electrons reside within a sub-shell we use a superscript after the secondary quantum number; if a 1s shell is filled by two electrons we can write this as $1s^2$.

The first shell, $n = 1$, contains only one s orbital, the second shell n = 2 contains one s and one p orbital, the third shell contains one s, one p, and one d orbital and the fourth shell similarly contains an s, one p, one d, and one f orbital. All of the shells with higher quantum numbers all contain s, p, d, and f orbitals.

6.2.1 The Aufbau principle and Hund's rule

The ***Aufbau principle*** (from the German word *Aufbau*—meaning building up) dictates that electrons sequentially fill vacant orbitals at the lowest energy levels available and this gives rise to a specified order, which may be predicted.

The energetic order and therefore the order in which the electrons fill orbitals is:

1s, 2s, 2p, 3s, 3p, 4s, 3d, 4p, 5s, 4d, 5p, 6s, 5d, 4f, 6p, ...

Hunds's rule tells us that in its ground state, an atom will adopt an electronic configuration with the greatest number of unpaired electrons filling the outermost partially occupied sub-shell of electrons; this means that electrons will only pair up within a sub-shell once each of the orbitals contains one electron. There are quantum mechanical considerations that can explain why this occurs, however, we can think of this rule as being a consequence of the repulsion that electrons have towards each other and by residing in different orbitals whenever possible, the space between them is maximized.

If we consider nitrogen, which has an atomic number of 7, we can write its electronic configuration as $1s^2$, $2s^2$, $2p^3$. We can be more specific, however, since there are three separate p orbitals which are known as the $2p_x$, $2p_y$, and $2p_z$ orbitals. Hund's rule dictates that for nitrogen the three 2p electrons will individually reside in each of the three p orbitals. We can therefore write the electronic configuration more accurately as being $1s^2$, $2s^2$, $2p^3$: ($2p_x^1$, $2p_y^1$, $2p_z^1$).

EXAMPLE 6.1 Write the electronic configuration for fluorine

Method

1. Determine the number of electrons fluorine possesses.
2. Write the electronic configuration according to the Aufbau principle.

Step 1: Fluorine has an atomic number of 9 and so possesses nine electrons.
Step 2: The electronic configuration for fluorine is $1s^2$, $2s^2$, $2p^5$.

The outermost shell of electrons are known as the ***valence electrons***. *Valence* literally means 'reactivity' or the ability to react. The valence electrons are named as such since it is electrons from this shell that are involved in the formation of either ionic or covalent bonds. It follows that the valency of an element is usually equal to the number of electrons that are required to either empty or fill the valence shell of electrons.

If one or more electrons within an atom are excited and relaxes back to its ground state via the emission of a photon, then a characteristic emission line will be observed for a particular element. If the electron relaxes back to the first electron state within the outermost shell, the emissions are said to belong to the ***Lyman*** series. Electrons that relax back to the position of the second, third, and fourth ground electronic states within the valence shell are said to give rise to ***Paschen, Brackett, and Pfund emissions***, respectively.

Whether an electronic transition is allowed or forbidden depends on the ***spin*** of the electrons involved in the process. The spin of the electrons within an atom is described by the ***spin quantum number***, s, and always possesses a value of $1/2$ for each electron. The spin of electrons can be thought of as being analogous to angular momentum. The direction of spin is denoted by the quantum number, m_s, and this is said to be either clockwise (+) or anticlockwise (−). The spin is therefore always either $+1/2$ or $-1/2$ for every electron. An electron with a spin of $+1/2$ is known as an **α electron** and is often written or denoted by the symbol ↑, while an electron with a spin of $-1/2$ is known as a **β electron** and is written as ↓. The electron spin describes the behaviour of the electron as if it were spinning as a tiny magnet and it is this behaviour that gives rise to the interaction of an electron with electromagnetic radiation.

Other fundamental sub-atomic particles also possess spin. Unlike momentum in classical mechanics, quantum mechanical angular momentum spin is quantized in units of $\pm 1/2$ or 1 and does not depend on the mass of the particle. Both protons and neutrons possess spins of $-1/2$, and so they have spin quantum numbers of $s = -1/2$. It should be remembered that electromagnetic radiation has a wave–particle duality and if we consider the particle light nature of radiation we find that photons possess a spin of $s = -1$. This is an important consideration since photons possess twice the angular momentum of electrons; it is this discrepancy that explains why electrons and electromagnetic radiation may, in some instances, interact yet at others times they cannot. When an electron is promoted from one quantum level to another, we can think of the incoming photon interacting and imparting energy to the surrounding electromagnetic field to give rise to the excitation. The sudden change in the spatial distribution of the electron as it moves from one orbital to another is associated with a large perturbation of the electromagnetic field associated with the atomic or molecular orbitals as the energy from the photon is absorbed. We may similarly think of the emission of a photon following the relaxation of an electron from a higher to a lower orbital as occurring as a result of the sudden jolt to the electromagnetic field as the electron moves from one orbital to another.

When a photon is absorbed or generated, the total angular momentum in the overall process must be conserved. This gives rise to ***selection rules*** for particular orbitals and describes which electronic transitions are permitted. It follows that an electron within a d orbital with $l = 2$ cannot make a transition to an s orbital with $l = 0$ because the electron cannot be given enough angular momentum. Another forbidden transition would be the transition of an s electron to another s orbital since there would be no change in angular momentum to accompany the absorption or emission of a photon.

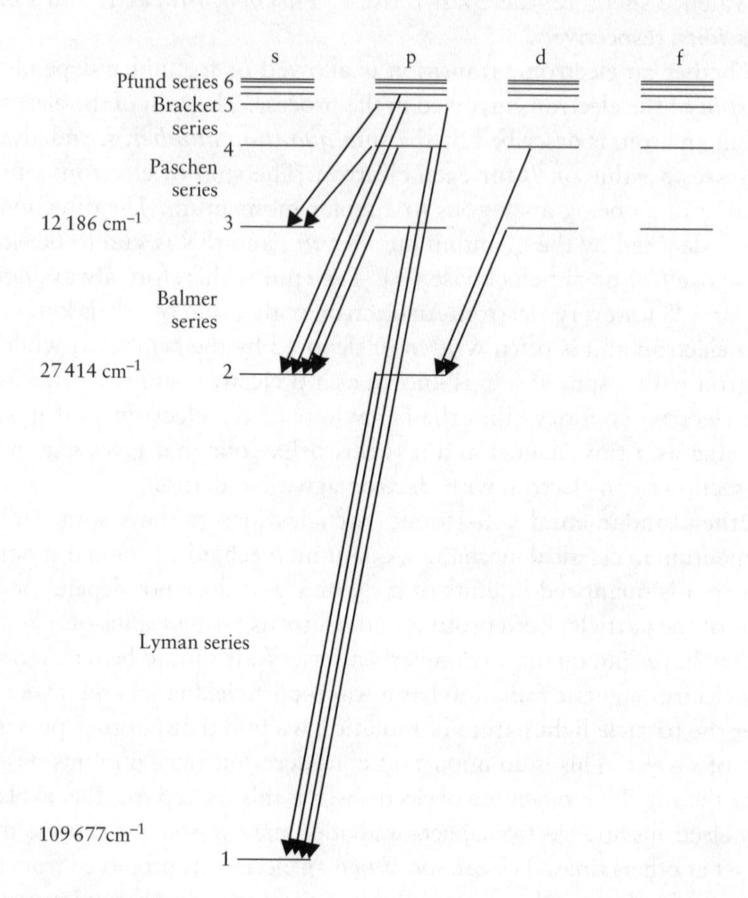

Figure 6.1 Simplified Grotian diagram for the atomic spectrum of hydrogen showing allowed electronic transitions.

If we ascertain which transitions are possible between the orbitals of an atom or molecule, we can summarize these within a *Grotian* diagram, an example of which is shown for hydrogen, in Fig. 6.1. In this case, due to the simple electronic structure of hydrogen, allowed electronic transitions only give rise to Lyman, Balmer, and Paschen emissions. We can even calculate the relative intensities of the spectra via quantum mechanics, although this is beyond the scope of this book.

6.3 UV–visible absorption spectra and deviations from the Beer–Lambert law

We already saw in Chapter 5 (Section 5.5) that many samples follow the Beer–Lambert (or absorption) law; i.e. the absorption of the sample increases linearly with the concentration of an analyte. If the sample fails to follow the Beer–Lambert law (or appears to) it is becomes rather more difficult to quantify the analyte.

Deviations in behaviour from the Beer–Lambert law are normally observed when calibration profiles fail to increase linearly with concentration; further investigations often reveal that absorption spectra significantly change in shape with concentration. The most common causes of this behaviour relates to interactions of the solute molecules with each other, interactions of solute and solvent molecules, or simply limitations in the instrumentation.

The change in colour of a potassium chromate, $K_2Cr_2O_7$, solution from an orange to yellow colour upon dilution is a good example of solvent effects and how this affects UV–visible spectrometry. The solute can reside in two forms: $Cr_2O_7^{2-}$ (orange) and $2CrO_4^{2-}$ (yellow), the relative concentrations of which are determined by pH, Eqn (6.1):

$$\underset{\text{(Orange)}}{Cr_2O_7^{2-}} + H_2O \rightarrow 2H^+ + \underset{\text{(Yellow)}}{2CrO_4^{2-}} \qquad (6.1)$$

The absorption spectra for $Cr_2O_7^{2-}$ and $2CrO_4^{2-}$ have different λ_{max} at ~370 and 350 nm, respectively.

Negative deviations from the Beer–Lambert law occur when the absorption observed is less than that predicted by the Beer–Lambert law for a given concentration of analyte.

Negative deviations from the Beer–Lambert will also always be seen to some extent if the illumination is not monochromatic. Let us consider an aqueous solution of, for example, $KMnO_4$, which has a λ_{max} at 520 nm. Now at a concentration of zero, light of all wavelengths will pass through the sample unimpeded and will therefore be picked up by the detector. As the concentration of $KMnO_4$ increases, more absorption (centred around a λ_{max} of 520 nm) will be increased. The light at all other wavelengths will, however, still pass through the sample and although less light will be monitored by the detector, the increase in absorbance will not be perfectly linear with concentration. All spectrophotometers are designed to irradiate the sample with monochromatic radiation, however, the quality of the monochromator and in particular its bandwidth should be considered carefully if a persistent negative deviation from the Beer–Lambert law is expected. The highest quality monochromatic radiation can be obtained using a laser, but since this approach would only allow a few wavelengths to be selected, very few UV–visible spectrophotometers adopt this approach.

Limitations in the instrumentation can also be related to concentration effects. If the analyte concentration is extremely low, then larger percentage errors will be introduced into estimating very small absorbances. Conversely, if the concentration of the analyte is sufficiently high to cause near total absorption of the light at λ_{max}, further increases in concentration will be extremely difficult to follow and the absorption calibration curve may appear to approach a plateau. It is therefore important that the concentration of analyte solutions are kept within a range of concentrations that yield absorbances in the range ~0.1–2.0. It is particularly important that if solutions show absorbances of 2 or greater that they are diluted by a known factor to bring their absorbances to values ideally below 1.0. We will consider briefly why it is so important to measure absorbance in a range of 0–2 and preferably 0–1. Remember that absorbance A, is a unit-less quantity since it is a $\log_{10}$ of a ratio of the intensity of the incident and transmitted light, that is, $\log_{10}(I_0/I)$.

With an absorbance of 2, we have:

$$\text{antilog}_{10}\, 2 = 10^2 = \frac{I_0}{I}$$

and so

$$100 = \frac{I_0}{I}$$

since I_0 may be assigned a value of unity or 1, then

$$100 = \frac{1}{I}$$

and so

$$I = \frac{1}{100}$$

and therefore ***it follows that an absorbance of 2 means that only 1% of the incident light is transmitted and that 99% is absorbed by the sample.***

If we follow by the same argument (and similar calculations), that an absorption, $A = 1$ corresponds to a 10% transmission of the incident light with 90% absorption occurring, while an absorbance of 0.1 means that ~79% of the incident light is transmitted and ~21% of the light is absorbed by the sample. If we consider, however, an absorbance of value of 0.01, then 97.9% of the incident light is transmitted and only 2.3% of the light is absorbed. It therefore follows that ideally samples should be prepared with concentrations so that absorbances ideally fall within the range 0.1–1 and certainly not more than 2.

6.4 Binary systems and the analysis of multi-component systems

We have already seen how a calibration curve can be constructed from the absorption measured at the λ_{max} across a range of concentrations so long as the Beer–Lambert law is followed—or at least approximated to. If two

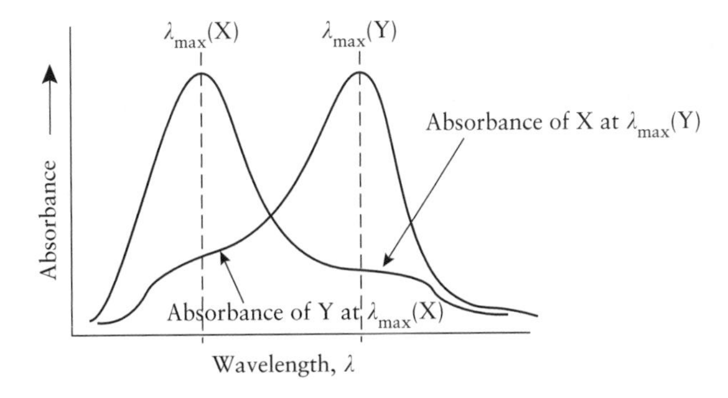

Figure 6.2 Absorption spectra for two analytes X and Y.

solutes, which possess overlapping absorption peaks are both present within one sample, it may be very difficult to quantitatively determine either analyte due to the interference possessed by the other.

Let us consider a solution that contains two analytes, X and Y, which possess absorption spectra that broadly overlap; Fig. 6.2. This behaviour normally prevents either analyte from being determined by measuring the absorbance at their respective λ_{max} without interference from the other solute. The Beer–Lambert law dictates that absorbances are additive since different solutes absorb irrespective of the presence of other solutes; this provides an approach to allow a determination of either analyte in the presence of the other (provided that they do not react with each other).

We must first know the molar absorptivity of both analytes at their λ_{max} and determine the absorption of the mixture at the λ_{max} for either analyte. There will inevitably be overlap between the absorption spectra of the two analytes or one would not be using this approach; however, the only requirement is that the λ_{max} do not coincide.

The total absorbance we shall observe at any one wavelength A_T will equal the sum of the absorbances of X (A_X) and Y (A_Y); Eqn (6.2):

$$A_T = A_X + A_Y \qquad (6.2)$$

The individual absorbances and therefore the total absorbance will vary with wavelength. The first step in determining the individual absorbances and therefore the concentration of each analyte is to determine the λ_{max} for the two individual analytes X and Y and we shall denote these $\lambda_{max}(X)$ and $\lambda_{max}(Y)$, respectively. Since both the analytes can contribute to the total absorbance seen at either wavelength, we next need to determine the molar absorptivities, ε, for the two analytes X and Y at their respective λ_{max} wavelengths $\varepsilon(X)$ and $\varepsilon(Y)$, and their molar absorptivities at each other's λ_{max} wavelength $\varepsilon(X)_Y$ and $\varepsilon(Y)_X$.

Finally, we need to measure the total absorbances at these two wavelengths $A_T(X)$ and $A_T(Y)$. The problem in determining the contribution to the absorbance of either analyte is that we have two unknowns. We can

PRACTICAL EXAMPLE 6.2

Determine the concentration of caffeine and aspirin in a commercial analgesic preparation.

Caffeine and aspirin are both found in many commercial analgesic preparations. Both compounds absorb in the UV–visible range and the spectra overlap to some extent, so that at any one wavelength the absorbance observed will be the sum of their individual absorbances.

Method

Step 1:

The first step is to measure the individual absorbance spectra of caffeine and aspirin. To do this, we need to make up standard solutions of both compounds.

Caffeine: (1 mmol dm^{-3}); Dissolve 38.8 mg caffeine in 50 cm^3 of methanol, add 10 drops of 4 mol dm^{-3} NaOH (bench dilute NaOH), and warm in a covered beaker for 15 min on a steam bath. Once the solution has cooled, transfer to a 200 cm^3 volumetric flask and top up to the mark with distilled or de-ionized water.

Aspirin (acetyl salicylic acid): (5 mmol dm^{-3}); Dissolve 180 mg aspirin in 50 cm^3 methanol, add 10 drops of bench NaOH and warm in a covered beaker on a steam bath for 15 min. Once the solution has cooled, transfer to a 200 cm^3 volumetric flask and make up to the mark with de-ionized or distilled water.

If we run a UV–visible spectrum of the two compounds between the wavelengths of 200 and 350 nm (ensuring the use of quartz cuvettes) we should obtain spectra similar to those in Figs 6.3 and 6.4.

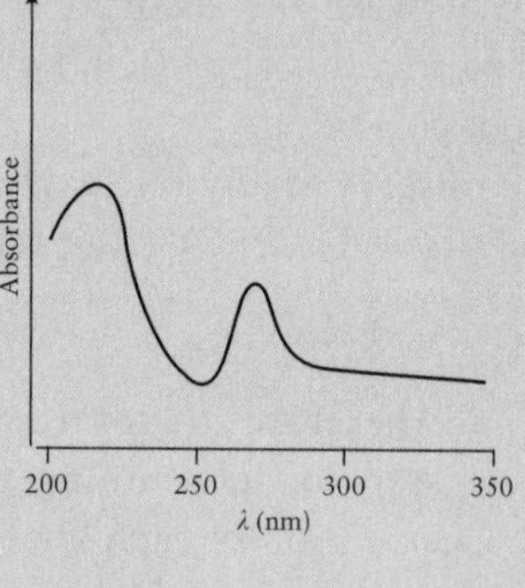

Figure 6.3 UV–visible spectrum for caffeine.

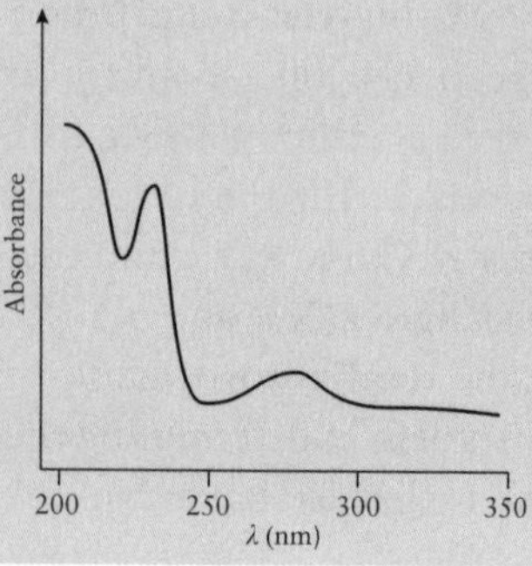

Figure 6.4 UV–visible spectrum for aspirin.

Step 2:

From the absorbance spectra determine the λ_{max} for the two compounds, which in this case are found to be approximately 210 nm for caffeine and 230 nm for aspirin.

Step 3:

Determine the absorbances for caffeine and aspirin across a range of concentrations at the two λ_{max} and plot calibration curves, as shown in Figs 6.5 and 6.6, respectively. Using the Beer–Lambert law determine the molar absorptivities of both aspirin and caffeine at both wavelengths.

Since $A = \varepsilon cl$, it follows that the slope of the caffeine calibration graph will be equal to εl. Since in this case a cuvette with a 1 cm pathlength was used:

- **The molar absorptivity ε of caffeine at 210 nm is 8510 $dm^3\,mol^{-1}\,cm^{-1}$.**
- **In a similar manner the molar absorptivity of caffeine at 230 nm is found to be 2120 $dm^3\,mol^{-1}\,cm^{-1}$.**

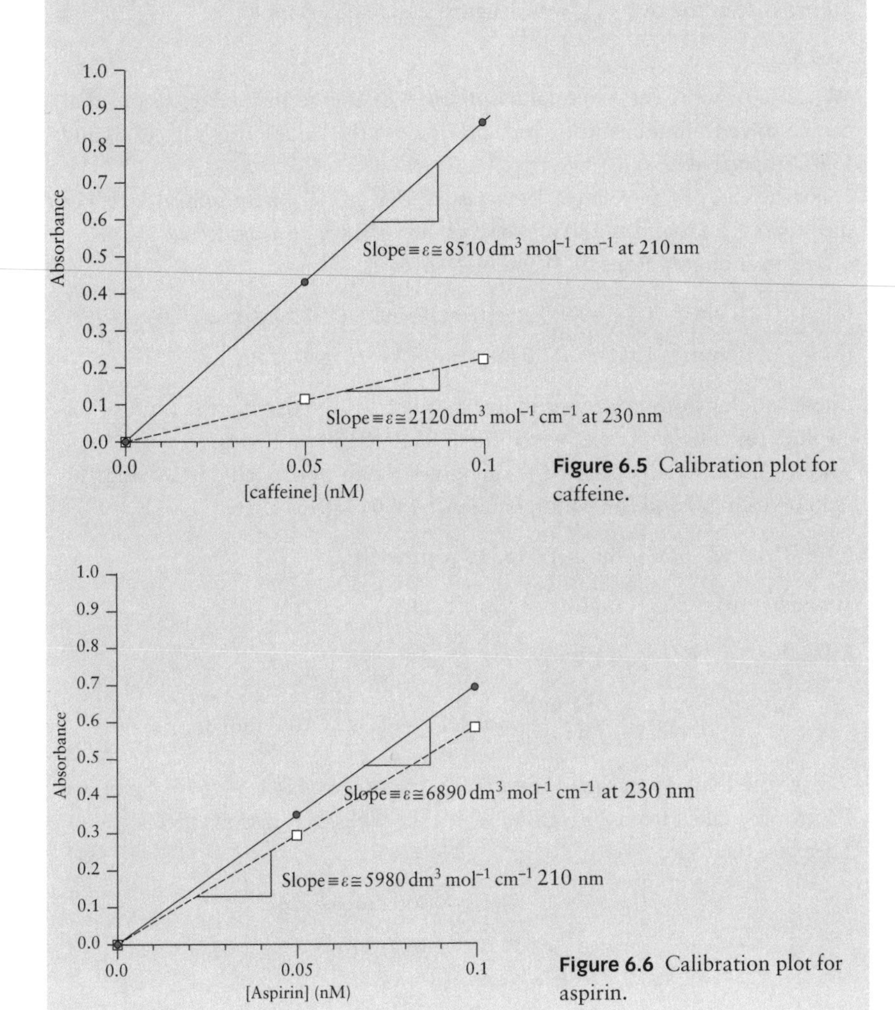

Figure 6.5 Calibration plot for caffeine.

Figure 6.6 Calibration plot for aspirin.

- **The aspirin calibration graph gives us a molar absorptivity for aspirin at 230 nm of 6890 $dm^3 mol^{-1} cm^{-1}$.**
- **The molar absorptivity, ε, for aspirin at 210 nm is found to be 5980 $dm^3 mol^{-1} cm^{-2}$.**

Step 4:

Prepare a solution of the analgesic by grinding the tablet and dissolving the powder in 50 cm^3 methanol, add 10 drops bench NaOH, and warm in a covered beaker on a steam bath for 15 min. Once the solution has cooled, transfer to a 200 cm^3 volumetric flask and make up to the mark with de-ionized or distilled water.

Ten cm^3 of this solution should be transferred to a 1000 cm^3 flask and diluted to the mark with distilled water.

Next, run a UV–visible spectrum of the solution containing both caffeine and aspirin between the wavelengths of 200 and 230 nm. Determine the total absorbance at the two λ_{max} wavelengths (210 and 230 nm).

Step 5:

We can now write the two total absorbances seen in terms of expressions that can be solved simultaneously, and substitute in the values into Eqns (6.3) and (6.4), respectively.

For the sake of this example, let us assume that at 210 nm the total absorbance measured = 1.11, and at 230 nm the total absorbance measured = 1.03.

Substitution into Eqns (6.3) and (6.4) gives us:

(a) A_T (210 nm) = 1.11 = (8510 [caffeine] × 1) + (5980 [aspirin] × 1)

(b) A_T (230 nm) = 1.03 = (6890 [aspirin] × 1) + (2120 [caffeine] × 1)

These simultaneous equations must be solved. If we consider the ratio of the ε values for caffeine, this gives a ratio of 8510 : 2120 or 4.014.

If we multiply (b) by 4.014, this gives us an approach for solving the unknowns in (a) and (b). Multiplying (b) by 4.014 gives (c).

(c) 4.014 = (27 656 [aspirin]) + (8510 [caffeine])

We may now use (c) − (a) to give (d)

(d) 2.904 = 21 676 × [aspirin]

$$\therefore \text{[aspirin]} = \frac{2.904}{21\,676} \text{mol dm}^{-3} = 1.34 \times 10^{-4} \text{ mol dm}^{-3}.$$

Since the solution was made up to 1000 cm^3, there was 1.34×10^{-4} mol but 10 cm^3 was taken from 200 cm^3 in which the tablet was dissolved.

There are therefore

$$[(1.34 \times 10^{-4})/10] \times 200 \text{ mol aspirin}$$
$$= 2.68 \times 10^{-3} \text{ mol aspirin.}$$

Relative molecular mass of aspirin = 180.16.
There are thus $2.68 \times 10^{-3} \times 180.16$ g aspirin in the tablet.
The tablet therefore contains 0.4828 g (~500 mg) aspirin.

Substituting the concentration of aspirin into (a) gives:

$$1.11 = 8510\,[\text{caffeine}] + [1.34 \times 10^{-4} \times 5980]$$

$$\therefore 8510\,[\text{caffeine}] = 1.11 - 0.83 = 0.31$$

$$[\text{caffeine}] = \frac{0.31}{8510} = 3.64 \times 10^{-5}\ \text{mol dm}^{-3}$$

Since the solution was again made up to 1000 cm^3, there are 3.17×10^{-5} mol but 10 cm^3 was taken from 200 cm^3 in which the tablet was dissolved.
There are therefore: $[(3.64 \times 10^{-5})/10\,] \times 200$ mol of caffeine $= 7.28 \times 10^{-4}$ mol of caffeine

Relative molecular mass of caffeine = 194.19.
There are therefore $7.28 \times 10^{-4} \times 194.19$ g caffeine in the tablet.
The tablet therefore contains 0.141 g (141 mg) caffeine.

solve this by describing the total absorbances we observe for the two analytes at $\lambda_{max}(X)$ and $\lambda_{max}(Y)$ in the form of two simultaneous equations which may then be solved.

Now the absorbance due to either analyte may be described by the Beer–Lambert law, and the absorbance at any wavelength will be equal to the absorbance due to X + Y.

It follows that at $\lambda_{max}(X)$:

$$A_T(X) = \varepsilon(X) \cdot \text{conc X} \cdot l + \varepsilon(Y)_X \cdot \text{conc Y} \cdot l \qquad (6.3)$$

and at $\lambda_{max}(Y)$:

$$A_T(Y) = \varepsilon(Y) \cdot \text{conc Y} \cdot l + \varepsilon(X)_Y \cdot \text{conc X} \cdot l \qquad (6.4)$$

These two expressions may then be solved as simultaneous equations to give us the concentrations of X and Y.

6.5 Analytical UV–visible fluorescence spectroscopy

In Chapter 5 we saw that in some instances photons that are absorbed can give rise to emissions of light due to the relaxation of the electron in an atom or molecule. Light is normally emitted over a short time-scale and we termed this effect *fluorescence*. Similar processes that give rise to the emission of light over extended periods of time are normally known as ***phosphorescent*** effects. Hot bodies that give rise to an extended and continual emission of light are said to give rise to ***incandescent*** radiation and the effect is known as ***incandescence***.

In this chapter, we shall consider a few further examples of how molecular fluorescence may be utilized for a range of practical analyses. Remember that the emitted (fluorescent) radiation will always be of a longer wavelength (and so less energetic) since some of the energy will be dissipated as heat and only the remaining energy will be available for the emission of a photon.

6.5.1 Fluorescent labelling

One of the biggest limitations of fluorescent spectroscopy is that very few molecules naturally fluoresce. One way of increasing the applicability of the technique is to ***fluorescently label*** the analyte with a compound or group that renders it fluorescent.

Molecules such as amino acids or proteins that contain NH_2 groups may, for example, be labelled with compounds such as fluorescamine to render them fluorescent. Fluorescent labelling has found a number of applications including, the 'fingerprinting' of complex mixtures and identifying their source of origin. In this way, oil spillages may often be traced to individual refineries or tankers since crude oils from different oil fields may—after suitable labelling—give rise to highly characteristic fluorescent spectra.

A number of chemical reactions can also give rise to the emission of UV or visible light and as such are known as ***chemifluorescent*** reactions. Light is emitted via the relaxation of an excited chemical state produced as a result of the chemical reaction. Chemifluorescent reactions may be exploited for the quantitative determination of a number of analytes at very low concentrations. Cobalt in this context may be determined down to ppm levels via the catalytic oxidation of luminol (*N*,*N*-3-aminophthaloylhydrazine) by hydrogen peroxide.

6.6 Raman spectroscopy

In Raman spectroscopy, high-intensity visible light is focused onto a sample. A laser is normally used for this purpose. The majority of light that is not absorbed will pass straight through the sample or be elastically scattered (i.e. scattered without any change in its wavelength); this scattering is known as ***Raleigh scattering*** and is of no analytical significance. A very small percentage of the light will, however, undergo ***inelastic***

scattering (typically <0.001%), and this gives rise to a change in frequency of the radiation to either longer or shorter wavelengths; this behaviour is known as the ***Raman effect***. The very small percentage of light that undergoes inelastic scattering explains why lasers are normally used for the incident light source. The inelastic scattering process involves photons either gaining or losing energy and this gives rise to ***Stokes*** or ***anti-Stokes*** lines, respectively.

This effect occurs as a result of an interaction between the incident photons and the vibrational (and possibly the rotational) levels of the molecules. The interaction cannot be described in terms of a simple absorption process but rather as a transfer of some of the energy of the incident photon to the molecule—or some energy from the molecule to an incoming photon. ***It is for this reason that we can think of the photons that emerge from the sample as being the same photons that comprised the incident radiation.***

In some respects, the ***Raman effect*** can be thought of as a phenomenon that bears some similarities to molecular fluorescence effects since the Raman effect also involves the emission of photons following excitation by incident radiation and the emission of photons. However, in this case, some of the energy of the incident photons is lost via the excitation of ground state electrons to higher vibrational states. Fluorescence must involve the absorption of radiation with the re-emission of photons at a different wavelength. By contrast, the incident radiation, *must not* be absorbed to any appreciable Raman effect.

Raman radiation is typically measured at 90° to the incident radiation. Many instrumental formats are manufactured and a simplified schematic of one popular arrangement is shown in Fig. 6.7. The emitted radiation is detected via a monochromator and photometer so as to record the wavelength corresponding to the Raman wavelength.

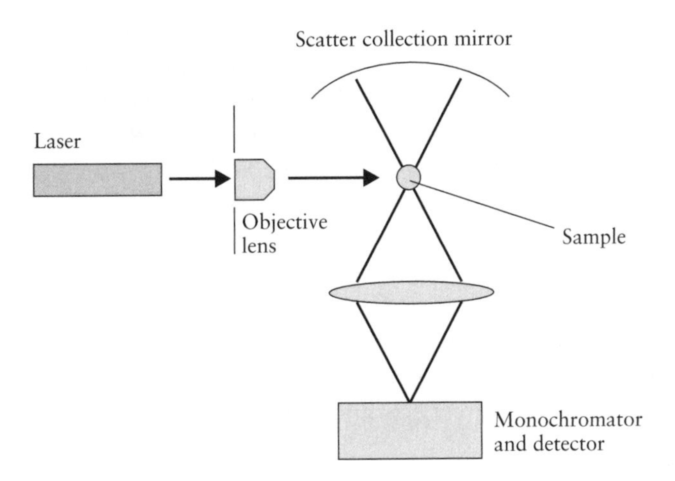

Figure 6.7 Schematic of simplified Raman instrumentation.

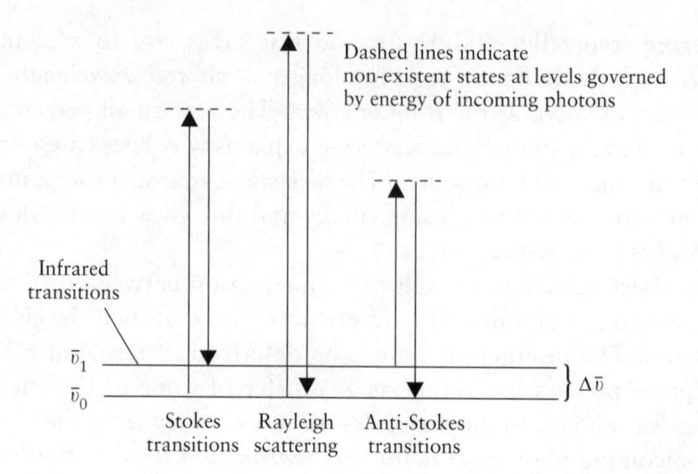

Figure 6.8 Origin of Raman Stokes and anti-Stokes lines.

The wavelength (expressed in wavenumbers, $\bar{\upsilon}$) of the scattered Raman radiation $\bar{\upsilon}_R$ can be related to the wavelength (again in wavenumbers, $\bar{\upsilon}$) of the incident radiation $\bar{\upsilon}_i$ and the absorbed shift, $\Delta\bar{\upsilon}_i$ in the scattered radiation, Eqn (6.7):

$$\bar{\upsilon}_R = \bar{\upsilon}_1 \pm \Delta\bar{\upsilon} \tag{6.7}$$

*where $\Delta\bar{\upsilon}$ is the shift in wavelength observed expressed in wavenumbers and is termed the **Raman shift**.*

We say that the Raman spectral lines are either due to ***Stokes transitions****. Stokes lines are associated with a shift to a lower frequency than the incident radiation*, that is, energy is taken from the sample to excite electrons to a more energetic level. *Anti-Stokes lines, by contrast, are associated with the relaxation of electrons to less energetic levels:* in this case, energy is imparted to the photons as they pass through the sample. The change in the frequency of the incident and transmitted radiation in either case is known as a ***Raman shift***.

Figure 6.8 shows how the Stokes and anti-Stokes transitions relate to the frequency of the incident radiation and the scattered Rayleigh radiation that does not undergo any shift in frequency. Rayleigh scattering is always far more intense than any scattering due to the Raman effect. For a molecule that exhibits Raman effects the Stokes lines will always be more intense than any anti-Stokes lines observed.

Raman effects are observed if the molecule is ***polarizable***, that is, if the shape of the molecule may be altered without the generation of a dipole moment. Absorption of IR radiation (Chapter 12) necessitates the generation of a dipole moment, so it follows that molecules that show absorption in the IR region of the electromagnetic spectrum will almost always fail to show Raman effects and vice versa.

Raman spectroscopy is always carried out in the visible region of the electromagnetic spectrum. The frequency of the incident radiation does not affect the observed Raman shift and so we can use any frequency that does not give rise to absorption and/or fluorescent effects that might complicate the measurement of the Raman effect. It is important to note that fluorescence effects, when present, are likely to swamp the faint Raman radiation. Before the advent of lasers it was common to utilize high-intensity mercury lamps in conjunction with band-pass filters to isolate the emission line at 435.8 nm. Lasers today, however, provide for higher intensity and highly monochromatic light sources that are inexpensive and moreover convenient to use. One of the most commonly used light sources is the helium–neon laser, which gives an intense line at 632.8 nm corresponding to a red light, that is unlikely to give rise to fluorescence. This wavelength may cause some problems since the Stokes lines will be seen at a longer wavelength and the sensitivity of most photo-multiplier tubes falls off above wavelengths of ~650 nm. Argon lasers that produce radiation at a wavelength of 488 nm are frequently used although caution should be exercised since these may give rise to fluorescence in some samples.

EXAMPLE 6.3 Carbon tetrachloride exhibits Stokes and anti-Stokes transitions at 218 cm^{-1} when irradiated with a helium laser with an incident wavelength of 632.8 nm. At what wavenumbers and wavelength will the Raman lines be observed?

Method

1. Calculate the wavenumbers for the incident radiation.
2. Calculate the wavelengths of the anti-Stokes and Stokes transitions.

Step 1: The incident radiation has a wavelength of 632.8 nm which is equivalent to 6.3×10^{-5} cm.

Expressed in wavenumbers this is equivalent to 15 873 cm^{-1}.

Step 2: The anti-Stokes transition is therefore observed at:

$$15\,873 + 218\ cm^{-1} = 16\,091\ cm^{-1}\ (1/16\,091\ cm^{-1} \equiv 621.5\ nm)$$

and the Stokes transition is observed at:

$$15\,873 - 218\ cm^{-1} = 15\,655\ cm^{-1}\ (1/15\,655\ cm^{-1} \equiv 638.8\ nm)$$

6.6.1 Applications of Raman spectroscopy

Raman spectroscopy should be thought of primarily as a qualitative rather than a quantitative analytical tool. Raman spectroscopy is particularly useful for the structural identification and 'fingerprinting' of an analyte, within a sample in a very similar manner to the way in which IR spectroscopy is often used, although it should be remembered that Raman active compounds may not normally yield an IR spectrum and vice versa.

Samples for Raman spectroscopy normally demand careful filtration or centrifugation to remove any suspended particulates that would increase Rayleigh scattering effects. It is also very important to ensure that mixed samples do not contain compounds capable of fluorescing, since as we have mentioned already, any trace of fluorescence will in most cases completely mask out any Raman wavelength effects.

6.7 Microwave spectroscopy

Microwave radiation can be adsorbed by a number of molecules in order to change its ***rotational quantum number***, J. For this reason, ***microwave spectroscopy*** is often also known as ***rotational spectroscopy***. For molecules to be able to interact with—and so absorb microwave radiation—they must first be polar. It follows that non-polar molecules, for example, those with a linear structure (e.g. CO_2), a homo-diatomic structure (N_2, O_2, etc.), a tetrahedral structure (such as CH_4), or an octahedral structures (e.g. SF_6) *will not* display any microwave spectrum. By contrast, heteronuclear diatomic molecules, (such as HCl), and other polar polyatomic molecules do display rotational spectra. The most useful frequency range for microwave spectroscopy is between approximately 8 and 40 GHz.

There are no absorption microwave characteristics that are characteristic of specific functional groups but, rather, it can be shown that microwave spectra consist of a series of lines separated by $2B$, where B is the ***rotational constant*** for a particular molecule. A representative microwave spectrum for *p*-chlorotoluene is shown in Fig. 6.9. If different isotopes within a molecule alter the dipole moment of a compound, separate peaks are shown for each; Fig. 6.9, for example, appears as a series of doublets due to the natural occurrence of the ^{35}Cl and ^{37}Cl isotopes

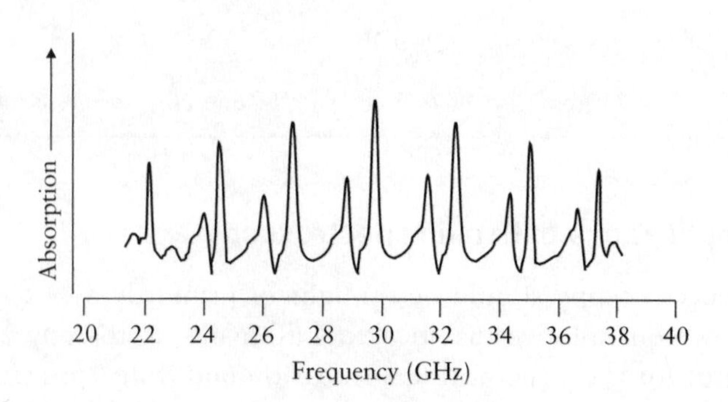

Figure 6.9 Microwave spectrum for *p*-chlorotoluene.

within the environment. The rotational constant, B, can in turn be used to calculate the moment of inertia, I, for that molecule, Eqn (6.8):

$$B = \frac{h}{8\pi^2 cI} \tag{6.8}$$

where h is the Planck constant, and c the speed of light.

If we can determine I, then we can also measure very accurately the bond lengths of polar molecules. Microwave spectroscopy is commonly used, for the fingerprint identification of polar molecules from the lines within the separation, since these are well catalogued and as we have already seen are intrinsically related to bond lengths.

6.8 X-ray fluorescence spectroscopy

X-ray fluorescence spectroscopy is one of the most powerful tools for identifying and quantifying heavy elements, either within compounds or in their elemental form. X-ray fluorescence spectroscopy is not *normally* capable of distinguishing between different compounds that might contain the same heavy element, although in exceptional cases this may be possible.

X-ray fluorescence spectroscopy, like UV–visible spectroscopy, is associated with electronic transitions (Chapter 6). X-rays are, however, more energetic than UV or visible radiation, so X-ray fluorescence spectroscopy is associated with the excitation and relation of electrons with larger energetic transitions. X-ray fluorescence occurs as a consequence of incident X-rays being absorbed and this may give rise to the promotion of inner-core electrons to the unoccupied valence orbitals of the atom followed by the relaxation of these electrons with the subsequent emission of photons in a similar manner to conventional UV–visible fluorescence spectroscopy. The incident radiation need not be of a specific frequency ***but must be sufficiently energetic*** to cause excitation of the innermost electrons to cause the X-ray fluorescence; it follows that for the identification of a particular element, the incident radiation must be greater than a characteristic threshold frequency.

Radiation in the X-ray region is normally generated either by the bombardment of a metal target with a beam of high speed electrons or via the use of a radioactive nuclide (see Section 13.2). A number of different X-ray tubes are commonly used, although tungsten target and iron-55 radioactive sources are amongst the most common since they both produce X-rays that are sufficiently energetic for most purposes.

The emitted fluorescent X-rays will always be of a lower frequency than the incident radiation. Again, this is similar to UV–visible fluorescence effects, since some of the energy of the incident photons will always be dissipated as heat upon the relaxation of excited electrons—and so the fluorescently emitted photon will be less energetic than the photons of the incident radiation. X-ray fluorescence spectroscopy may only be used to *identify* elements with atomic weights greater than approximately 20 and may only really be used for *quantitative* determinations for elements with atomic weights greater than about 40. The reason X-ray fluorescence effects are only seen with the heavier elements is that it is only these elements that possess sufficient electrons to be able to give rise to the electronic transitions from the inner shells to the outer valence electron orbitals that are sufficiently large to correspond to the absorption/ emission of X-ray radiation. In many cases, the relaxation of the excited electrons may occur via a series of stepwise electronic transitions, which will give rise to the emission of a number of quantised photons that will be seen as discrete lines within the characteristic X-ray fluorescence spectrum for a given element.

X-ray fluorescence spectra may be measured by instruments based on either (a) energy dispersive or (b) non-dispersive principles. We shall consider the principles of both instruments in turn.

6.8.1 Energy-dispersive X-ray fluorescence spectrophotometers

A schematic diagram of the components of an energy dispersive X-ray fluorescence spectrophotometer is shown in Fig. 6.10. Energy-dispersive X-ray spectrophotometers utilize polychromatic X-ray sources comprising either a radioactive nuclide, or X-ray electron bombardment tube. Dispersive X-ray fluorescence spectrophotometers offer obvious advantages in that they do not possess any moving parts since all of the frequencies that go to make up the fluorescence spectrum are measured simultaneously. In instruments of this type, the photo-multiplier detectors

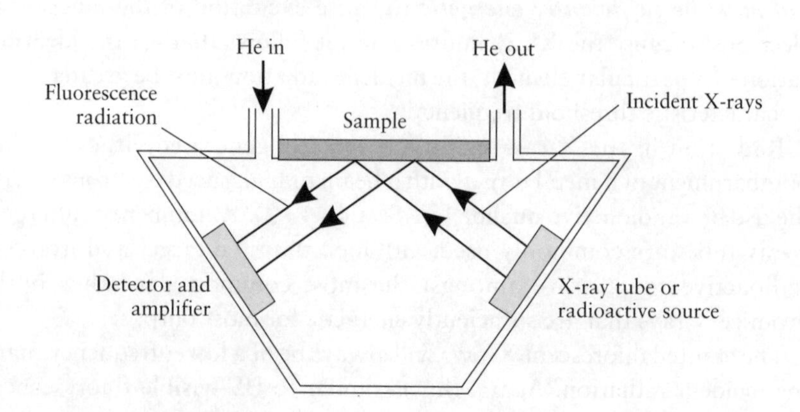

Figure 6.10 Schematic of an energy dispersive X-ray fluorescence spectrometer.

are coupled to the appropriate instrumentation to allow for energy and therefore wavelength discrimination. Energy dispersive instruments also offer some advantages in comparison to non-dispersive instruments in terms of their superior signal to noise ratios.

The principal drawback associated with energy-dispersive instruments is that some loss of resolution occurs below wavelengths longer than approximately 1 Å. Since energy-dispersive instruments measure all of the wavelengths simultaneously, they can provide continuous spectra in which individual lines correspond to the characteristic X-ray fluorescence lines of different elements, as we shall see later.

The area under each peak of spectra of this type may be integrated and related to the concentration of the element to be quantified by comparison to spectra of standard reference materials of known concentrations.

6.8.2 Non-dispersive X-ray fluorescence spectrophotometers

The principal features of a non-dispersive X-ray fluorescence spectrophotometer are shown in Fig. 6.11. The X-ray source is often ^{55}Fe, although other sources are sometimes used. The fluorescence radiation passes through a filter and onto a photo-multiplier based counter; in some instances a pair of matched and twin detectors may be employed as shown in Fig. 6.11. Instruments of this type are normally used for the quantitative determination of individual elements and therefore allow for spectra to be taken across a range of wavelengths.

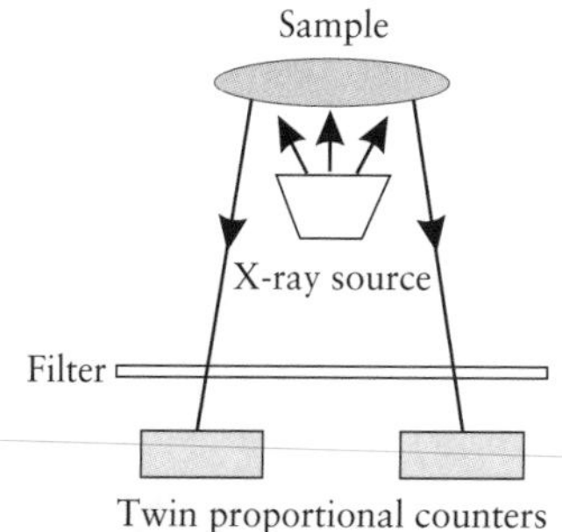

Figure 6.11 Schematic of a non-dispersive X-ray fluorescence spectrometer.

6.8.3 Applications, advantages, and limitations of X-ray fluorescence spectroscopy

X-ray fluorescence spectroscopy unquestionably represents one of the most powerful techniques for the elemental analysis for complicated samples. X-ray fluorescence spectroscopy is normally non-destructive and this means that valuable items such as jewellery, paintings, or archaeological treasures can be analysed with confidence and also lends itself to the examination of very small through to massively large objects. A representative X-ray fluorescence spectrum is shown for a paper sample with print (i.e. ink) on its surface in Fig. 6.12.

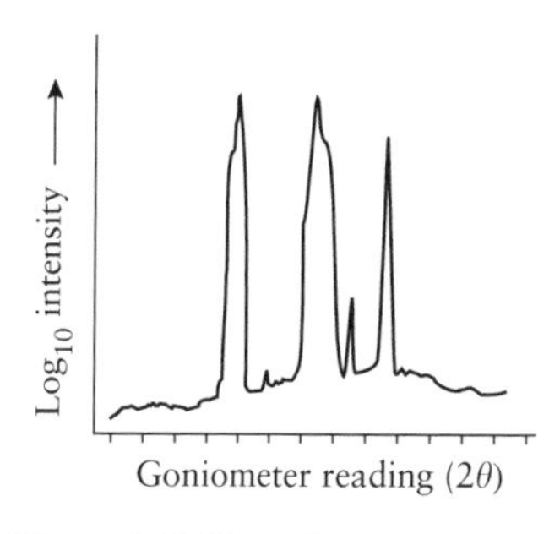

Figure 6.12 X-ray fluorescence spectrum for paper sample with print.

Rapid sampling times greatly facilitate routine sampling, which, with the lack of sample preparation requirements, greatly contributes to the versatility of the technique. There are many examples where X-ray fluorescence techniques may be used in difficult or harsh conditions: the elemental composition of alloys may, for example, be determined during the melt process to allow for the correction of mixture during manufacture.

X-ray fluorescence techniques are also widely used for the analysis of liquid samples, with applications including the determination of pigments within paints as well as heavy (poisonous) metals such as cadmium and barium within petroleum.

X-ray fluorescence spectroscopy has also been widely used for the determination of atmospheric pollutants. For analyses of this type, air is normally drawn through filters to collect the particulate matter, which may then be analysed within the X-ray fluorescence spectrophotometer.

As with any other technique, there are a number of limitations and disadvantages associated with X-ray fluorescence spectroscopy. First, X-ray fluorescence spectroscopy is often not as sensitive as other optical methods with lower limits of detection typically being of the order of ppm levels at best. To counteract this, however, the accuracy and precision of X-ray fluorescence spectroscopy are often equal to—or may even exceed—that offered by other competing techniques such as atomic absorption spectroscopy or UV–visible spectroscopy. Moreover, it is common to have a working concentration range of a fraction of 1–100% for the analyte of interest since the area under the fluorescence peak normally remains proportional to the concentration of the analyte under almost all conditions. Unfortunately, the instrumentation is always laboratory based and tends to be relatively expensive in capital expenditure terms. The cost of instruments varies enormously, however, and the resolution and performance of the machine is normally directly related to the money you are prepared to pay.

X-ray fluorescence spectroscopy cannot be used for the determination of the lighter elements and in cases where rather more complex samples require a more complete stoichiometric analysis, X-ray fluorescence spectroscopy may have to be used in conjunction with other analytical techniques. ***Matrix effects*** represent one of the most serious potential drawbacks with using X-ray fluorescence spectroscopy and may considerably affect the quantitative analysis of a number of elements. There are two main types of matrix effects that we need to consider and these are ***absorption effects*** and the so-called ***enhancement effect***.

Absorption matrix effects occur as a result of the bulk sample matrix absorption of X-rays. It is clear that the X-ray beam will penetrate the sample to some extent. It therefore follows that both the incident X-ray beam and the fluorescent X-ray spectrum will traverse some distance through the sample. It is therefore inevitable that both the incident and fluorescent X-rays will be attenuated to some extent due to absorption of the X-rays by the sample. The instrument will have been calibrated using standard reference materials. However, it does not follow that the absorption spectrum of the sample and the reference material will be the same. If the sample displays a greater absorption of either the incident or fluorescent X-rays, lower than expected results will be observed. Conversely, if the reference material displays greater absorption characteristics, then measured intensities will be falsely elevated. It is important that erroneous increases in the signal are not confused with matrix enhancement effects, which shall be considered next.

Enhancement effects occur when the sample contains an element which, when excited by the incident X-ray beam gives rise to an X-ray fluorescence line and which, in turn, causes further excitation of the element being determined. The intensity of the fluorescence line will clearly be related to the power of the incident radiation and if this is increased we will see a falsely enhanced (and so erroneous) signal.

6.9 Mössbauer spectroscopy

Mössbauer spectroscopy is a fluorescence technique utilizing γ-radiation. So far we have considered: (a) UV–visible fluorescence spectroscopy, which is associated with the excitation and relaxation of atomic and molecular valence electrons; and (b) X-ray fluorescence spectroscopy, which involves the excitation and relaxation of the innermost electrons of heavier elements. Mössbauer spectroscopy is associated with even more energetic processes, and follows the excitation and relaxation of *intra-nuclear* rather than electronic energy levels. A simplified schematic of a Mössbauer spectrophotometer is shown in Fig. 6.13.

Mössbauer spectra yield extremely sharp lines in comparison to those associated with X-ray fluorescence spectroscopy. Iron represents the most widely studied element since the nuclear energy levels of one of its isotopes ^{57}Fe are readily accessible by excitation with γ-radiation. Radioactive ^{57}Co is commonly used as a γ-radiation source for the excitation of ^{57}Fe to a metastable energy level approximately 14.4 keV above the stable ground state. The γ-rays emitted during relaxation to the ground state give rise to Mössbauer spectra that are characteristic for a given nuclide. In the case of iron, the nuclide must be ^{57}Fe and not the more

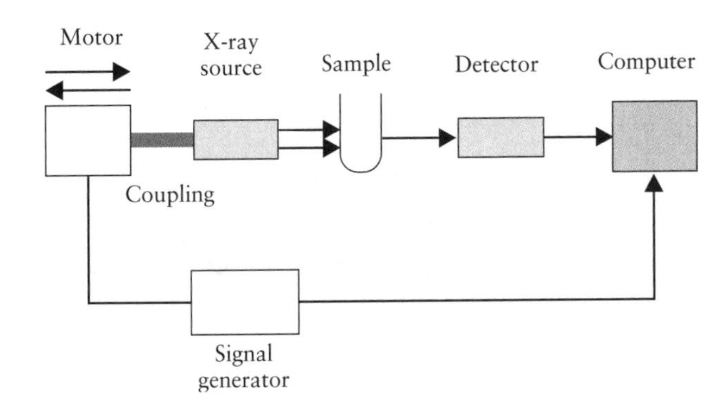

Figure 6.13 Schematic of a Mössbauer spectrometer.

common ^{55}Fe and, even though ^{57}Fe is only found in 2.2% abundance, this is sufficient to allow an analysis to be performed. The sample may, however, be a component within a compound (i.e. not in its elemental form). Different chemical states can give rise to chemical shifts and these must be addressed if spectra are to be interpreted correctly. Chemical shifts occur as a consequence of extremely small changes in nuclear energy levels upon chemical reaction, and the formation of compounds. Since the Mössbauer lines are so narrow, very small changes in nuclear energy states can give rise to chemical shifts in both γ-ray absorption characteristics and the subsequent emission of the Mössbauer spectrum. These problems may be overcome by moving the emitter of the incident radiation relative to the analyte sample and making use of a Doppler shift for the excitation of the sample. Consideration of the frequency shift shows that the sample and incident radiation source should be moved at velocities of the order of a few millimetres per second, which is easily obtainable. Spectrophotometers are therefore designed to allow for movement of the radiation source at a *constant acceleration* relative to the sample. In this way, a range of differing velocities may be scanned in order to ensure that both the initial absorption and subsequent emission of γ-rays are recorded.

The spectra are normally expressed as a function of velocity of the radiation source relative to the sample in millimetres per second. Note that peaks are recorded as a decrease in transmission and therefore point downwards in the convention normally adopted. The spectrum normally consists of one or two peaks due to the major different energy states and these often show hyperfine splitting by the electric and magnetic fields existing within the nucleus. ^{57}Fe, for example, causes six peaks to be seen originating from two main energy levels $I = \frac{1}{2}$ and $I = \frac{3}{2}$, with each being split into a triplet, Fig. 6.14.

Mössbauer spectroscopy may be used to analyse any sample that contains the appropriate element, and may give additional information relating to the crystal structure and valance states of the compounds in

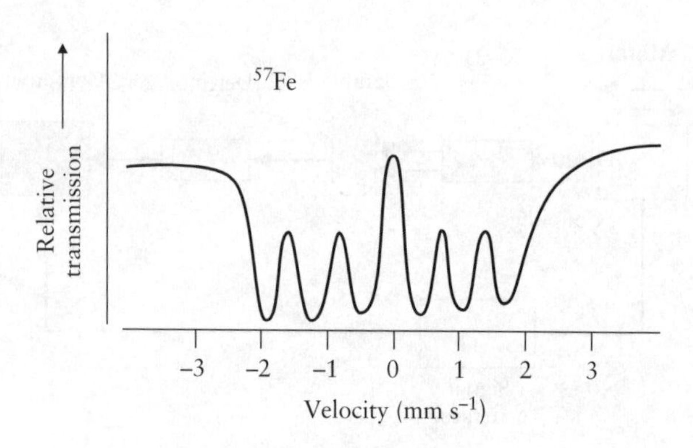

Figure 6.14 Mössbauer spectrum for ^{57}Fe.

which the element resides. Unfortunately, Mössbauer spectroscopy may only be used to analyse samples containing a relatively small number of elements and the majority of applications are based on the detection of ^{57}Fe, ^{61}Ni, and ^{119}Sn. Although there are a number of other elements that can give rise to a Mössbauer effect, these often depend on isotopes, which are in such small natural abundance so as not to be of use for routine analysis.

6.10 UV photo-electron spectroscopy

UV photo-electron spectroscopy (**UPS**) is a technique whereby a sample is irradiated with far UV (high frequency) radiation to initiate the ionization of atoms within the sample by the promotion and ejection of *valence electrons*. The wavelength at which radiation is absorbed corresponds to the energy required to cause ionization of a particular element, which may be exploited for fingerprinting identification. A UPS spectrum for N_2 is shown in Fig. 6.15. The ionization energies are often expressed as *ionization potentials*, in which case they are quoted in keV. The energy levels of valence electrons may of course alter on the formation of compounds and it follows that the UPS ionization energies will in these cases be altered in comparison to their elemental states. Alternatively, photo-electron spectroscopy may also be used to provide information relating to the chemical nature of a compound.

A closely related and more widely used technique is ***X-ray photo-electron spectroscopy*** (**XPS**), which is used to excite and remove electrons from the inner electronic shells lying below the valence shell. X-rays are more energetic than UV radiation and hence may excite and cause ejection of electrons from inner shells.

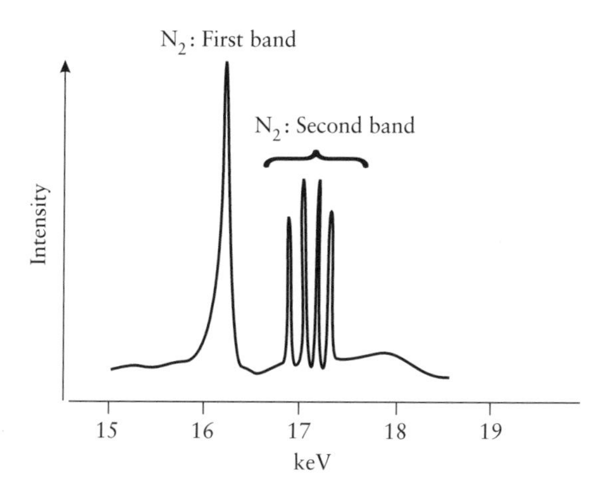

Figure 6.15 UPS spectrum for N_2.

Exercises and problems

6.1. Write the electronic configuration for oxygen.

6.2. An element has an electronic configuration of: $1s^2$, $2s^2$, $2p^6$, $3s^2$, $3p^6$, $4s^1$. Identify the element.

6.3. Describe what is meant by the fluorescent labelling of compounds.

6.4. Explain why some compounds fluoresce while others do not.

6.5. Explain why some analyses fail to obey the Beer–Lambert law.

6.6. Explain why some compounds absorb infrared radiation, others are Raman active but these two phenomena are not observed within simple molecules.

6.7. Explain what is meant by: (i) Stokes; and (ii) anti-Stokes transitions.

6.8. A Raman spectrum for a compound is obtained following excitation with a frequency of 435.8 nm (from mercury). Raman lines are observed at 443 nm and 463 nm. Calculate the Raman shift for both of these lines.

6.9. Manganese concentrations within rock samples are to be determined via X-ray fluorescence. Fluorescence intensity for different concentrations of standards were determined as below:

% Mn	Counts s^{-1}
0.05	100
0.1	200
0.2	405
0.3	598
0.4	810
0.5	998

A rock sample was found to have a count rate of 152 s^{-1}; calculate the concentration in the rock sample.

6.10. A solution is known to contain solely aspirin and caffeine but in unknown concentrations; your task as an analytical chemist is to determine the concentrations of both compounds in this solution.

The UV–visible spectrum of the solution is recorded between the wavelengths of 205 and 300 nm. Caffeine is known to exhibit a λ_{max} at 210 nm with a molar absorptivity of 8510 $dm^3\ mol^{-1}\ cm^{-1}$. Aspirin is known to exhibit a λ_{max} at 230 nm of 6890 $dm^3\ mol^{-1}\ cm^{-1}$. (The molar absorptivity, ε, of caffeine at 230 nm is 2120 $dm^3\ mol^{-1}\ cm^{-1}$. For aspirin, ε at 210 nm is 5980 $dm^3\ mol^{-1}\ cm^{-1}$.) The total absorbance for the caffeine and aspirin containing solution at 210 nm is found to be 1.15 and the total absorbance at 230 nm is found to be 1.02.

From this information, determine the concentrations of caffeine and aspirin within this solution.

6.11. A Raman shift is seen associated with a relaxation of electrons to a less energetic state; will a Stokes line or an anti-Stokes line be observed?

6.12. Carbon tetrachloride is irradiated with an argon laser with a wavelength of 488 nm (20 492 cm^{-1}). Three Stokes lines are observed corresponding to shifts of –459, –314, and –218 cm^{-1}, respectively. Three anti-Stokes lines are observed with shifts of 218, 314, and 459 cm^{-1}, respectively. Calculate the wavelengths and wavenumbers associated with these transitions.

Summary

1. The uncertainty principle states that at any one instance we cannot define exactly where an electron will be.

2. The orbitals and electrons that comprise a principal quantum level are collectively known as a shell and are denoted by the letters k, l, m, and n.

3. The sub-shells of an atom are labelled by both the principal and secondary quantum numbers.

4. An orbital is filled by two electrons and cannot, according to the Pauli exclusion principle accommodate any more electrons.

5. The Aufbau principle dictates that electrons fill orbitals in the order: 1s, 2s, 2p, 3s, 3p, 4s, 3d, 4p, 5s, 4d, 5p, 6s, 5d, 4f, 6p,....

6. Electrons relaxing back to the position of the first electronic ground state are said to give rise to Lyman transitions.

7. Electrons relaxing back to the second, third, and fourth ground electronic ground states within the valence shell are said to give rise to *Paschen, Brackett*, and *Pfund* emissions, respectively.

8. The spin of electrons is described by the spin quantum number, *s*, and this is said to be either clockwise (+) or anticlockwise (−).

9. Diagrams showing which electronic transitions are possible are known as Grotian diagrams.

10. Deviations from the Beer–Lambert law may be due to a number of effects such as interactions of solute molecules with each other or non-monochromatic light sources.

11. The concentration of solutions should be adjusted so that measured absorbance falls between values of 0 and 2 (and preferably 0 and 1).

12. UV–visible determinations of mixtures containing two or more absorbing species may sometimes be achieved via the use of simultaneous equation based resolution of absorbances.

13. Non-fluorescent compounds may sometimes be rendered fluorescent by fluorescent labelling approaches.

14. Raman spectroscopy is based upon the absorption of radiation and the subsequent re-emission of photons of lower energy following excitation of ground state electrons to higher vibrational states.

15. Raman shifts to lower frequencies are known as Stokes transitions. Raman shifts to higher frequencies are known as anti-Stokes frequencies.

16. The absorption of microwave radiation can give rise to changes in the resonance quantum number, *J*, and is so sometimes known as rotational spectroscopy.

17. X-ray fluorescence spectroscopy relies on the absorption of X-rays and the associated promotion of their electrons to the unoccupied valence orbitals of the atom by the relaxation of these electrons and the subsequent emission of photons.

Further reading

Atkins, P. W. and Friedman, R. S. (2005). *Molecular quantum mechanics* (4th edn). Oxford University Press.

Colthup, N. B. (1989). *Introduction to infrared and Raman spectroscopy*. Academic Press.

Lakowicz, J. R. (1999). *Principles of fluorescence spectroscopy*. Kluwer Academic.

Mayo, D. W., Miller, F. A., and Hannah, R. W. (2004). *Course notes on the interpretation of infrared and Raman spectra*. Wiley.

Valeur, B. (2002). *Molecular fluorescence: an introduction—principles and applications*. Wiley-VCH.

Atomic spectroscopy in analytical chemistry

7

Skills and concepts

This chapter will help you to understand:

- The origins of atomic spectral absorption and emission transitions.
- The factors that affect atomic spectra and how spectral lines may be broadened by Doppler, collision, pressure, and temperature effects.
- The principles underpinning flame atomic absorption spectroscopy.
- The operation of a hollow cathode lamp.
- Some applications of flame absorption spectroscopy.
- The principles of flame absorption spectroscopy and the key components of a flame absorption spectrometer.
- The advantages and disadvantages of flame absorption spectroscopy in comparison to alternative approaches.
- What is meant by a Delves cup and how one may be used.
- The principles of graphite furnace spectrometry.
- The operation of electric arc and spark sources and the use of these in atomic emission spectroscopy.
- The operation of a flame photometer for flame emission spectroscopy.
- The scope of applications for flame photometry.
- The principles underlying inductively coupled plasma (ICP) emission spectroscopy.
- Analytical applications for ICP techniques including fluorescence approaches.

7.1 The origins of atomic spectral transitions—an introduction

In Chapters 4 and 6, we saw how the promotion and relaxation of valence electrons can give rise to UV–visible and fluorescence spectroscopy.

With few exceptions, applications involve the determination of compounds since very few elements exist in the free elemental state, and it is the valence electrons which are involved with the bonding whether it be ionic or covalent in nature. The bonding orbitals and therefore the electrons are primarily associated with the compound and not an individual atom *per se*.

In this chapter, we shall consider ***atomic spectroscopic techniques***, which also depend upon and exploit electronic transitions. Atomic spectroscopic approaches involve and exploit the electronic transitions of electrons not involved in bonding, that is, either non-valence electrons within atoms or compounds (or the valence electrons of elemental atoms or ions).

Atomic spectroscopic techniques may broadly be classified as either being based upon ***atomic emission*** or ***atomic absorption*** processes. Atomic emission spectroscopy normally involves the emission of photons as electrons relax from excited states back to their ground states. ***Atomic absorption*** techniques, by contrast, are based upon the capture of photons, as electrons are promoted or even lost in the formation of an ion.

7.2 The nature of atomic absorption spectra

Atomic spectra originate from electronic transitions between *atomic* orbitals and give rise to extremely narrow absorption lines, with wavelength band widths typically of the order of 0.1 nm or so. Remember that electronic transitions to or from bonding orbitals cause far broader absorption peaks and this type of spectrum is observed within UV–visible molecular spectroscopy. Atomic absorption peaks are very much narrower than those seen in molecular UV–visible spectroscopy since there are no bonding orbitals within the outer valence electronic shell. Atomic absorption spectra typically occur in the UV, visible, and IR regions of the electromagnetic spectrum.

7.3 Factors that affect atomic spectra

There are a number of factors that can effect the width, intensity, or even frequency of an atomic emission or absorption line. Generally, the factors that affect atomic absorptions also affect atomic emissions.

7.3.1 Lifetimes of transition states, the Heisenberg uncertainty principle, and line widths

Quantum theory tells us that if the lifetime of the transition state approaches infinity, then the line width of the spectral line width will approach zero. Similarly, as the transition time increases, the line width will also increase. The Heisenberg uncertainty principle states that we cannot predict the exact position and momentum of an electron at any given time. It follows that if we cannot predict these properties for an electron at any given moment before or during an electronic transition, then we are left with uncertainties in the transition times and therefore the widths in the spectral lines. Theoretical line widths as described by quantum mechanics are sometimes known as ***natural line widths*** and are often in the order of 10^{-4} Å.

7.3.2 Pressure or collision broadening

Collisions between the emitting or absorbing species with other atoms or ions lead to small changes in ground state energy levels, and hence a spread in the wavelengths of the absorbed or emitted radiation. Within a flame, the collisions occur largely between the atoms of the analyte and various combustion products of the fuel and these may result in broadening of two or three orders of magnitude over the natural line widths. Similar effects are often observed as a consequence of collisions within the plasma excitation media of inductively coupled plasma (ICP) spectroscopy. Line broadening in hollow cathode and electrode discharge lamps is largely caused by collisions between the emitting atoms themselves.

7.3.3 Doppler broadening

The Doppler effect, whereby an apparent wavelength change occurs on moving quickly towards or away from a sound source, is well known. Doppler shifts are also often observed in atomic spectroscopy and can lead to apparent shifts in the wavelengths at which atomic absorption or emission lines are observed and in this way lead to their broadening. This phenomenon occurs due to the emitting or absorbing species travelling at high velocities as a result of thermal excitation from the flame or plasma. Atoms moving directly towards the detector will emit radiation that will be observed by the detector as being of a slightly shorter wavelength than radiation emitted by atoms travelling at right angles to the detector. In a similar manner, radiation emitted by atoms moving away from the detector will be observed as being of a longer wavelength than that emitted by atoms moving at right angles to the detector.

A range of atomic velocities relative to the detector will be observed due to: (a) the natural variation in the speeds of the atomic population; and

(b) the different directions in which atoms will travel relative to the detector. These factors will together cause the detector to record a number of different wavelengths in the form of a Gaussian distribution curve centred in intensity around the natural line width.

This spread in observed wavelengths due to the Doppler effect causes a net broadening of the emission lines.

The Doppler effect causes line broadening in atomic absorption spectroscopy for exactly the same reasons. The Doppler broadening of atomic absorption or emission lines becomes more pronounced in flames or plasma of higher temperatures due to the increased net velocity of the atoms. Doppler broadening typically leads to a doubling of natural line widths.

7.3.4 Temperature effects and atomic spectra

Temperature effects can alter the nature of atomic spectra in several ways. We have already seen that heating may, via the Doppler effect, lead both to the broadening of atomic spectral lines accompanied by a decrease in peak heights.

Heating (whether it be via a flame, furnace, or plasma) is often used to complete the atomization of a sample for atomic absorption spectroscopy. Generally, the more heat that is supplied, the greater will be the efficiency of the atomization process and this in turn leads to an increase in the atom population and so the intensity of the atomic spectrum. Whilst it should be noted that as the temperature is increased a greater number of atoms will be ionized, this effect is normally insignificant in comparison to the increase in the efficiency of the atomization process.

Atomic emission spectroscopy is further affected by the temperature of the flame or plasma as heat is either used directly or as a contributing factor to electronically excite the atoms to a state from where they emit radiation upon relaxation. It follows that the more heat that is supplied, then the greater will be the proportion of atoms that will be excited; and this, in turn, will lead to an emission line of greater intensity being recorded. Since the temperature has a more pronounced effect on atomic emission spectroscopy, it follows that it is more important that the temperature of the flame in emission spectroscopy be more closely controlled than in atomic absorption techniques, especially if quantitative analyses are to be relied upon.

Atomic absorption techniques typically rely on a larger population of atoms in the sample than emission-based techniques since the sample has only to be in a ground electronic state to be monitored; remember that in atomic emission spectroscopy, only the proportion of the sample thermally excited may give rise to atomic spectral lines. The ratio of unexcited to excited atoms may range from 10^3 to 10^{10} or more. In theory, absorption-based techniques should therefore offer a considerably greater sensitivity over emission-based approaches. There are in fact so many other variables that influence the sensitivity of the two approaches, that atomic emission- and absorption-based techniques typically offer comparable sensitivities and lower limits of detection.

7.4 An introduction to atomic absorption spectroscopy

Atomic absorption spectroscopic techniques involve the quantification of the energy (via wavelength and intensity monitoring) absorbed from an incident radiation source for the promotion of elemental electrons from the ground state. The wavelength and absorption may then be monitored and recorded in the form of a ***spectrum***. We will discuss different forms of absorption spectroscopy in the following sections of this chapter.

7.5 Flame atomic absorption spectroscopy

Flame atomic absorption spectroscopy is the most widely used form of atomic spectroscopy. Parts per million (ppm) levels of many metal ions may be readily determined by means of what has now become a relatively simple experimental procedure. In practice, the technique relies on a supply of free elemental atoms or ions being electronically excited by monochromatic light; the absorption that occurs is then measured by the instrumentation.

The first problem that must be overcome, involves supplying a source of atoms in the elemental (or free elemental ionic) form. This is achieved with a ***nebulizer***, Fig. 7.1, in conjunction with an air/acetylene flame. The first stage is to form a micro-droplet aerosol of the analyte solution by means of a nebulizer. In this process, a peristaltic pump passes a continual supply of the analyte solution into the path of a jet of compressed air to produce a fine mist of tiny droplets. The spray may then be directed into the path of a long and thin air/acetylene flame to give rise to the atomization of the analyte.

The combustible acetylene fuel gas for the flame must be supplied with either air or pure oxygen as an oxidant; the resulting flames are known as air/acetylene or oxy/acetylene flames, respectively.

The combustion gases are mixed prior to combustion; Fig. 7.2. The flame is typically directed through a gas jet slit approximately 10 cm in length and 2–3 mm wide. Acetylene burns to provide temperatures of around 2000°C to 2200°C. If higher temperatures are required, acetylene/nitrous oxide fuel gas mixtures may be used (see Section 7.5.5).

The acetylene and air for flame atomic spectroscopy are mixed prior to being passed through the gas jets and it is at this point the mixture is

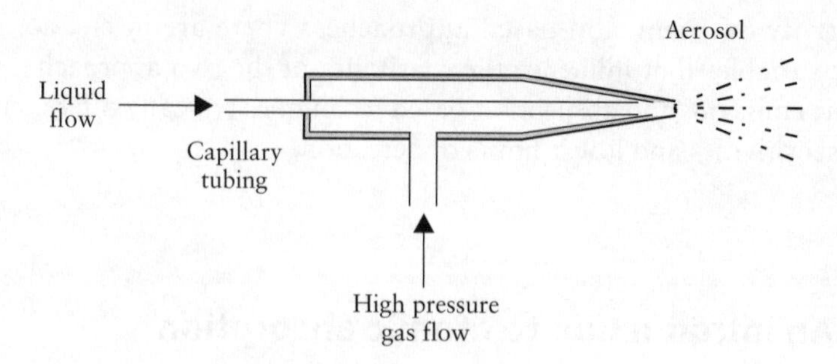

Figure 7.1 Schematic of a jet nebulizer.

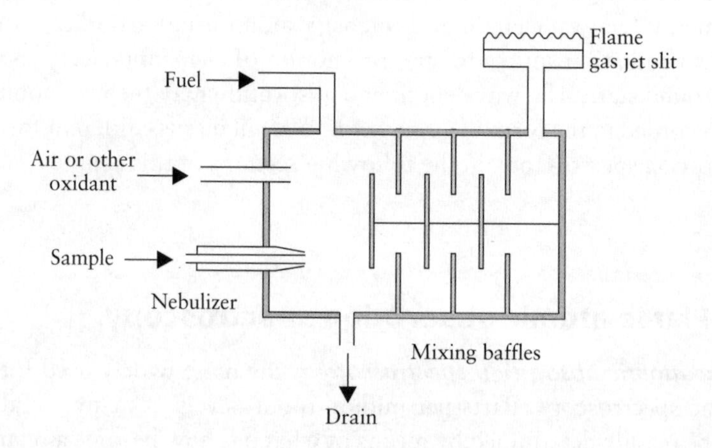

Figure 7.2 Mixing of fuel, sample, and oxidant prior to combustion.

ignited. It is important to vent away the exhaust gases and it is normal practice to place an exhaust hood directly above the exit chimney of the spectrophotometer.

The solvent within the fine droplets of the analyte (which is almost always water) evaporates extremely quickly at these temperatures. The metal salt in turn vaporizes and this is reduced within the high temperatures experienced within the flame to complete the atomization process.

The flame is shaped to allow incident radiation to be passed through a continual supply of the atomized sample; Fig. 7.3. A detector (which is normally a photo-multiplier tube) can then monitor the intensity of the radiation and thus the absorption.

We saw in Chapter 4 that when using UV–visible spectroscopy, the bandwidth of the incident radiation source should ideally be considerably narrower than that of the absorption peak. The same rule applies to atomic absorption spectroscopy if we wish to avoid unacceptable errors. We should recall that the absorption bandwidths of many UV–visible spectra often span several tens of nanometres and in this situation a monochromator with a bandwidth of less than ~0.5 nm is perfectly acceptable. Atomic absorption bandwidths are, however, extremely narrow,

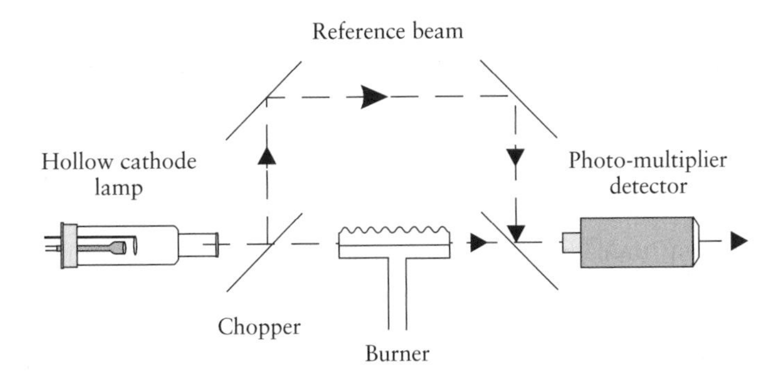

Figure 7.3 Schematic of a flame atomic absorption spectrometer.

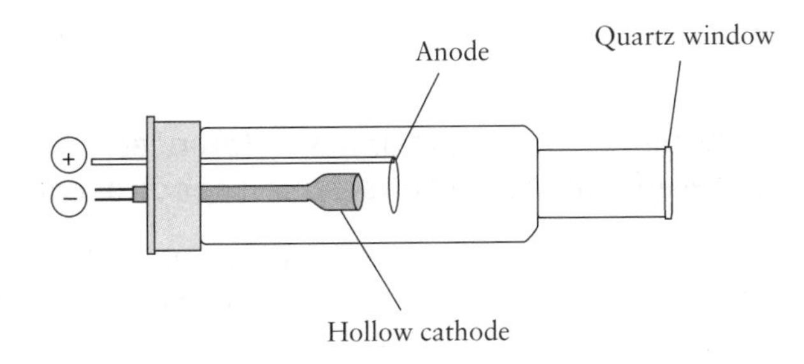

Figure 7.4 Schematic of a hollow cathode lamp.

typically being of the order of <0.01 nm. For this reason, it is especially important that the light source used is capable of producing radiation centred at exactly the correct wavelength and with an extremely narrow bandwidth. This can be achieved by use of a ***hollow cathode lamp***, Fig. 7.4, which is a gas discharge lamp that exploits the emission characteristics of the same element that we wish to monitor. It follows that separate and individual hollow cathode lamps are required for each element to be monitored. Elements may also only be analysed one at a time. Modern atomic absorption spectrophotometers often possess a carousel containing a number of different lamps so that different elements may be determined at will by simply rotating a different lamp into position.

The hollow cathode lamp (Fig. 7.4) consists of a tubular shaped hollow cathode—from which it derives its name, together with a small ring shaped anode as depicted in Fig. 7.4. Both electrodes are encapsulated within a glass envelope that is filled with neon at low pressure. Each lamp is rendered specific for a particular metal by coating the cathode's surface with the element to be analysed. When a high voltage is applied across the electrodes, the neon ionizes and the impact of positively charged ions at the cathode surface vaporizes some of the cathode lining. Further collisions give rise to electronic excitation in these ions and, upon relaxation, the light emitted corresponds to the exact absorption characteristics of a particular element. Remember that, in this context, the electronic configurations of the ions within the hollow cathode lamp will be identical to the atomized analyte within the air/acetylene flame. Upon relaxation the ion returns to the cathode and once more forms part of the coating. In this way, the lining is not consumed but rather is dynamically stripped and re-formed with time.

It should be remembered that the emission wavelength bandwidth should be narrower than that the absorption bandwidth of the analyte. The hollow cathode lamp will indeed always provide a narrower emission bandwidth than the absorption bandwidth of the analyte in the air/acetylene flame. In this context, the absorption and emission bandwidths of an element both broaden with increasing temperature. The analyte is heated to a temperature of 2000°C or so in the air/acetylene. The hollow cathode lamp, by contrast, operates at temperatures only slightly elevated from room temperature.

7.5.1 Absorption measurements and interferences encountered with atomic absorption spectroscopy

The photo-multiplier tube detector continually monitors the intensity of the transmitted light, and hence the absorption. The signal therefore depends on a constant supply of the analyte sample being fed into the flame via the nebulizer. Most spectrophometers are designed so that a piece of peristaltic polypropylene pump tubing can be fed into the sample for a few seconds as seen within the representative output for Ni; Fig. 7.5. The spectrophotometer normally then outputs a near steady state signal for a few seconds (Fig. 7.5); most modern spectrophotometers automatically take a mean value for the absorption over a few seconds, which is saved as a computer file or hard copy output. As with many analytical techniques, the quantitative nature of the procedure can be compromised by a number of interferences. In the case of atomic absorption spectroscopy, interferences normally arise from errors in determining the intensity of the transmitted radiation, or the atomization of the analyte.

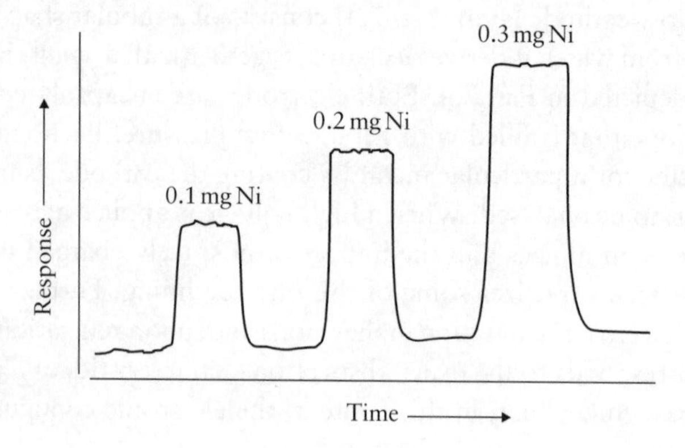

Figure 7.5 Atomic absorption output response of different concentrations of Ni.

The air/acetylene flame emits light like any other flame. It is clearly not possible to place the flame within a light proof enclosure, which means that unless we are to work within a completely darkened room (which again is not feasible) then background light can act as an interference. If we are to monitor the absorption of the analyte within the flame, we must therefore have some way of differentiating the intensity of the transmitted light from the hollow cathode lamp and background light sources. In practice, an atomic absorption spectrophotometer utilizes two approaches in conjunction with each other to separate the desired signal from the background light interference.

The first step is to place a monochromator between the flame and the photo-multiplier tube detector. In this way, most of the light from the flame that is from chemiluminescent reactions is excluded. Notice how the placement of the monochromator in the path of the *transmitted* light beam compares with UV–visible spectroscopic applications in which we place a monochromator in the path of the *incident* radiation source. It is of course possible that the flame emits light at the same wavelength as the analyte, and so we must take precautions to measure only the contributions from the analyte emission.

Another factor that we should remember is that however carefully the combustion jets are designed, the shape of the flame will be dynamic and this will introduce random fluctuations into the effective optical path via turbulence.

The hollow cathode lamp is normally stimulated via a pulsed staircase voltage supply, that gives rise to a discontinuous pulsed output from the lamp. The frequency at which the lamp is pulsed is normally 50 Hz. The photo-multiplier detector may be synchronized with the voltage supply profile by means of a lock-in amplifier or phase sensitive detector. Together these methods allow an extremely simple and effective approach for determining, in almost any light conditions, the intensity of the transmitted radiation from the hollow cathode lamp and hence the absorption that occurs. An analogy to this approach is to think of how we can determine the speed of rotation of a wheel by means of a stroboscope.

The other principal types of interference that must be guarded against are those that are caused by chemical effects. Atomic absorption spectroscopy relies on the thermal atomization of the analyte within the flame. It is sometimes possible for the analyte to form thermally stable compounds in the flame, which do not display atomic absorption effects and therefore will not contribute to the analytical signal. Chemical interferences of this type will normally lead to erroneously low estimations of the analyte concentration. Aluminium is particularly notorious for giving rise to problems of this type since it can form thermally stable aluminates with a number of metals such as calcium. The aluminates have variable stoichiometries, and are normally written in the form of 'Ca–O–Al'. The normal approach for overcoming these problems involves introducing a nebulized stream of another element that has a higher affinity for the interferent than the analyte, while also being capable of forming compounds that are sufficiently thermally stable to withstand the temperatures of the air/acetylene flame. In the case of aluminium, we can add lanthanum, which will very readily form compounds of the form Ca–O–La, again of variable stoichiometry.

7.5.2 Applications of flame atomic absorption spectroscopy

Flame atomic absorption spectroscopy is most widely used to determine trace levels of metals within different types of sample. We have already seen that the sample must be in solution so that it may be passed into the flame in the form of a nebulized spray and atomized. This does not preclude the analysis of solid samples providing they can first be dissolved. Metallic alloys such as steels can often be dissolved within nitric acid and in this way the cobalt concentration within steels may, for example, be determined with great accuracy within different grades of stainless steel—either for comparative purposes or as a quality control measure at the site of manufacture.

Environmental samples for the monitoring of river or esturine waters downfield of sites of industrial activity require little or no pre-treatment. Soil and rock samples may often also be analysed with ease. Other industrial applications include the analysis of the metallic content of paints and polymers, both for safety and quality control purposes. Flame atomic absorption spectroscopy is also widely used in medicine to determine metal content of a range of pharmaceutical preparations and even of whole blood, blood serum, or urine samples. The calcium and iron contents of fortified foods such as cereals and children's foods are also regularly analysed via atomic absorption spectroscopy. Although atomic absorption spectroscopy is used primarily to determine metals, the technique can be exploited for a number of different analyses via combinational chemistry. A good example in this context is the analysis of aldehydes via Tollen's reagent that releases two moles of elemental Ag upon the oxidation of each mole of aldehyde.

7.5.3 The relative advantages and disadvantages of flame absorption spectroscopy

Flame absorption spectroscopy is the most used form of atomic spectroscopy due to the relative ease by which determinations down to ppm level concentrations, or even lower, may be readily obtained. There are a number of disadvantages with this approach, however, some of which can be addressed by relatively minor modifications in the technique. One of the principal drawbacks associated with the use of the flame is that it must firstly evaporate the solvent within the analyte sample and then atomize the analyte within a short time period. It is estimated that as little as 0.1% of the sample is atomized and passes through the path of the light beam. Methods that permit a greater efficiency of the atomization process such as the graphite furnace (which we describe in the next section) potentially offer several further orders of enhanced sensitivity. These limitations can largely be overcome by firstly calibrating the instrument on a regular basis and secondly via the use of certified reference materials.

7.5.4 The calibration of flame atomic absorption spectrophotometers and the use of certified reference materials

The absorption observed by flame spectrophotometers may vary considerably from one instrument to another with the same type of sample. This behaviour differs significantly from UV–visible spectroscopy where we might expect the readings from one instrument to concur with another. There are several reasons why atomic absorption spectrophotometers cannot be treated in the same manner. In essence, it is not practical to either cause total atomization of the sample or monitor all of the absorption that occurs. If we calibrate the instrument, however, with the use of suitable certified reference materials, we can normally obtain reliable and quantitative determinations even at very low concentrations, since the behaviour of the instrument should be reproducible. The controlling influences may include any factors that affect the temperature of the flame, the sensitivity of the photo-multiplier tube, and any chemical interferents.

Certified reference materials can often be purchased that are formulated to have very similar compositions as the samples for analysis (e.g. river water, metallic alloys, etc.). The rationale here is that the reference samples will contain very similar quantities of interferents, and will be subjected to the same pre-treatment regimes as the real samples. The calibration readings are normally performed with dedicated software within the instrument for this purpose. An alternative approach may be to add known quantities of a certified reference analyte to the sample in the form of a standard additions calibration. This may be preferable if the sample to be analysed is rather complex or if suitable standardized certified reference materials are not available and this could, for example, include the analysis of soil or paint samples.

7.5.5 The use of nitrous oxide/acetylene versus air/acetylene flames

While the lower detection limits of flame atomic absorption spectroscopy are favourable in comparison with many other techniques, the use of an air/acetylene flame does impose some limitations. The flame is merely used to atomize the sample and for this reason the control of its temperature is crucial. Ideally, we would wish to atomize the entire sample without the creation of ions that absorb light at different wavelengths and it follows that the design of the flame burners and the gaseous fuel mixture are therefore crucial to ensure a constant, uniform, and ideally controllable temperature. An explosion risk exists if the flame burns back to the mixing chamber. To prevent this occurring the gases are pre-mixed prior to combustion. The design of the gas jets to produce a high velocity gas flow and the use of air instead of pure oxygen both minimize the risk of explosion although every precaution should always be taken. The use of air limits the maximum temperature, although if temperatures greater than around 2200°C are required, nitrous oxide can be used instead of air to allow temperatures of up to approximately 3000°C to be achieved.

7.5.6 Delves heating cups

An improvement in the sensitivity and lower limits of detection obtainable can be achieved by using the air/acetylene flame *solely* as a heat source for the atomization of the sample. In this arrangement, the solvent is evaporated prior to the sample being introduced to the flame. The dissolved sample is placed within a metal receptacle called a ***Delves cup***. Delves cups are often made of nickel, so cannot be used with acidified samples without prior neutralization. Alternatively, the cups may be made of other suitable materials such as tantalum.

The Delves cup is gently heated on a hot plate or other heat source to evaporate the solvent. Only at this stage should the cup be placed into the hottest part of the flame to vaporize the sample. The Delves cup is often used in conjunction with a quartz tube which resides in the hottest part of the flame into which the vaporized sample is passed. In this way, the quartz tube acts as a holding tank that appreciably lengthens the time that atoms display absorption and this often greatly increases the sensitivity of the technique.

7.6 Graphite furnace atomic absorption spectrophotometers

Graphite furnaces totally negate the need for a flame by utilizing a graphite electrical resistive heating element shaped in the form of a hollow tube. Samples are therefore atomized by an ***electrothermal atomization*** approach. The graphite furnace approach has become the most widely adopted form of electrothermal atomization. Electrothermal atomization approaches can achieve a near 100% atomization efficiency, and if we remember that air/acetylene methods frequently only permit as low as 0.1% of the sample to be atomized, we can readily see that the graphite furnace approach can potentially offer a 1000-fold increase in sensitivity. The graphite tube surrounds the sample with a very small volume and (when combined with electrical heating) allows a far greater control of the temperature than is obtainable by means of a flame. There are, moreover, no fluctuations in the optical path length due to thermal convection within the flame. The unwanted generation of light by the combustion of the fuel gases within the flame is of course also eliminated.

Samples are injected through a window in the roof of the graphite tube furnace via a micro-pipette. Graphite furnace tubes are typically 5 cm long and have diameters of approximately 1 cm. The tubes are designed to be easily interchangeable for ease of cleaning and replacement. Electrical contacts for the heating element are made at either end of the tube. The entire tube is normally surrounded by a water-cooled metallic jacket. An external noble gas stream (normally argon) prevents the ingress of oxygen from the atmosphere, which would otherwise result in incineration of the tube. Inert gas is also passed in through both the ends of the graphite furnace tube and leaves through the sample injection window. This gas not only serves to exclude oxygen but also helps carry away vapour generated during the initial heating stages. The temperature is normally firstly gently raised to evaporate the solvent. Once the sample is completely dried, the temperature may be rapidly ramped to vaporize the analyte. The incident light source from the hollow cathode lamp passes in through one end of the graphite tube and through the path of the vaporized sample. The vaporized sample typically resides in the light path for around 1 s or more and this also helps enhance the sensitivity of this approach in comparison to flame absorption spectroscopy. The transmitted light may then be monitored as it passes through and out of the other end of the graphite tube.

Since the temperature of the sample is ramped it is vaporized over some period of time even though the desired temperature is reached as quickly as possible. In this case the output signal of a graphite furnace atomic absorption spectrophotometer rises to a maximal value and then drops again to the baseline value.

Interferences arising from inter-element effects can unfortunately often be far more pronounced in graphite furnace based instruments in comparison to their effects within flame spectrophotometry. Background absorption also tends to be more pronounced in graphite furnace based determinations, which can be especially significant in samples that contain high concentrations of either organic or inorganic salts such as, for example, within biologically derived or environmental samples. Standard addition type calibrations normally overcome these problems and should therefore be regarded as normal good experimental practice.

Graphite furnace tubes, Fig. 7.6(a), are often used in conjunction with ***L'vov*** platforms, Fig. 7.6(b). The platforms are also made of graphite and are located within the graphite tube just beneath the sample entrance port; Fig. 7.6(b). In this way, the sample lands on the L'vov platform and not the internal walls of the graphite furnace. The sample is firstly gently heated to evaporate off the solvent in the normal way. When the sample is heated for atomization, the temperature of the internal section changes less rapidly, however, than the walls of the furnace and this tends to lead to more reproducible results.

The porosity of the graphite used for the construction of the graphite furnace and indeed a L'vov platform may lead to the erroneous lowering of sample signals since the sample may be absorbed into the graphite pores. This problem may be largely overcome by coating all graphite surfaces with a thin layer of pyrolytic carbon in order to seal the pores of the graphite surface. Pyroltyic carbon is deposited by passing an inert mixture of hydrocarbon gas such as methane through the tube at a highly elevated temperature. Coatings of this type can be deposited layer by layer to form extremely homogeneous and impervious coatings. Some graphite furnaces are designed so that the coatings may be replenished during their lifetimes.

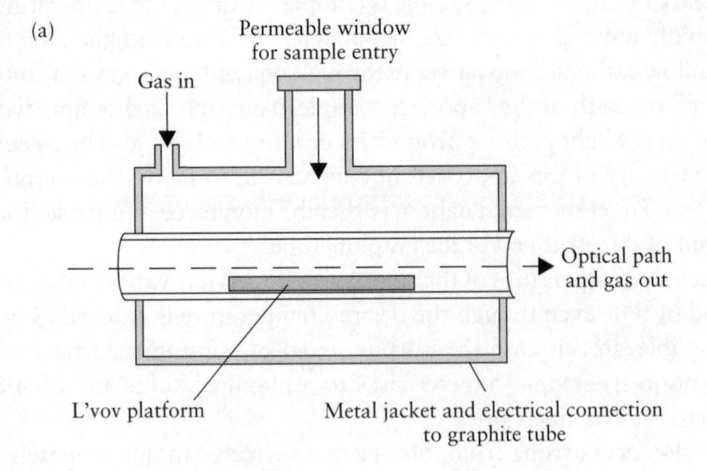

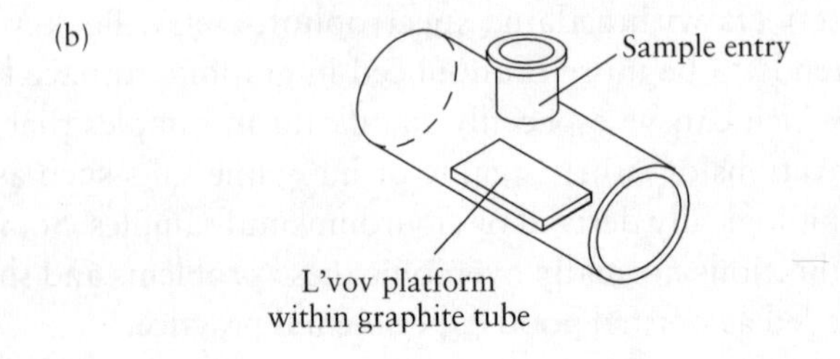

Figure 7.6 Schematic of (a) a graphite furnace tube and (b) a L'vov platform.

7.6.1 Applications of graphite furnace atomic absorption spectroscopy

Graphite furnace atomic absorption spectroscopy is rapidly gaining popularity due to the simplicity, enhanced safety, and inherently greater sensitivity that the technique offers in comparison to flame atomic absorption spectroscopic approaches. Lower limits of detection are frequently in the order of parts per billion levels or lower, and indeed represent some of the most sensitive analytical techniques available. The graphite furnace also allows the analysis of samples with volumes as little as 10 μl or less and so has become a powerful tool for micro-analysis applications.

Graphite furnace atomic absorption spectroscopy may be used for the determination of many different forms of sample ranging from solutions through to solids and even gaseous vapours if one is prepared to perform some form of sample pre-treatment procedure. Mercury, for example, may be analysed from environmental air, water, or even soil samples following its collection in a potassium permanganate solution. The permanganate oxidizes organomercury compounds as well as elemental mercury to form a solution of mercuric ions. Excess permanganate may be removed with hydroxylamine and a reducing agent such as $SnCl_2$ is then added to reduce the mercuric ions back to elemental mercury. Liquid mercury has a sufficiently high mercury vapour pressure, that if an inert gas such as N_2 is bubbled through an ***Erlenmeyer flask***, mercury vapour can be swept through to a gas sample tube and this in turn can be subsequently discharged into the graphite furnace sample entry port.

7.7 An introduction to atomic emission spectroscopy

Each of the different forms of atomic emission spectroscopy are based upon the emission of light upon the relaxation of elemental electrons from excited states. There are several ways in which we may cause the initial excitation and these give rise to the different names by which these different forms of spectroscopy are known. The most common forms of atomic emission spectroscopy include electrical arc or spark emission, flame emission, and plasma emission based techniques.

7.8 Electric arc or spark emission spectroscopy

Electrical arc or 'spark' based forms of atomic emission spectroscopy are often known simply just as 'atomic emission spectroscopy', and care should therefore be taken not to confuse these techniques with other forms of atomic emission spectroscopy.

Electrical arc or spark emission spectroscopies are often used to determine the inorganic content of many samples whether they be solids, liquids, or gases. A schematic diagram of a spark emission spectrometer is shown in Fig. 7.7(a). A solid sample is normally packed as powder into a hollow thimble-shaped electrode and a high-voltage discharge arc or spark is passed between this and a counter electrode; Fig. 7.7(b). The electrodes must be made out of a material that will not interfere with the analysis; carbon is often used since it is a good conductor, is thermally stable, and may be easily shaped. Silver and copper electrodes are sometimes used when these elements cannot interfere with the analysis. Porous graphite electrodes are also used for the uptake of liquid samples to facilitate their handling and analysis. An alternative approach, however, involves evaporating the analyte solution onto the surface of a supporting electrode. The counter electrode is often shaped into the form of a tapered cone, since this normally gives rise to the most stable and reproducible spark. Some instruments employ two identical sample supporting electrodes, between which the spark or arc is passed. If only one sample-supporting electrode is to be used this is normally polarized to act as the anode. Metallic samples for analysis may sometimes be directly used as one of the electrodes since some instruments are equipped to be able to hold and make electrical contact with irregularly shaped samples. In some instances, it may be preferable to machine the sample into the form of a regularly shaped electrode. Powdered solid samples may sometimes be compressed under high pressure with powdered carbon or copper to form 'briquettes' or pellets for ease of handling. The heating effect of the spark or arc causes the sample to evaporate. Elemental electrons that are thermally excited quickly relax and give rise to the emission of photons—the energy of which is equivalent to $h\nu$. Each element gives rise to a number of different emissions since an excited electron can fall via a number of different transitions before it returns to its ground state.

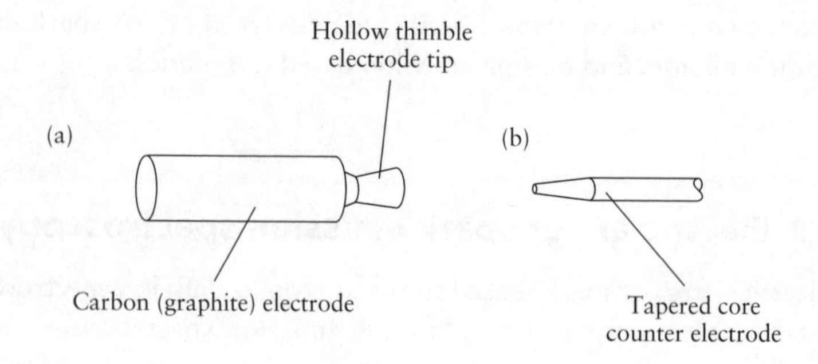

Figure 7.7 Hollow thimble shaped electrode and high voltage discharge.

7.8.1 Spark sources

Periodic sparks induced by regular pulsed dc potentials give rise to the most reproducible and reliable atomic emission spectra. The frequency of these sparks is typically in the range of 180–220 Hz. A periodic burst of sparks for 20 s or so is normally required for the measurement of each spectrum. The *net* current flow over a period of time is normally far smaller with spark emission sources than with an arc, due to the discontinuous nature of the spark, although momentary currents may be in the order of 1000 Å or more. The overall temperature experienced with spark sources will be considerably lower than when an electrical arc is used. Charge, however, is carried within a very narrow pathway or *streamer* through the centre of the spark where temperatures of 40 000–45 000 K are reached. It is in this region where the analyte is stimulated to give rise to atomic emissions and, for this reason, atomic spectra from spark sources tend to be more intense than those arising from arc sources.

7.8.2 Spark sources and laser microprobe analyses

Lasers may, in some cases, be used in conjunction with spark sources for ***microprobe atomic emission*** spectroscopic determinations. The laser (e.g. ruby) is pulsed to periodically vaporize very small samples of material from a spot surface area of 5×10^{-3} mm^2 or less into the gap between a pair of graphite electrodes, which are again excited by means of a synchronized spark source to generate the emission of the atomic spectrum.

7.8.3 Arc sources

The arc is normally initiated either by means of a spark or, alternatively, by bringing together the two electrodes until they almost touch, and then separating them to the required gap. The current which flows between the two electrodes may vary from 1 to 40 Å or more and may be stimulated by either dc or ac potentials ranging from 20 to 300 V or more. The arc is often applied for approximately 20 s to allow time for each spectrum to be recorded.

Charge is passed through the arc by means of electrons and ions within a plasma. The ions are formed via thermal excitation as a result of the flow of current. The process is normally self-sustaining once the arc has been formed between the two electrodes. The atomized analyte enters the plasma as it is volatized from the sample supporting electrode(s) with the temperature of the plasma typically being in the range of 4000–5000 K. It is common to experience the emission of intense interferent lines from –CN containing molecules from the atmosphere unless a controlled gas stream of CO_2, He, and Ar is used to continually flush the space between the electrodes.

It is clear that the intensity of the atomic spectrum will be dependent on the concentration of the analyte within the arc at any given time. However, it is important to realize that the intensity of the atomic spectrum will also be a function of time since different samples will atomize at different rates. In some cases, spectra will gradually reach a maximal intensity which decays slowly with time while others are most easily observed at shorter time intervals.

Arc atomic spectroscopy does not offer the same degree of precision as, for example, ICP or flame atomic emission spectroscopy, although the level of sensitivity may be greater and this can offer advantages if the determination of trace elements is required. Chemical interferences tend moreover to be less significant with arc atomic spectroscopy due to the high temperature within the arc.

7.8.4 Instrumentation for the measurement of atomic emission spectra

The light corresponding to the atomic emission is passed through either a dispersive grating or prism where it is differentiated into its respective wavelengths. The spectrum may then be recorded either by means of a photographic film, or more commonly by electronic means. The electronic recording of atomic spectra certainly facilitates the determination of the intensity of the spectral lines, which in turn helps us perform quantitative determinations. In many situations, it is, however, often easier to record the spectrum photographically especially if the primary purpose of the analysis is for the identification of one or more elements within a sample, that is, a qualitative determination. The camera may often be lowered or raised in a stepwise fashion to allow a number of spectra to be sequentially recorded one above another and this clearly helps comparisons to be made between, for example, a certified reference material and samples for analysis.

A schematic of a spark emission atomic emission spectrometer is shown in Fig. 7.8. The final spectrum consists of a series of lines each corresponding to a particular electronic transition. 'Real' samples will of course contain two or more elements capable of giving rise to atomic emissions and in these cases the spectra will be superimposed. It is therefore important that before we conclude that a particular element has been identified, a number of different lines are matched between the standard reference sample and the sample for analysis. A discharge emission spectrum for a Ni and Cd containing sample is shown in Fig. 7.9. Quantitative determinations often prove difficult via discharge emission spectroscopy although the intensity of the spectral lines reflects the concentrations of individual elements within the sample. The intensity of these lines may, moreover, be affected by factors such as the sensitivity of, for example, photographic film and other environmental ambient influences. Despite these difficulties, discharge

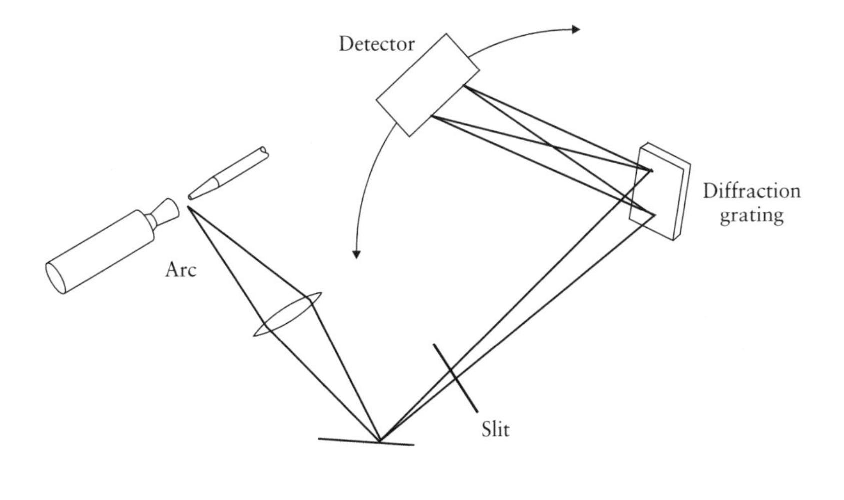

Figure 7.8 Schematic of apparatus for measurement of atomic emission spectra.

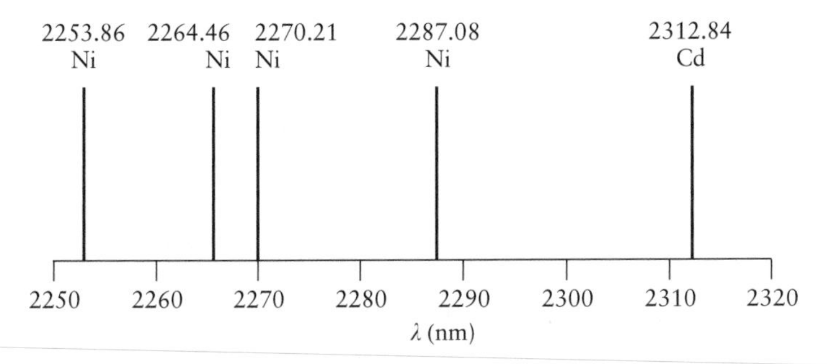

Figure 7.9 Atomic spark emission spectrum for Ni and Cd containing sample.

emission spectroscopy is often routinely used for quantitative determinations for a number of applications.

7.8.5 Quantitative applications of arc and spark emission spectroscopy and the use of standards

In favourable conditions, both arc and spark forms of atomic emission spectroscopy are capable of parts per billion levels of sensitivity. The greatest problems associated with performing quantitative determinations are associated with: (a) overcoming the effects of the many variables that can affect the sensitivity of the analyses; and (b) erroneous measurements caused by effects of chemical interferents. In essence, if quantitative determinations are to be performed, then it is necessary to use a series of internal standards, certified reference materials and/or possibly standard additions based calibrations.

We can recall that internal standards (Chapter 2) include additions of different species other than the analyte and are used for reference purposes if the sensitivity of a process is prone to fluctuation. In the case of arc or spark atomic emission spectroscopy, we should introduce an element with as similar chemical properties to the analyte element as possible. Ideally, the reference element should therefore have very similar ionization energies and excitation levels to the analyte element so that temperature fluctuations affect two species as closely as possible. If these conditions are met, then the ratio of the spectral lines for this element should increase or decrease if environmental or instrumental factors modulate the magnitude of all responses. Compromises have in practice sometimes to be made, especially if we are to use an internal standard to monitor a multi-element analysis.

The analytical approach must always be calibrated. This is more easily achieved by preparing a series of standard samples and constructing a calibration curve. It is important that these samples resemble as closely as possible the samples to be analysed both in their physical form and their chemical composition (except of course in the case of the analyte element itself).

It essential that any arc or spark atomic emission analysis be verified with certified reference materials (Chapter 2). Samples containing known concentrations of the analyte in question can be obtained from organizations such as NAMAS or the LGC (see Chapter 2) and determined using the calibration curve; an even better approach could be to perform an analysis on a certified reference material prior to being informed as to the concentration of the analyte and to incorporate this approach as part of an accreditation process (Chapter 2).

7.9 Flame emission spectroscopy (flame photometry)

Flame emission spectroscopy is also known as ***flame photometry***, and is widely used in clinical laboratories to quantify potassium and sodium as electrolytes in blood, although many other metallic elements may also be determined by this technique.

The analyte sample should normally be prepared in the form of a solution. The solution is then normally continually passed via a peristaltic pump through a nebulizer and then into the path of a flame in the form of a fine micro-droplet mist. Some of the sample will be atomized within the flame and then excited, leading to the promotion of electrons; upon relaxation the electronic transitions gives rise to the characteristic wavelength emissions for the constituent elements within the sample. A relatively low temperature flame of methane/air or natural gas/air is usually used, since this imparts sufficient thermal energy for the determination of sodium and potassium whilst minimizing the effects of possible interference from other species. A typical atomic spectrum for potassium is shown in Fig. 7.10.

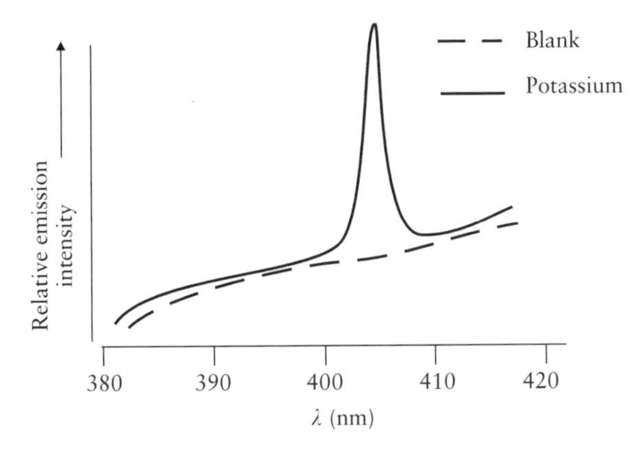

Figure 7.10 Atomic emission spectrum for potassium in an air/methane flame.

Instruments are normally designed to record either a complete spectrum for the simultaneous qualitative determination of a number of differing elements or, alternatively, to allow the quantitative measurement of individual elements. Alternatively, some instruments are designed to permit either type of analysis.

If a complete spectrum is recorded, individual elements may be identified by the λ_{max} within the emission spectra. If individual elements are to be measured quantitatively then the intensity of the output spectrum may be measured over a narrow wavelength range of 0.05 nm or less corresponding to one of the emission λ_{max} for that particular element.

Complete emission spectra are typically recorded by passing the output radiation through an interference filter (to remove the contribution from the combustion of the gaseous fuel for the flame) and then onto a dispersive prism or filter and photo-multiplier tube. The prism or filter is normally rotated so that individual wavelengths are monitored. In this way, the entire spectrum is recorded.

Instruments for the quantitative determination of individual elements, such as sodium or potassium, contain filters designed to only allow the passage of a very narrow bandwidth of radiation (typically 0.05 nm or less) corresponding to one of the characteristic emission lines of the element in question. Some instruments are designed with two or more complete optical measuring systems to permit for the simultaneous determination of different elements within the same sample. In these instruments, the light being emitted from the flame is split into a number of separate beams (one for each element) and these are directed through different filters and onto separate photo-multiplier tubes. An alternative design of the instrument involves having only one photo-multiplier tube but having multiple filters for the determination of different elements. It follows, however, that only one element can be analysed at a time in instruments of this type.

EXAMPLE 7.1 The potassium content of a blood serum sample is to be determined by flame emission spectroscopy using a standard additions approach. Two 1 cm^3 additions are added to 10 cm^3 water aliquots and these are labelled A and B. Ten μl of a 0.05 M KCl solution is added to water sample A. The emission signals in arbitrary units of the two water samples A and B are determined to be 144.0 and 78.9, respectively. Calculate the concentration of K^+ within the serum.

Method

1. Calculate the molar quantity of KCl added to the sample.
2. Calculate the signal due to the addition of KCl.
3. Calculate the signal due to K^+ in the blood serum sample.
4. Calculate the concentration of K^+ in the blood serum sample.

Step 1: Calculate the molar quantity of KCl added to the sample.

$$\text{The quantity of KCl added} = 1 \times 10^{-6} \times 0.05 \text{ moles of KCl} = 5 \times 10^{-8} \text{ moles of KCl}$$

Step 2: Calculate the signal due to the addition of KCl.

$$\text{Signal due to KCl addition} = 144.0 - 78.9 \text{ arbitrary units} = 65.1 \text{ arbitrary units.}$$

Step 3: Calculate the signal due to K^+ in the blood serum sample. The number of moles of K^+ in the serum sample is then:

$$5 \times 10^{-8} \times \frac{78.9}{65.1} \text{ moles of } K^+ = 6 \times 10^{-8} \text{ moles of } K^+$$

Step 4: Calculate the concentration of K^+ in the blood serum sample. The volume of the original sample was 1 cm^3 and so:

$$[K^+] = \frac{6x10^{-8}}{1} \times 1000 \text{ mol dm}^{-3} = 6.06 \times 10^{-5} \text{ mol dm}^{-3}$$

$$\text{or} \sim 0.06 \text{ mmol dm}^{-3} K^+$$

7.9.1 Interferences and calibration in flame emission spectroscopy

The interferences encountered in flame emission spectroscopy are typically similar to those in flame atomic absorption methods. Fluctuations in the intensity of the light emitted by the flame may also cause problems. It is therefore necessary to calibrate the instrument at regular intervals with standard reference materials or to spike samples via the standard additions approach (Chapter 2). Fluctuations in the intensity of the flame should affect the determination of two elements by a proportionate amount. An alternative approach involves spiking samples with an element that is known to be absent and to use this as an ***internal standard*** by which fluctuations may be determined and thus corrected for. Lithium is widely used in this context as an internal standard for the clinical determination of sodium and potassium in blood or serum.

7.10 Plasma emission spectroscopy

Plasma emission spectroscopy utilizes a plasma as the excitation source for an atomic emission. A plasma is an electrical conducting gaseous mixture that contains a significant number of cations and electrons. Plasmas may be formed: (a) via the use of microwave field sources; (b) by passing a dc current between electrodes; or (c) by inducing a current flow via the use of a high powered radio-frequency electromagnetic field. DC plasmas are typically formed within instruments employing ***plasma jet*** sources whereas the plasmas that are formed via radio-frequency coils are known as ***inductively coupled plasmas or*** ICPs. In this section, we shall consider dc plasma and ICP-based approaches, since these are the two most widely used forms of plasma atomic spectroscopy. Microwave-induced plasma atomic emission spectroscopy is not so widely used and will not be considered further here, although the interested reader is referred to other works.

7.10.1 Inductively coupled plasma atomic emission spectroscopy

Inductively coupled plasma spectroscopy has now become the most widely used form of plasma emission spectroscopy today. The ICP is produced within a device known as a ***torch***, a schematic for which is shown in Fig. 7.11. The principal components of the torch are: (i) a central quartz tube with a tapered tip in the form of a jet through which a vaporized or nebulized sample is passed within a stream of argon; (ii) a concentric outer quartz jacket through which a stream of argon flows in a circular and upwards direction; and (iii) a radio-frequency coil that surrounds the nozzle of the torch. The nozzle is normally also made of quartz and may take the form of an extension of the outermost sleeve of this quartz argon-carrying jacket.

Ionization of the argon emerging from the central sample tube is initiated by a spark from a Tesla coil and this causes a localized but very rapid heating of the gas. A copper induction coil surrounds the quartz tube to provide an alternating electromagnetic field within the flowing stream of

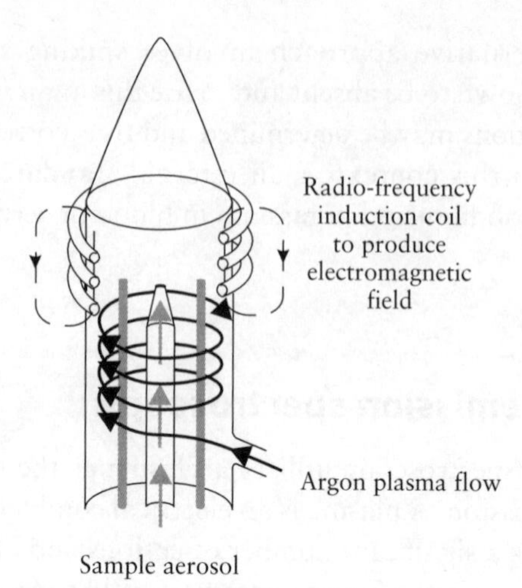

Figure 7.11 Schematic of an ICP torch.

argon gas. Ions and electrons formed from the initial spark interact with this field. Their movement gives rise to ohmic heating which sustains the plasma. The core temperature of the plasma is typically in the order of 10 000 K. Argon flowing up and out from the outer quartz sleeve mixes with the plasma being formed from the argon emerging from the sample quartz tube. The effect is to: (a) cool the outer perimeter of the plasma; and (b) direct its flow in order to prevent thermal damage to the outer quartz nozzle tip. Argon would normally be considered an electrical insulator, however, eddy currents within the gas lead to its heating, which, in turn, increases its electrical conductivity. The coil is excited by a 5–80 MHz radio-frequency generator and typically imparts 1–2 kW to the plasma. The stream of ionized argon once formed is capable of absorbing sufficient power from the induction coil to sustain the plasma. Induction coils are typically able to impart 2 kW of power at a frequency of 27 MHz to the plasma. The continuous heat of the plasma within the quartz tube in this way acts as the excitation source, which, in turn, gives rise to atomic emission.

Plasmas possess flame-like shapes with brilliant white cores that are topped by a tapered tail. The core is not optically transparent and may extend a few millimetres above the sample jet, forming a continuum in which argon and other ions recombine with electrons. Directly above the core, the density of continuum diminishes. In this region the plasma becomes optically transparent. Spectral measurements are generally made at a height of 15–20 mm above the nozzle. This region of the plasma is largely free from the argon lines and so is ideal for analytical purposes.

The sample is supplied in the form of a nebulized or vaporized spray into the path of the argon gas stream as it passes through the quartz tube. These approaches offer an obvious advantage over conventional electrical spark or arc emission spectroscopy since the analyte may be supplied in the form of a solution as opposed to a powered solid sample.

The emission spectrum is once again resolved into its component wavelengths via dispersive prisms or diffraction gratings. Spectra may be recorded either via photographic films or, more commonly now, multiple photo-multiplier tubes and monochromatic filters matched for the determination of individual elements.

The analyte ions will have been heated within an environment of 4000–8000 K for 2–3 ms by the time they reach the optically transparent region (observation point) of the plasma. These temperatures are considerably greater than those encountered within even the hottest nitrous oxide/acetylene flame which leads both to a greater efficiency in the atomization of the sample as well as minimizing the oxidation and so loss of the atomized analyte. The temperature profiles within the observation region of the plasma are typically extremely uniform which helps maintain the atomized analyte within a state that gives rise to the atomic emission spectrum. The high temperatures of ICPs make this approach suitable for the determination of analytes requiring considerable excitation such as zinc, cadmium, manganese, and calcium as well as elements which readily form oxides such as those of boron, phosphorus, uranium, and tungsten. Linear calibration profiles are often observed that may extend over several orders of concentration. Lower limits of detection are typically in the order of parts per million concentrations and may in some cases even be as low as parts per billion concentrations or less. It is clear, therefore, that ICP detection limits typically exceed those obtainable by flame photometry and are certainly competitive with atomic absorption spectroscopy.

7.10.2 DC Plasma jet atomic emission spectroscopy

The plasma is formed by passing a dc current between matched electrodes through a flowing stream of argon. Three electrodes (two anodes and one cathode) are normally arranged in a Y configuration; Fig. 7.12. Current flows from the anodes to the cathode. The anodes take the form of hollow graphite rods through which argon is passed. The plasma is initiated by bringing the cathode close enough to the anodes so that an arc is established. Currents of 10–15 A typically flow once the plasma has been established with the temperature of the plasma typically ranging from 5000 to 10 000 K, which is cooler than most plasmas found within ICP instruments. The nebulized sample is sprayed into the path of the plasma between the two anodes cause the atomic emission spectra that may then be recorded as with other forms of atomic emission spectroscopy.

The advantages of using dc plasma spectroscopy include the consumption of less argon than inductively coupled plasma spectroscopy requires. Sensitivities of dc plasma spectroscopy are, however, typically one order of magnitude less than those obtainable via ICP spectroscopy. Spectra obtained via dc plasma spectroscopy tend to be rather simpler and contain

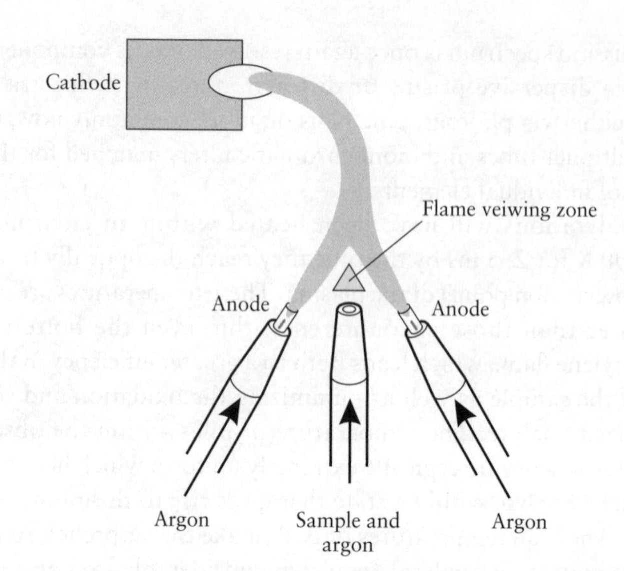

Figure 7.12 Schematic of a three-electrode dc plasma jet.

fewer lines than those seen via ICP, since dc spectra are primarily caused by atoms and not ions; we have already discussed that the plasma in a dc jet source at the observation point is considerably cooler than that encountered within an ICP and so less ionization of the sample occurs. The instrumentation associated with dc plasma spectroscopy is considerably less expensive than that of ICP. There are, however, some disadvantages associated with the use of dc plasma jet instrumentation including, for example, the need to replace the graphite electrodes in dc plasma instruments every few hours of use, thereby contributing significantly to the running costs of the instrument.

7.11 Atomic fluorescence methods based on ICP

Atomic fluorescent measurements can be determined within samples atomized in an ICP upon excitation by a hollow cathode lamp. In this way, only analyte atoms corresponding to the appropriate elemental cathode lamp will be excited. Fluorescent emissions may then be determined within a dispersive arrangement (Section 7.5) so that differing elemental fluorescent measurements can be made by simply using different cathode lamps. In some instruments, an optical filter may be placed in front of the photo-multiplier tube to remove background interference from the plasma source and this in turn often helps lower the practically achievable lower limits of detection.

Exercises and problems

7.1. Explain the basic differences between atomic emission and atomic absorption spectroscopy.

7.2. Why are hollow cathode lamps used in preference to other radiation sources?

7.3. Explain what is meant by: (i) Doppler; and (ii) pressure broadening.

7.4. Why are high temperature nitrous oxide/acetylene flames sometimes required within atomic absorption spectroscopy?

7.5. What are the relative advantages and disadvantages of using ICP torches?

7.6. A 10 ppm solution of lithium gives an atomic absorption signal of 12% absorption; what is the atomic absorption sensitivity?

7.7. Explain why atomic emission spectra consist of discrete lines rather than broad bands.

7.8. A drinking water supply is suspected of being contaminated with lead. Samples of water aspirated directly into an air/acetylene flame gave an absorbance of 0.68 at 283.3 nm. Standard solutions containing 0.5 and 1.0 ppm were found to exhibit absorbances of 0.43 and 0.86, respectively. Assuming the Beer–Lambert law is obeyed calculate the concentration of lead within the water sample.

7.9. Explain why sharp line radiation sources are desirable for atomic absorption spectroscopy.

7.10. Describe the principles underlying flame emission atomic spectroscopy.

7.11. Explain how and why reference standard materials are often used within both atomic absorption and emission spectroscopic techniques.

7.12. A serum sample is analysed by flame emission spectroscopy for potassium using a standard additions approach. Two 1 cm^3 additions are added to 10 cm^3 water aliquots and labelled A and B. Twenty μl of a 0.025 M KCl solution is added to water sample A. The emission signals in arbitrary units of the two water samples A and B are determined to be 88.5 and 58.9, respectively. Calculate the concentration of K^+ within the serum.

7.13. The sodium (Na^+) content of a brine solution is to be analysed by flame emission spectroscopy. Two 5 cm^3 additions are added to 10 cm^3 distilled water. The first of these samples exhibits an emission of 3310 (arbitrary units). The second sample has 50 μl of a 0.1 M NaCl solution added prior to analysis; this exhibits an emission of 3550 arbitrary units. Calculate the concentration of NaCl within the brine solution.

Summary

1. Atomic spectroscopy techniques involve exploitation of non-valence electronic transitions.

2. Atomic spectroscopy techniques may be broadly classified as being based upon either atomic emission or atomic absorption processes.

3. Atomic emission spectroscopy involves the emission of photons as electrons relax from excited states back to ground states.

4. Atomic absorption spectroscopy is based on the capture of photons as electrons are excited.

5. Atomic spectra originate from electronic transitions between atomic or elemental ionic orbitals.

6. Atomic spectral lines may be broadened by a number of phenomena such as pressure or collision broadening effects.

7. Flame absorption spectroscopy is the most widely used form of atomic absorption spectroscopy.

8. Two of the most widely used gaseous mixtures for flames are air/acetylene and nitrous oxide/acetylene.

9. Hollow cathode lamps for specific elements are used to provide extremely monochromatic radiation for excitation.

10. Samples are sometimes placed within Delves cups to allow lower limits of detection.

11. Graphite furnaces can be used in conjunction with L'vov platforms for the electrothermal atomization of samples.

12. Atomic emission spectroscopy involves approaches such as electric arc, spark, laser microprobe, flame emission, or plasma emission approaches.

13. Plasmas may be formed: (a) via the use of microwave field sources; (b) by passing a dc current between electrodes; or (c) by inducing a current flow via the use of a high powered radio-frequency electromagnetic field.

14. DC plasmas are typically formed within instruments employing plasma jet sources whereas the plasmas that are formed via radio-frequency coils are known as inductively coupled plasmas or ICPs.

15. Atomic fluorescent spectroscopic techniques may be performed using ICP approaches in conjunction with a hollow cathode lamp.

Further reading

Broekaert, J. A. C. (2001). *Analytic atomic spectroscopy with flames and plasmas*. Wiley-VCH.

Cullen, M. (2003). *Atomic spectroscopy in elemental analysis*. Sheffield Analytical Chemistry. Blackwell Publishing.

Golightly, D. W. (1992). *Inductively induced coupled plasmas in analytic atomic spectra*. Wiley.

Hollas, J. M. (2002). *Basic atomic and molecular spectroscopy*. Tutorial Chemistry Texts, Royal Society of Chemistry.

Schlemmer, G. and Radzuik, B. (1999). *Analytical graphite furnace atomic absorption spectrometry: a laboratory guide*. Birkhauser Boston.

Softley, T. P. (1994). *Atomic spectra*. Oxford Chemistry Primers, Oxford University Press.

Separatory methods and chromatography

8

Skills and concepts

This chapter will help you to understand:

- How to use distribution coefficients for calculating how analytes may separate themselves between different phases.
- How to perform solvent extractions.
- What is meant by a solid-phase extraction and how these may be used.
- The principles of chromatography including the use of stationary and mobile phases and the calculation of R_f values.
- What is meant by the capacitive factor and the selectivity factor and how these are used in calculations.
- The factors contributing to chromatographic broadening.
- How to perform paper and thin layer chromatography for the separation of simple mixtures.
- The principles underlying both gas chromatography (GC) and liquid chromatography (LC).
- The operation and use of injection ports for use with GC and LC separations.
- Factors affecting the choice and use of differing columns with GC and LC techniques.
- The operation and suitable use of photodiode, fluorescent, electrochemical (amperometric and conductivity), infrared, differential refractive index, evaporative light scattering, and mass spectroscopy based devices for LC-based analyses.
- The operation and suitable application for silica (and its derivatives), styrene-divinyl benzene, alumina, controlled pore size, hydroxyapatite, and agarose packed columns.
- Principles of zone capillary electrophoresis and the apparatus used.

8.1 Mixtures and the need for separatory methods

Samples frequently contain a complicated mixture of different components. Liquefied food, environmental river water, and clinical blood samples can all, for example, contain a mixture of solutes, micelles, colloidal, and even suspended particulate materials. While it is true that many analytical techniques are designed to be inherently selective, many 'real' samples contain such a bewildering cocktail of components that it is often necessary to perform separations as part of the routine analytical procedure. Hence, *Separatory Science* forms a major cornerstone of analytical chemistry.

In some instances it might be necessary to exclude interfering components from the detection system by, for example, some form of chromatography or solvent extraction procedure. In other situations, we may wish to concentrate the analyte by removing some or all of the supporting solvent from the mixture. Examples here could include precipitation reactions (e.g. as used within gravimetric analyses) or the evaporation of a solvent. In each of these situations, some form of phase change is used to separate the components since their solubility will differ within solvents and across phase-change boundaries. Sodium chloride is soluble in water, for example, but not gaseous air; if we evaporate some of the water, the concentration of the salt increases until it eventually precipitates. Again, if we shake two immiscible liquids such as tetrachloromethane and water, the solubility of, for example, a carboxylic acid will be different between the two phases. In this way we can perform a separation via a ***solvent-based extraction*** and we shall consider this approach in the next section.

8.2 Solvent extraction methods

Solvent extraction methods normally rely on mixing two immiscible solvents and are, thus, known as ***liquid–liquid***-based extractions. Organic chemists often use liquid–liquid-based extractions to separate a newly synthesized molecular product whereas analytical chemists extract an analyte from a complicated mixture in order to simplify its analysis. The most common approach is to extract the species of interest from an aqueous system into an organic solvent. In general, large un-ionized and non-polar solutes will be more soluble in organic solvents than polar solvents such as water.

It is generally possible using this approach to extract and concentrate covalent molecular species, uncharged metal chelates, and ion-association complexes. The separation and selectivity are normally governed by a number of processes in dynamic equilibria with each other that can be influenced by a number of factors, such as pH. The efficiency of the extraction will ultimately be determined by the relative solubility of the extractant in the two solvents under a set of pre-defined conditions. The ***distribution or partition coefficient***, K_D, is numerically equal to the ratio of concentrations of the solute within the two phases at equilibrium (Eqn (8.1)), and this allows us to quantitatively predict how much of the solute will be extracted. Note that K_D is a unitless parameter since it is a ratio of two concentrations and hence the units of the numerator and denominator cancel each other out.

$$K_D = \frac{[S]_{org}}{[S]_{aq}} \tag{8.1}$$

where K_D is the distribution coefficient, and $[S]_{org}$ and $[S]_{aq}$ are the concentrations of solute S in the organic and aqueous phases, respectively.

A liquid–liquid extraction of this type can be performed by vigorously shaking the two immiscible liquids together for some time in a separating funnel, and then allowing the two phases to separate, Fig. 8.1. The lower layer can then be run off through the tap at the base of the funnel. The solute will always distribute itself between the two phases so that the ratio of the concentrations remains constant as predicted by the partition coefficient. It follows that if we have 100 cm^3 of the organic phase to extract the solute from the aqueous phase, it will be more efficient to extract the aqueous phase with four 25 cm^3 aliquots rather than using all of the 100 cm^3 in one process. This point is illustrated in Example 8.1.

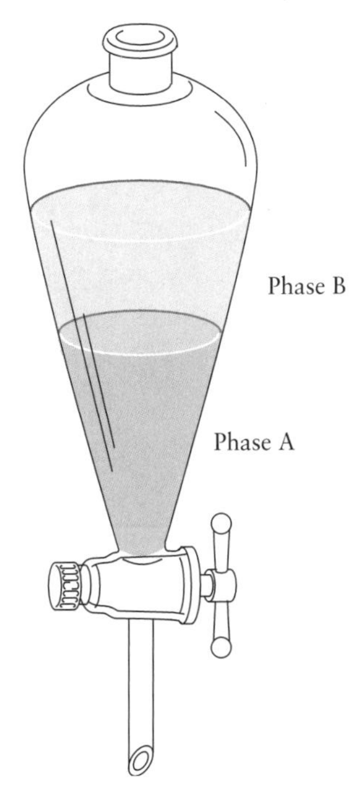

Figure 8.1 Liquid–liquid extraction between two immiscible phases.

EXAMPLE 8.1

The distribution coefficient K_D of an organic salt between hexane and water is 90. A quantity of 0.1 mol of the salt is dissolved in 100 cm^3 of water. Predict how many moles of the salt will remain within the aqueous phase following extraction by (a) using 100 cm^3 and (b) sequentially using four 25 cm^3 aliquots of hexane to extract the salt from the aqueous phase.

Method

1. Using the partition coefficient K_D, calculate the concentration of the organic salt at equilibrium in both phases when 100 cm^3 of hexane is used to extract the aqueous phase.
2. Again using the partition coefficient K_D calculate the concentration of the salt remaining in the aqueous phase following extraction with 4×25 cm^3 of hexane.

Step 1:

$$K_D = \frac{[\text{salt}]_{org}}{[\text{salt}]_{aq}} = 90$$

It follows that

$$[\text{salt}]_{aq} \text{ after extraction} = \frac{\text{vol. of aqueous phase}}{(\text{vol. of organic extractant} \times K_D) + \text{vol. of aqueous phase}}$$

$$\times \text{ no. of moles of salt}$$

Therefore, no. of moles of salt in the aqueous layer

$$\left(\frac{100}{(100 \times 90) + 100}\right) \times 0.1 = 3.28 \times 10^{-6}$$

Thus, 3.28×10^{-6} M concentration of the salt remains in the aqueous phase following extraction with 100 cm^3 of hexane.

Step 2:

For calculation of the concentration using 4×25 cm^3 aliquots of hexane a similar approach may be adopted. Therefore,

$$[\text{salt}]_{aq} = \left(\frac{100}{(25 \times 90) + 100}\right)^4 \times 0.1$$

Thus, 3.4×10^{-8} M concentration of the salt remains in the aqueous phase following extraction with 4×25 cm^3 of hexane.

8.3 Solid-phase extractions

Solid-phase extraction methods may be used as an alternative to solvent extraction for separating mixtures when the use of multiple solvents might prove excessively cumbersome and/or expensive. A solid partitioning phase normally takes the form of either a powdered silica or a polymer powder support packed into a custom-fabricated cartridge. In some cases the separation may be accelerated by pressure exerted by means of a syringe barrel as shown in Fig. 8.2. Analytes may be separated or extracted from a liquid mixture by interaction and partitioning with the solid support by means of van der Waals, electrostatic, or hydrogen bonding as well as, in some cases, size exclusion or retention type interactions.

Another popular approach involves using functionalized organic hydrophobic groupings such as C_{18} moieties bonded to the solid support for the uptake of organics from aqueous samples via van der Waals interactions. Trace hydrophobic organics are pre-concentrated on the column as the sample is introduced and drawn through either under vacuum or by means of a plunger. Solutes are said to be ***eluted*** as they leave a solid-phase extractant cartridge (or indeed any other chromatography column). Solutes may be eluted from solid-phase extractant cartridges with an appropriate solvent. In some cases partial evaporation of the solvent may be used for further pre-concentration of the sample prior to its analysis.

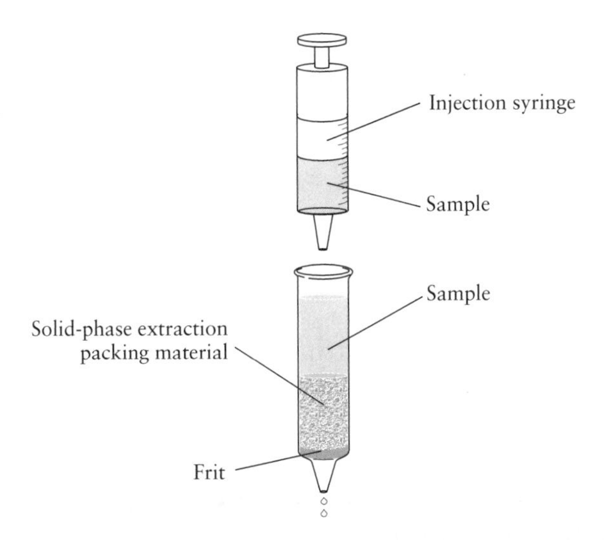

Figure 8.2 Solid-phase extraction cartridge.

Solid-phase extraction supports are also sometimes used in the form of filter-type extraction disks that may be placed within Büchner-type funnels for the vacuum-assisted separation of mixtures. Membranes are sometimes fabricated in the form of functionalized silica disks although these are often brittle or fragile and for this reason are often strengthened by a surrounding outer polymer (e.g. PTFE) support to aid mechanical strength, Fig. 8.3. Functionalized powdered silica is often commonly entrapped within an inert fibrous polymer support membrane made from, for example, PTFE to produce a flexible extraction phase; that is, free from the brittleness and fragility of pure silica.

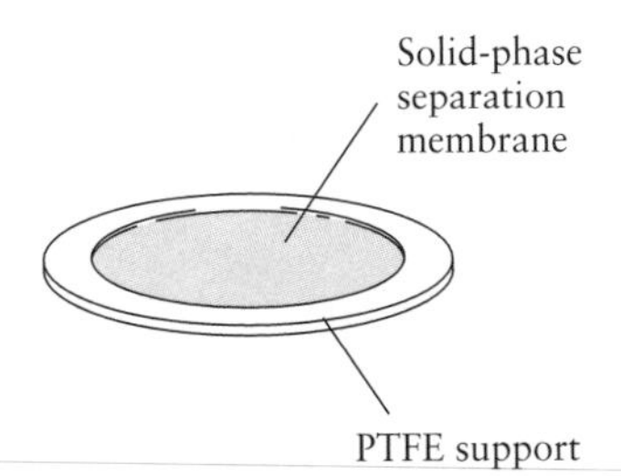

Figure 8.3 Solid-phase separation membrane.

8.4 An introduction to chromatographic methods

Chromatography is the generic name assigned to the many different separatory techniques, and owes its name to the Russian botanist Mikhail Tswett who first coined the term in the early 1900s. Tswett separated a number of pigments including xanthophylls and chlorophylls by passing solutions of these mixtures through glass columns packed with finely divided calcium carbonate. Each pigment travelled through the columns at different rates and ultimately appeared as coloured bands. This gave rise to the name chromatography from *Chroma* the Greek for colour and *graphein* meaning 'to write'.

All forms of chromatography rely on a mixture coming into contact with two phases and then moving one phase relative to the other. The two phases are known as the ***stationary phase*** and the ***mobile phase***. The components of the sample distribute themselves between the two phases as dictated by their relative solubilities (or ***affinities***) towards the two phases.

Components that do not interact with the stationary phase pass quickly through within the mobile phase. Conversely, components that interact strongly with the stationary phase travel very slowly.

Components will move with speeds that are dictated by the relative times spent in the mobile and stationary phases. These times are, in turn, governed by the partition coefficients for each component towards the two phases. In this way, different components can be separated by the relative speeds at which they pass through the stationary phase.

8.5 Elution chromatography—chromatography with two liquid phases

Elution chromatography can be performed in several different ways. The stationary phase in all cases is a solvent (e.g. water) adsorbed and, therefore, immobilized on a solid support such as cellulose-based paper or silica within a packed column. The sample to be ***resolved*** into its component parts is dissolved in a small volume of a second solvent that will act as the mobile phase.

The sample solution is then added to the solid support (and immobilized stationary phase); as the mobile phase travels along the stationary phase, solutes distribute themselves between the two phases according to their appropriate partition coefficients. Additional mobile phases may be added to ***elute*** the components from the stationary phase at different rates and one way of thinking of this type of chromatography is as a continuous series of liquid–liquid extractions. Eventually, all of the components will elute from the column and this way the mixture is resolved.

8.6 Theory of chromatographic separations

8.6.1 Partition coefficients

All chromatographic separations are governed by partition coefficients, K_D, for solutes between the stationary and mobile phases. For a solute S, a dynamic equilibrium is set up, that is, Eqn (8.2):

$$S_{mobile} \rightleftharpoons S_{stationary} \quad (8.2)$$

and the partition coefficient, K_D, is equal to the ratio of the concentration of the solute within the two phases, Eqn (8.3):

$$K_D = \frac{[S]_{stat}}{[S]_{mob}} \tag{8.3}$$

where K_D is the partition coefficient, and $[S]_{stat}$ and $[S]_{mob}$ are the concentrations of solute S in the stationary and mobile phases, respectively.

Ideally, the value of K_D should remain constant across a wide range of solute concentrations to ensure that the ratio of $[S]_{stat}$ and $[S]_{mob}$ itself remains constant. Chromatography performed under these conditions may be assumed to be linear in its behaviour and this allows quantitative determinations to be made from the chromatogram. In practice, under most normal working conditions, K_D can be taken as being a constant, although at very high concentrations the stationary phase can become either partially or totally saturated.

8.6.2 Retention times

Many forms of chromatography use some form of column. A mixture for separation is introduced at one end of the column, and at different time intervals, solutes are eluted. Figure 8.4 demonstrates this principle for two solutes A and B that travel through the column at different rates. If A travels faster than B, then A may be collected from the end of the column before B is eluted and in this way the two solutes can be separated.

The time taken for a solute to elute from a column is known as its ***retention time***, t_R. If a solute within the mobile phase totally fails to interact with the stationary phase then it will travel at the same speed as the

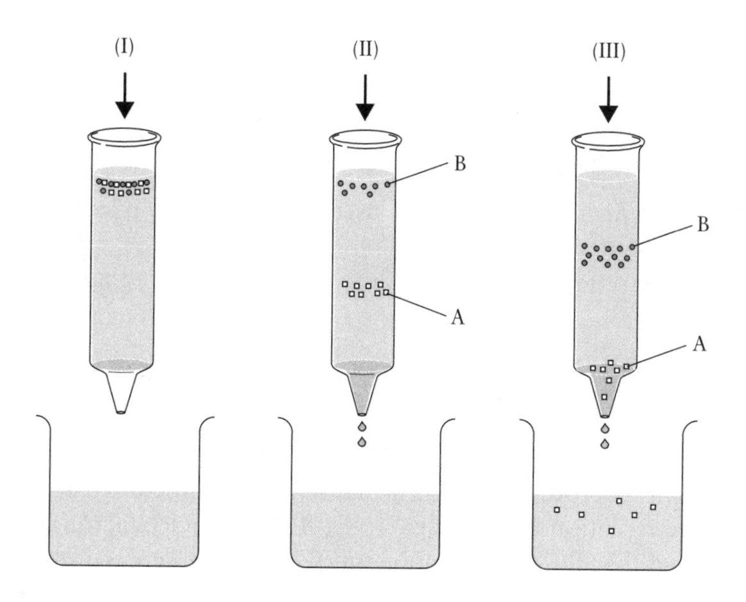

Figure 8.4 Separation principle for two solutes A and B: (I) a mixture containing A + B is introduced at the top of the chromatography column; (II) A travelling through column faster than B; (III) solute A is collected before B is eluted from the column and so separation is achieved.

mobile phase—we can in this case denote this time as t_{mob}. If the solute spends part of the time within the mobile phase and part of the time within the stationary phase, then its rate of progress will be governed by the partition coefficient, K_D, and it follows that different solutes will be eluted from the column at different times depending on their partition coefficients with the column.

The average linear rate of movement, u, of the mobile phase may be expressed by, Eqn (8.4),

$$u = \frac{L}{t_{mob}} \tag{8.4}$$

where L is the length of the column.

In a similar manner, for any chromatographic peak we can express the average linear rate $\overline{\nu}$ of solute migration (Eqn (8.5) as

$$\overline{\nu} = \frac{L}{t_R} \tag{8.5}$$

The retention time, t_M, for a solute can be related to its partition coefficient, K_D, by expressing the migration rate of the solute $\overline{\nu}$ in terms of a fraction of the velocity of the mobile phase, Eqn (8.6):

$$\overline{\nu} = u \times \text{fraction of time spent in the mobile phase} \tag{8.6}$$

We must, however, take into account the volumes of the stationary and mobile phases, which we can call V_{stat} and V_{mob}, respectively.

In this way, the rate of solute migration for a particular chromatographic peak can be expressed as:

$$\overline{\nu} = u \times \frac{1}{1 + K_D V_{stat}/V_{mob}} \tag{8.7}$$

8.6.3 The capacity factor

The ***capacity factor*** is a parameter used to compare the relative rates of solute migration along columns.

The capacity factor, k' for a solute may be calculated according to Eqn (8.8):

$$k' = \frac{t_R - t_{mob}}{t_{mob}} \tag{8.8}$$

k' should ideally be in the range of 1–5. If k' falls very much below 1, then elution will be proceeding so fast that it will be difficult to accurately determine retention times. Conversely, if the capacity factor becomes much larger than 20, then retention times will become excessively long.

EXAMPLE 8.2 If the retention time for a chromatic peak, t_R, is 65 s and t_{mob} is 30 s, calculate the capacity factor, k′.

Method

Calculate k' according to Eqn (8.8) that is: $k' = (t_R - t_{mob})/t_{mob}$

$$k' = \frac{65 - 30}{30} = 1.17\,s^{-1}$$

8.6.4 The selectivity factor

The selectivity factor α for two solutes is defined as the ratio of the larger (l) partition coefficient K_l' as the numerator and the smaller (s) coefficient K_s' as the denominator for the two phases, Eqn (8.9). This is equal to the ratio of the larger capacity factor K_l' (numerator) and the smaller capacity factor, K_s'.

$$\alpha = \frac{K_l'}{K_s'} \tag{8.9}$$

Note that in this arrangement, α will always be greater than unity.

It follows from Equations 8.8 and 8.9 that the selectivity factor α for two solutes may be readily calculated from a chromatogram, since:

$$\alpha = \frac{(t_R)_l - t_{mob}}{(t_R)_s - t_{mob}} \tag{8.10}$$

EXAMPLE 8.3 If two chromatographic peaks have capacity factors of 2.4 s^{-1} and 3.8 s^{-1}, calculate the selectivity factor.

Method

Calculate α according to: $\alpha = k'_l/k'_s$

$$\alpha = \frac{3.8}{2.4} = 1.58$$

8.6.5 The efficiency of chromatographic columns

The efficiency of a chromatographic column can be described in terms of either the ***number of theoretical plates*** N, or the ***plate height*** *H* (sometimes known as the ***height equivalent of a theoretical plate, HETP***). The number of theoretical plates and the plate height are related by Eqn (8.11) the ***Van Deemter*** equation named after the first scientist to theoretically describe an expression for quantifying the efficiency of chromatographic columns:

$$H = \frac{L}{N} = A + \frac{B}{\overline{\mu}} + C\overline{\mu} \quad (8.11)$$

where H is the HETP, L is the length of the column (normally in cm) and A, B, and C are constants relating to a particular system. $\overline{\mu}$ is the average linear velocity of the mobile phase.

The three constants A, B, and C are related to different parameters that affect the efficiency of the chromatographic separation. The parameter A essentially describes the effects of eddy currents caused by the variability of diffusional pathlengths that occur between irregularly spaced particles within a column, and is, therefore, independent of the velocity of the mobile phase. B is related to longitudinal or molecular diffusion of the analyte within the mobile phase. C represents the rate of mass transfer of the analyte between the stationary phase and the mobile phase.

The efficiency of the column improves as (a) the number of plates increases, and (b) as the plate height becomes smaller. The number of theoretical plates can vary from a few hundred through to several hundred thousand, while the plate height may vary from a few millimetres to a few tens of micrometres (thousandths of a millimetre).

The terminology of using theoretical plates and plate heights to describe the efficiency of chromatography columns is a historical legacy from a theoretical model of chromatography that has largely been superseded. The use of the term 'theoretical plate' *should not* be thought of as describing any physical representation of the column or its operation—but, rather, as an arbitrary parameter to describe its efficiency.

We have already seen that a chromatographic peak will broaden as its retention time increases. The retention time of a peak will increase if the column length increases. It, therefore, follows that as the column length increases, chromatographic peaks will broaden. Since the peak will take the form of a Gaussian distribution curve we can express its shape in terms of the width encompassing plus or minus one standard deviation σ. It follows that we can express the column efficiency in terms of the variance (σ^2), Eqn (8.12):

$$H = \frac{\sigma^2}{L} \quad (8.12)$$

Note that L has units of cm, σ^2 has units of cm^2, and this dictates that H also has units of cm and can, in fact, be thought of as being related to the length of the column that contains $(L - \sigma)$ proportion of the analyte.

The number of theoretical plates may also be obtained directly from a chromatogram, since it can be shown that

$$N = 16\,(t_R/w)^2 \tag{8.13}$$

where t_R is the retention time and w is the width of the base of the chromatographic peak.

We have already seen that the retention time of a solute is related to the length of the column. In practice, it is easier to measure retention times directly from chromatograms and to use these to express the efficiency of the column.

EXAMPLE 8.4 A chromatographic peak is found to have a retention time of 52 s. The base width of the peak is equivalent to 3.2 s by intersection of the sides of the peak with the baseline. If the column is 500 cm in length, calculate the HETP in terms of centimetres per plate.

Method

Calculate N from the expression $N = 16\,(t_R/w)^2$ and then calculate HETP using the expression $\text{HETP} = \frac{L}{N}$.

$$N = 16\,(t_R/w)^2 = 16\left(\frac{52}{3.2}\right)^2 = 16 \times (16.25)^2 = 4225$$

Therefore,

$$\text{HETP} = \frac{50}{4225} = 0.012 \text{ cm per plate}$$

8.6.6 The shapes of chromatographic peaks

The separation of two solutes A and B is shown in Fig. 8.5 and it can be seen that chromatographic peaks typically take the shape of a normal Gaussian distribution curve. The shapes of these peaks can be attributed to the random motion of solute particles as the solution moves through the column. There will be a retention time that corresponds to the greatest number of solute molecules being eluted from the column; some solute molecules are eluted a little earlier than this and some a little later. The solute molecules undergo many thousands of transfers between the mobile and stationary phases they pass through the column, however, the length of time a molecule spends in each phase is random and highly unpredictable. The solute can only move through the column while it is in the mobile phase and so it follows that if the solute molecule spends

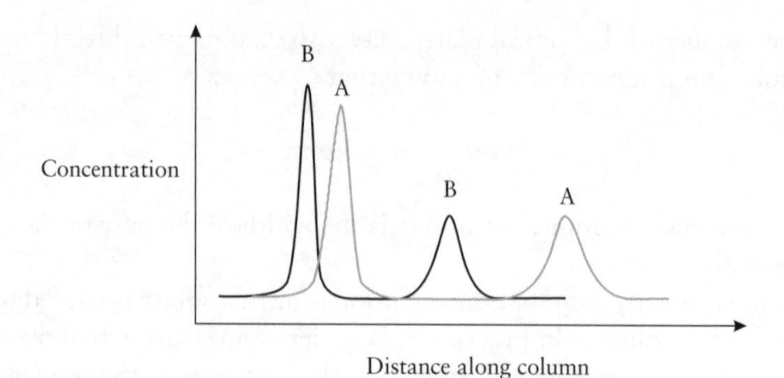

Figure 8.5 Simple chromatographic separation.

more time within the stationary phase it passes through the column more slowly. Conversely, if the molecule spends a greater proportion of its time in the mobile phase, it will pass through the column more rapidly. The interchange of a solute from one phase to another requires the expenditure of energy and, as with any system in which energy transfer occurs, the process is random in nature. The result of these random interchanges between the two phases gives rise to the broadening of chromatographic peaks. The breadth of a peak is related to the mean length of time a solute takes to elute from the column, that is, the retention time. For this reason chromatographic peaks with longer retention times tend also to be broader.

EXAMPLE 8.5 If the width of a solute peak within a chromatogram at one-half its height is 5.2 mm, what will be its base width by extrapolation?

Method

If we assume that a chromatographic peak to Gaussian in shape, then the width at half height will be equal to $\pm 1\sigma = 2\sigma$. Similarly, the width of the base of the peak can be taken by extrapolation of the tangents to give a peak base width equal to $\pm 2\sigma = 4\sigma$.

If $2\sigma = 5.2$ mm, then $4\sigma = 2 \times 10.4$ mm.

8.6.7 Band or peak broadening

The broadening of chromatographic peaks is a phenomenon normally referred to as ***band broadening***. Band broadening is caused by a number of effects. A major cause of band broadening originates from flow distribution effects of the mobile phase within columns. The flow is fastest in the centre of a column due to rheological (or frictional) drag effects at the walls and since this effect continues throughout the length of the column, chromatographic bands continue to broaden the longer the column is.

A second cause is the diffusion of molecules within the mobile phase as it passes along the stationary phase. As we have seen, the concentration of a species will be higher in the centre of the lumen of a chromatographic column than at the edges due to flow distribution effects. Molecules therefore also tend to diffuse towards the edges, that is, down a concentration gradient. This process will continue as long as the chromatographic separation occurs and it follows that the longer a chromatographic column is, the greater will be the broadening of its bands.

A third effect contributing to band broadening is known as the *Eddy diffusion of solutes*. Many columns contain particulate packing materials as the stationary phase. Some solute molecules will by chance travel in a more direct line than others through the column whereas others will undergo more diversions along the way. This motion will, in turn, allow some solute molecules to be eluted faster from the column than others and so the chromatographic band broadens.

The presence of small areas within columns containing 'stagnant' mobile phase can give rise to a further reason for band broadening effects. Some columns are packed with particulate stationary phases that give rise to pores through which the mobile phase may not readily pass and in these areas the mobile phase can 'stagnate'. In this case, a solute molecule ceases to be carried along the column and will only leave this area of stagnation by diffusion.

8.7 Paper chromatography

Most people first encounter the use of paper chromatography at school. Paper chromatography probably represents the simplest form of chromatography available and yet is still widely used.

The cellulose fibres of paper can either act directly as the stationary phase or can provide a support for the adsorption of a liquid stationary phase such as water (see Section 8.5). One end of the paper is typically immersed a short distance into a reservoir of the mobile phase. The solution mixture is then spotted on the starting line and allowed to dry. It is important to make this spot as concentrated yet as small as possible to prevent premature spreading of the sample. It is, therefore, normal practice to add several very small spot additions, one at a time—on top of each other—once each spot has completely dried.

Melting point tubes may be used to make spot addition of the solution; better results can be obtained by drawing out the capillary tubes in a micro-burner flame and snapping the tubes in half once cool. The extra-fine tips produced in this way allow spots of 0.5-mm diameter or less to

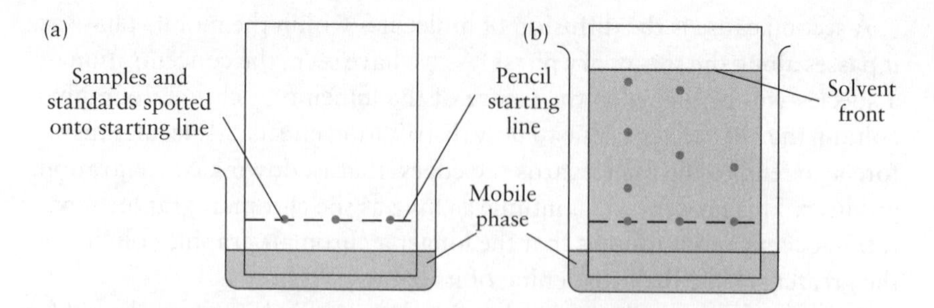

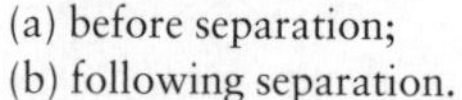

Figure 8.6 Paper chromatography demonstrating calculation of R_f values: (a) before separation; (b) following separation.

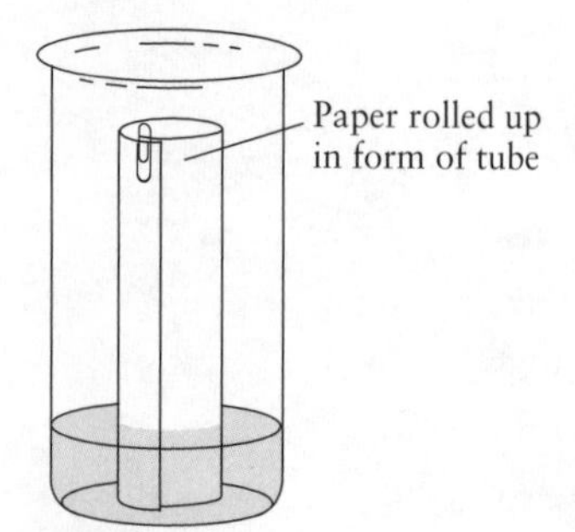

Figure 8.7 Paper chromatography with paper rolled into the form of a tube.

be concentrated and this produces finer resolution of the components with less smearing of the individual component spots.

A thin *pencil* line is then drawn at a distance of a centimetre or two from the bottom of the paper to act as a starting reference line for the calculation of R_f values. *Note*: a pencil must be used since inks contain pigments that will themselves be soluble and, hence, separate when the chromatogram is run.

The paper should be hung from a support to allow the bottom end of the paper to be immersed into a chromatography tank, Fig. 8.6; the mobile phase will then travel up the paper with time by capillary action. An alternative arrangement is to curl the paper into the form of a tube that can be secured by a paper clip and stood upright within the tank, Fig. 8.7. Care should be taken that the pencil line and the spots are a centimetre or so above the surface of the solvent mobile phase.

If the paper is cut so as to be 3–4 cm or more in width, a number of different spots may be placed on the staring line—as long as sufficient space is left between each spot. A lid should be placed on the tank thereby ensuring the atmosphere surrounding the paper is saturated with the vapour of the mobile phase. Just before the solvent front reaches the end of the paper another pencil line is drawn to mark the distance it has travelled. The paper should then be removed from the tank and allowed to dry.

The R_f values for each component are then calculated as the ratio of the distance travelled by the spot and the distance travelled by the solvent front from the starting line, that is, Eqn (8.14):

$$R_f = \frac{\text{Distance travelled by centre of spot}}{\text{Distance travelled by solvent front}} \qquad (8.14)$$

The R_f of a particular component should be the same whether it forms part of a complicated mixture to be resolved or as a single compound travelling along the paper. It is, therefore, possible to identify spots by their R_f values by spotting concentrated drops of compounds that are suspected to be found within the mixture.

It should be noted that the R_f values are highly dependent on experimental conditions. It is, therefore, essential for identification purposes to run chromatograms of mixtures and single spots of different possible components at the same time and on the same piece of chromatography paper—if we hope to identify components within a mixture with any certainty.

If the different components are coloured, then it is easy to see where the spots have travelled. A number of commonly used dyes and inks can be qualitatively analysed into their component constituents, and, in this way, we may find out how many different coloured pigments are in blue or black writing inks—or a number of commercially available food dyes.

Non-coloured constituents can sometimes also be resolved from mixtures of different components. Amino acids, for example, may be resolved with a mobile phase formed from a 4 : 1 : 5 mixture by volume of 1-butanol, glacial ethanoic (acetic) acid, and water. The paper can then be sprayed with ninhydrin (indane-1,2,3-trione), which stains the amino acids purple and allows the relative positions of different acids to be identified.

We shall describe a simple experimental procedure that can be performed within any laboratory to demonstrate how paper chromatography can be used:

PRACTICAL EXAMPLE 8.6

Many commercial food dyes contain three or more edible pigments such as Tartrazine, Sunset Yellow FCS, Indigo Carmine, or Amaranth. These components can be separated and identified by paper chromatography if one uses a mobile phase consisting of a 1 : 100 ammonia/water mixture.

Method

Step 1: Pour the mobile phase into the chromatography tank to a depth of 1 cm. Cut pieces of filter paper to a length 2–3 cm less than the height of the chromatography tank. Draw a thin pencil line 2.5 cm from the base of the chromatography paper. Some designs allow the papers to be suspended from a glass rod or similar support. An alternative approach is to ensure the paper is cut to a width of 10 cm or more so that it may be rolled into the form of a paper tube that is secured at the top with a paper clip.

Step 2: Apply small spots of each dye using drawn out melting point tubes on the starting pencil line. Prepare spots of the dye and two or three commercial food dyes (preferably choose colours such as deep purple, red, or blue, which are likely to contain a number of different pigments). Deposit as little of the dye as possible with each spot addition. Add several small spots to concentrate the sample.

Step 3: Fill the tank with the mobile phase to a depth of 1 cm. Hold the filter paper up to the side of the tank to ensure that the height of the solvent lies below that of the pencil line.

Step 4: Carefully place the chromatography paper in the tank so that the solvent can travel up the chromatography paper by capillary action.

Step 5: When the solvent front reaches approximately 3–4 cm from the end of the paper, remove it and carefully draw another pencil line at this point. The paper should now be allowed to dry.

Step 6: The R_f values for each of the spots may be calculated by comparing the distance moved by the centre of each spot with the distance moved by the solvent front from the starting line, that is,

$$R_f = \frac{\text{Distance travelled by centre of spot}}{\text{Distance travelled by solvent front}}$$

Step 7: See if any of the pigments within the food dyes can be identified by comparison with the R_f values of any of the individual pigments.

8.8 Thin-layer chromatography

Thin-layer chromatography (TLC) is conceptually very similar to paper chromatography, but often offers better separations and tends, moreover, to yield a more reproducible behaviour. TLC utilizes a finely divided solid such as alumina immobilized on a glass or polymer backing plate. The mobile phase may be water, aqueous ammonia solution, or some other mixture such as an alcohol/water/ethanoic acid solution. Alumina is highly polar and separation between the stationary and mobile phases may involve adsorption, partition, and/or ion exchange processes.

For details of how quenching agents operate refer to Chapter 5, Section 5.13 and Chapter 6, Section 6.5.

TLC plates are developed in an almost identical manner to that used for paper chromatography. Thin pencil lines are drawn at a height so as to just clear the mobile phase when placed upright in a chromatography tank, see Fig. 8.8. Care should be taken that the surface of the TLC plate does not become scratched since this would spoil the separation.

The alumina is often coated with a fluorescent material in order to help visualize components as they are separated on the plate. The compounds act as ***quenching agents*** and are seen as dark patches when viewed under UV light.

The spots can then be identified and labelled using pencil to allow R_f values to be calculated. Alternatively, organics can be stained using iodine staining.

TLC is normally used for a qualitative analysis of mixtures of non-volatile compounds such as pharmaceuticals or dyes. Organic chemists frequently use TLC to determine whether or not synthetic samples contain impurities. A single spot on a TLC plate indicates the presence of only one compound, whereas the presence of two or more spots must indicate that the sample contains a mixture of compounds.

Like paper chromatography, TLC is normally used as a qualitative technique for the identification of components within a mixture since it is hard to quantitatively deposit known quantities of the mixture on the plate. There are, however, instances where TLC is used *quantitatively*—especially if no other technique is available that readily lends itself to the quantitative determination of trace components within a mixture. Many examples of this type are found within analytical biochemical enzyme based analyses and assays. Quantitative analyses often involve the radioactive labelling fluorescent analytes. The chromatographic bands may first be identified under UV light and then collected by scraping the alumina from the plate to allow quantification by radioactive counting. It is clear that approaches of this type are extremely difficult to perform without incurring large errors due to difficulties in (a) reproducibly applying aliquots of the mixture to the plate and then (b) recovering all of the separated analytes from the plate.

TLC plates are often treated with reagents such as iodine or derivatizing agents to help visualize components that cannot be seen by the naked eye.

Iodine can be achieved by placing the TLC plate within a glass developing chamber with a small amount of iodine. For safety reasons this procedure should always be performed within a fume cupboard.

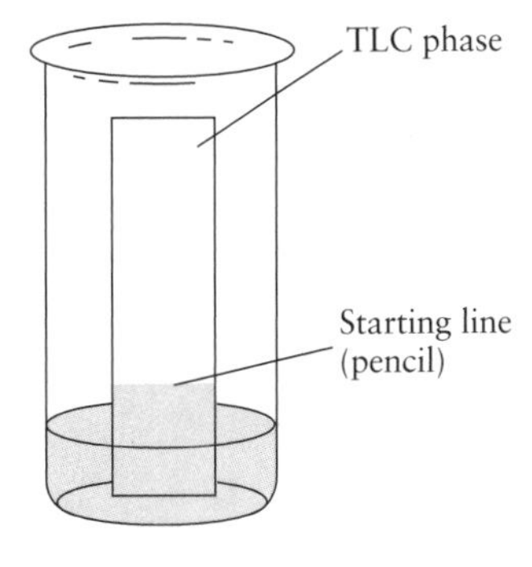

Figure 8.8 Thin layer chromatography.

8.9 Gas chromatography and gas–liquid chromatography

8.9.1 Introduction and principles of gas chromatographic separations

Gas chromatography (GC), as the name suggests, utilizes a carrier gas as the mobile phase together with a stationary phase within a packed or open-tubular (capillary) column. If a liquid stationary phase is used, the technique is known as gas–liquid chromatography or GLC. A schematic for a gas chromatograph is shown in Fig. 8.9. GLC was first described by Martin and Synge in 1941 and since this time GLC has become one of the most extensively used and powerful analytical tools for the separation and identification of components in complicated mixtures.

Separation occurs by partitioning gaseous samples between a carrier gas and the stationary phase. The sample must either already be in the gaseous phase (or be transformed into the gaseous phase by heating), so

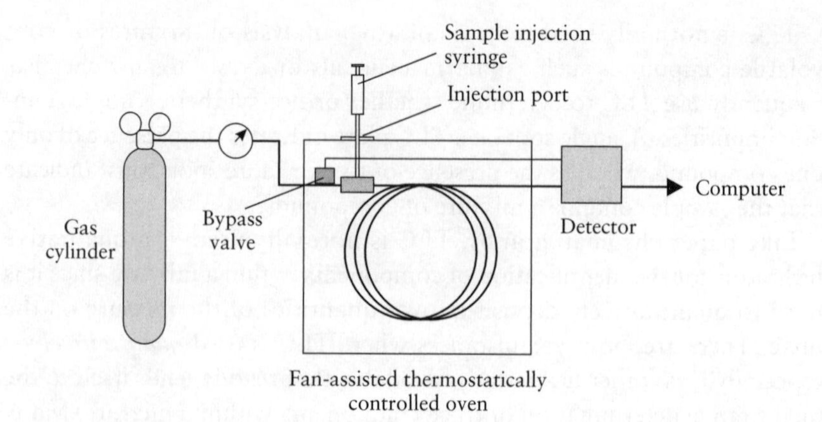

Figure 8.9 Schematic for simple gas chromatography.

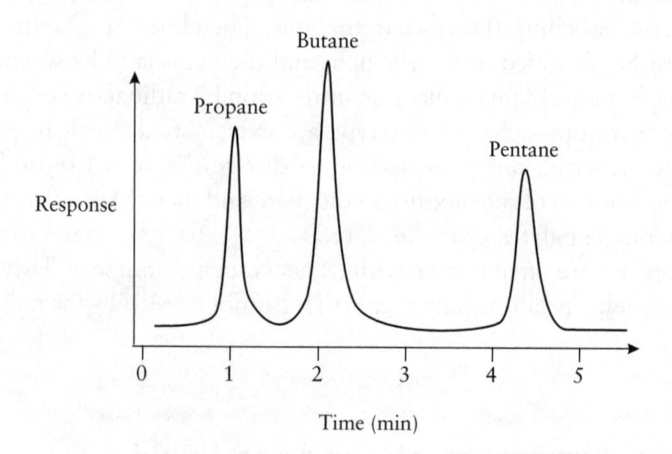

Figure 8.10 GC separation of simple alkanes.

that it may be passed into the carrier gas stream to be carried along the column. A high-purity chemically inert carrier gas such as nitrogen, carbon dioxide, helium, or argon is used although the choice is normally determined by the type of detector to be used. Higher-density gases give rise to slower but more efficient separations while lower-density gases give faster separations if not so well defined. A typical GC separation for pentane, butane, and propane is shown in Fig. 8.10.

Packed columns Within GLC packed columns the liquid phase is adsorbed onto an inert solid phase such as firebrick or diatomaceous earth or fused silica that will be finely divided to provide a large surface area for adsorption of the liquid. Particle sizes typically range from 60 mesh (250 μm mean particle diameter) through to 100 mesh (150 μm mean particle diameter). Solid supports are normally chosen so as to provide surface areas in excess of $1\,m^2\,g^{-1}$. Some solid supports are also pre-treated or *de-activated* (e.g. via silanization) to prevent the irreversible adsorption of polar analytes such as alcohols.

The liquid phase must be chemically inert, thermally stable, and possess a boiling point at least 100°C higher than the maximal operating temperature of the column. Columns may be packed in a number of ways and the choice of the liquid phase is crucial in determining the separational properties of the column. A dispersion of the finely divided support is frequently prepared, for example, in a volatile solvent that also contains the liquid support as a dissolved solute component. The slurry may then be introduced into the column to leave a packed column and liquid support upon evaporation of the solvent with the stationary liquid film typically varying from 0.1 to 1 μm in thickness.

Open-tubular capillary columns Open-tubular or capillary columns are now the most widely used types of columns and offer superior separational performance over packed counterparts in terms of both separational speed and the number of theoretical plates possible. Capillary columns are fabricated of either glass or fused silica having internal diameters ranging from ~0.25 to 0.5 mm and lengths of 25 to 50 m. The walls are much thinner than their packed column counterparts and possess outer diameters typically of 0.3 mm but are given extra strength by exterior polymer coatings. The resulting columns, being both strong and flexible, are coiled to permit their placement within thermostated oven enclosures as depicted within the schematic of Fig. 8.9. The inner surface of the capillary column is coated with the liquid stationary phase. Fused silica columns offer particular advantages by their ability to resist adsorption of analyte.

8.9.2 Injection ports, columns, and thermostating

Let us consider the schematic for a simple GC/GLC chromatograph as shown in Fig. 8.9. The supply of the carrier gas is typically regulated to maintain a constant gas pressure throughout the chromatograph irrespective of variations in the supply cylinder pressure. The flow of the carrier gas may normally be monitored via an in-line flow meter. Inlet pressures are typically in the range of 10–50 psi and yield carrier gas flow rates of between 25 and 150 $cm^3 min^{-1}$. Columns are typically packed within a length of glass or polymeric tubing. Columns are often coiled to allow columns of up to several metres in length to be housed within a fan-assisted thermostatically temperature controlled oven. The temperature of the column should ideally be controlled to within a few tenths of a Kelvin and this helps identification of chromatographic peaks in complicated mixtures by comparison of closely spaced retention times. Oven temperatures should normally be chosen so as to be equal to, or slightly greater than, the average boiling points of the components within the mixture. The choice of temperatures is often a compromise since lower temperatures permit optimal separations but elevated temperatures shorten retention and thus separation times. If the sample is known to contain a mixture of compounds with widely differing boiling points, then an optimal separation can often be achieved by increasing the temperature of the column with time—either linearly or in a step-wise manner with time.

The sample injection port is normally also thermostatically heated in the housing of the oven to (a) facilitate rapid vaporization of liquid samples and (b) allow thermal equilibration of the carrier gas, column, and analyte sample before the chemical separation commences. In practice, the sample injection is often maintained at a higher temperature than the bulk oven temperature to prevent sample condensation within the port assembly. Samples are normally injected through rubber septa using specialist micro-litre volume GC syringes.

A two-channel detector then differentially monitors the emergent eluent in comparison to a pure gas stream by means of a by-pass valve (see Fig. 8.9); in this way, background signals can be continually subtracted from the total measured response so that chromatograms relate solely to the analytes within the mixture being analysed.

It is important that samples are introduced into the flowing gas stream quickly and in as small volumes as possible to obtain good separational performance. Liquid sample injection ports typically allow a volume of 0.1–20 μl samples to be injected with excess volumes being passed to waste via an overflow by-pass. Gaseous samples are normally injected using gas syringes and a special gas sampling valve.

8.9.3 Detectors for GC and GLC

Detectors monitor and allow quantification of analytes as they are sequentially eluted from the chromatographic column. Ideally, they should (a) respond to any compound other than the carrier eluent gas and (b) the response should increase linearly with increasing analyte concentration over as wide a range of concentration as possible. It should be appreciated that detectors impart no selectivity to the system and that it is solely the separational capability of the chromatograph that allows identification and quantification of individual analytes within a mixture. There are a number of different detectors that are commonly used and we shall, in the following sections, describe the operation of some of the most widely used devices.

Flame ionization detectors

Flame ionization detectors measure the current that can pass between a pair of oppositely polarized electrodes positioned either side of a hydrogen/air flame. A schematic of a flame ionization detector is shown in

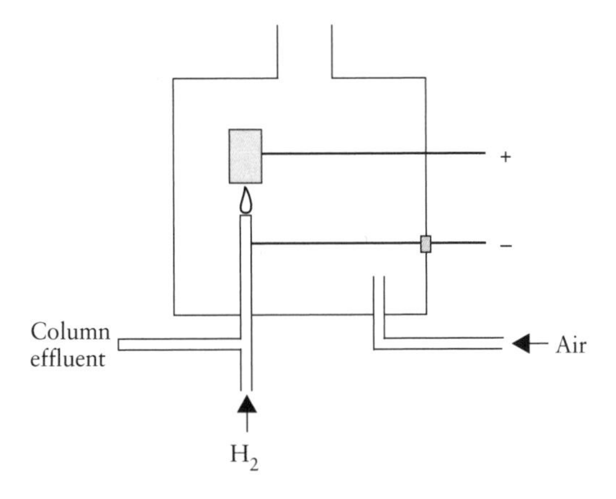

Figure 8.11 Schematic of flame ionization detector.

Fig. 8.11. The flame gives rise to a gas plasma that possesses a high electrical resistivity in the absence of ions. However, the temperature of the flame pyrolyses most organic compounds to produce cationic intermediates and electrons that act as charge carriers between the two electrodes. Ions are collected at the anode that is known as the ***collector***. The current that flows may then be amplified and recorded as chromatographic peaks, as components are eluted from the column. The response depends upon the number of carbon atoms within the analyte molecule as well, of course, as its concentration. The redox state of the carbon can also to some extent affect the sensitivity of this detector to some extent with fully oxidized carbon sometimes failing to ionize within the flame; in these cases responses will either be lowered or may even be completely absent.

The carrier gas is normally helium, nitrogen, or argon, due to their thermal stabilities, non-combustibility, and chemical inertness.

Flame ionization detectors, Fig. 8.11, are more widely used than any other form of detector due to their simplicity, general robustness, high sensitivity (which may allow determinations down to 10^{-13} g cm^{-3}), as well as their capability for responding across a wide range of concentrations.

Flame photometric detectors

Flame photometric detectors are primarily used for the determination of compounds containing phosphorus and/or sulphur and include, for example, pesticides and air- or water-borne pollutants. The eluent within this detector is sprayed into the path of a low-temperature hydrogen/air flame and the UV/visible emissions photometrically recorded, Fig. 8.12. The flame will, in the case of phosphorus, give rise to a short-lived H–P–O species that emits light at approximately 510 and 526 nm. In a similar manner, sulphur may be converted to S_2, which emits light at 394 nm. A number of different compounds containing the halogens, phosphorus, and a number of metals may be detected following emissions at characteristic wavelengths.

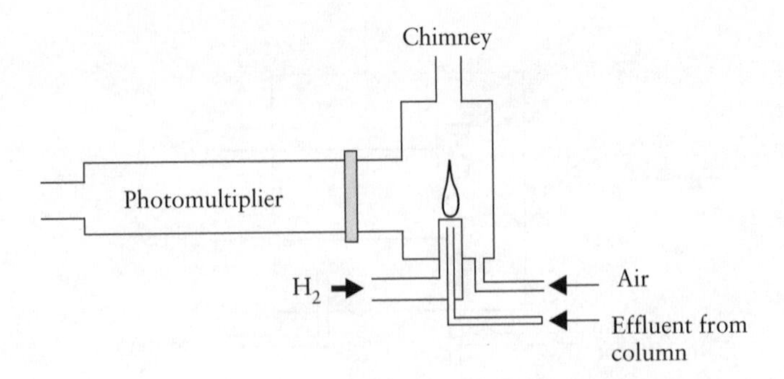

Figure 8.12 Schematic of flame photometric detector.

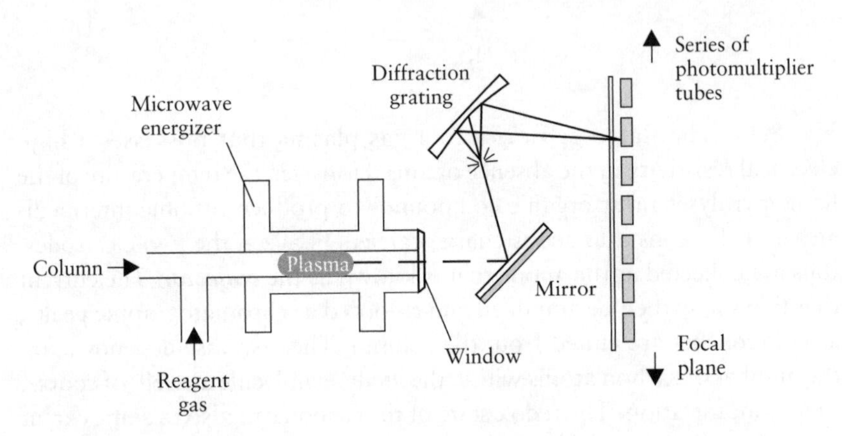

Figure 8.13 Schematic of atomic emission detector.

Atomic emission detectors

In this type of detector, the eluent is first passed into the path of a microwave-induced helium plasma (Fig. 8.13), which is sufficiently energized to both allow atomization of the sample as well as give rise to the appropriate atomic emissions. Emissions are dispersed via a diffraction wavelength to allow differentiation and thus monitoring of the individual atomic emissions via a series of movable photo-multiplier tubes.

Electron capture detectors

A β-emitter such as tritium or ^{63}Ni is used to irradiate the eluent as it emerges from the chromatographic column. ^{63}Ni sources may be used with column temperatures up to 350°C, whereas tritium sources may only be used with temperatures up to 220°C—since above these temperatures rates of tritium loss becomes unacceptable. The β-particles (electrons) cause ionization within the carrier gas (e.g. nitrogen) and this gives rise to the further release of electrons. A pair of oppositely polarized electrodes are positioned either side of the eleuent gas stream and the current that flows between these electrodes is monitored to give rise to the response of the detector, Fig. 8.14. Organic compounds (analytes) tend, however, to capture electrons and this decreases the current between the paired electrodes. A decrease in current, therefore, corresponds to a chromatographic peak as an analyte is eluted from the column. It follows that the *measured* response of the detector is inversely related to the current flowing between the electrodes.

The greater the electronegativity of a molecule, the greater will be its efficiency towards capturing electrons and this, in turn, will lead to variable sensitivity towards differing groups of organic compounds. Electron capture detectors tend to exhibit the greatest sensitivity towards nitro, carbonyl, halogen, and certain metallo group containing compounds although compounds with low electronegativities may also sometimes be determined via derivatization with, for example, chloroacetates.

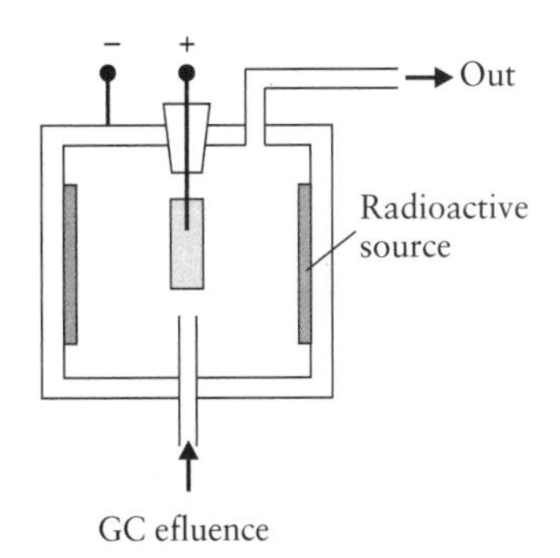

Figure 8.14 Schematic of an electron capture detector.

Thermal conductivity detectors

Thermal conductivity or hot wire detectors (Fig. 8.15) operate by monitoring the thermal conductivity of the carrier gas stream in the presence or absence of analyte molecules. As the carrier gas elutes from the column, it passes over an electrically heated filament wire. The temperature of the wire, and so its resistance, change depending on the thermal conductivity of the gas and this, in turn, will be modulated by presence and concentration of any analyte within the gas stream as it is eluted from the column. Most organic compounds possess thermal conductivities six to seven times lower than nitrogen or helium, which are most commonly used with this type of detector. The gas stream cools the filament so even small quantities of an analyte may significantly heat the filament. Two matched detectors are normally placed in the steam of the pure carrier gas (via a bypass loop) and the eluent gas stream, respectively, for comparison. The two detectors are connected so as to form opposite arms of a Wheatstone bridge, and in this way, *differences* in the resistance of the filament with analytes are monitored as they elute from the column, irrespective of ambient temperature fluctuations or, for example, in situations where temperature ramping regimes are employed.

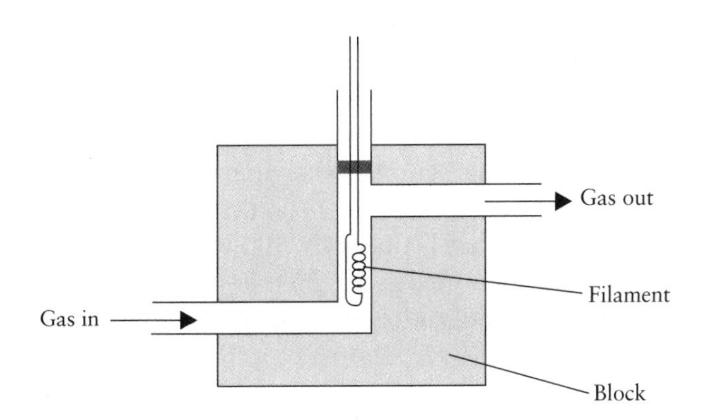

Figure 8.15 Schematic of a thermal conductivity detector.

8.9.4 GC–GLC coupled to other instrumental approaches for detection purposes

GC and GLC are often coupled to additional analytical techniques to further enhance analytical sensitivity. Approaches such as these are known as hyphenated methods and normally exploit the separatory capability of the GC or GLC to first fractionate the mixture into its component parts to allow quantification by, for example, mass spectroscopy (GC-MS), infrared spectroscopy (GC-IR), or nuclear magnetic resonance spectroscopy (GC-NMR).

Earlier approaches involved collecting eluent fractions to be analysed separately by a second analytical technique. Today, modern computerized instrumentation normally allows the direct online coupling of the two instruments to allow a real-time quantification of the eleuent as it emerges from the column. Not only is this approach less time-consuming but frequently allows an enhancement in resolution since any collected eluent fraction will necessarily have been collected over some period of time and will, therefore, represent, to some extent, a recombined mixture.

8.10 High-performance liquid chromatography (HPLC)

8.10.1 Introduction to HPLC

HPLC represents one of the most widely used forms of chromatography today. The term HPLC was originally coined as an abbreviation for *High-Pressure Liquid Chromatography*; however, as the performance of the technique improved, the acronym was retained and was progressively used as a shorthand for *High-Performance Liquid Chromatography*. The 'high performance' relates to the ability of HPLC for offering very selective and, therefore, high-quality separations in the minimum of time. This separatory capability is achieved by passing a mobile liquid phase through a finely divided (typical particle size of a few μm diameter) stationary support under high pressure. A schematic of a typical high-performance liquid chromatograph is shown in Fig. 8.16. The stationary support phase is uniformly packed in a stainless steel 3–4 mm bore and 10–30 cm long column that is normally housed within a thermostated oven. HPLC is most widely used as a highly sensitive and selective analytical technique for the identification and quantification of analytes within complex mixtures. In some preparative procedures, HPLC may be also used for the purification of some products or compounds.

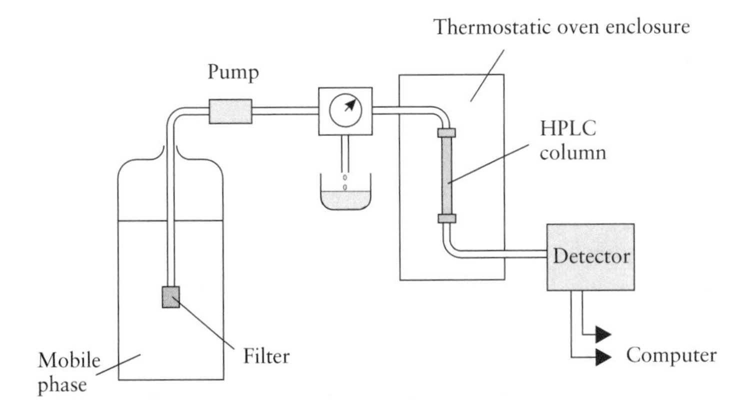

Figure 8.16 Schematic of a simplified HPLC apparatus.

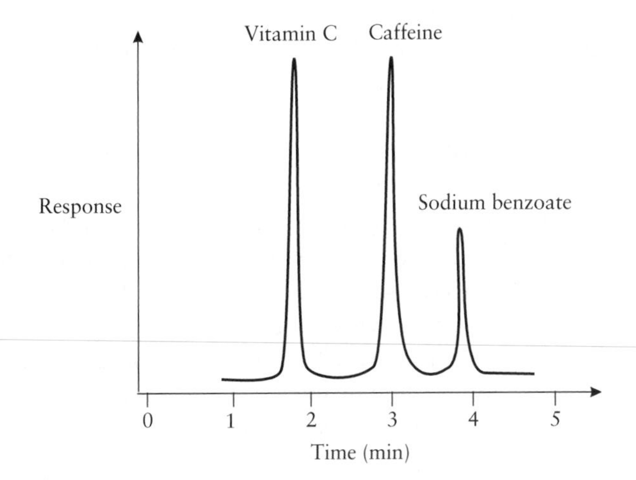

Figure 8.17 HPLC separation of vitamin C, caffeine, and sodium benzoate.

Analytes are normally dissolved within a solvent, or mixture of solvents. The mobile phase is chosen to allow the most efficient separation in the minimum of time; in practice, the mobile phase is often chosen to be the same (or a similar mixture of) solvents as found in the analyte sample. In some cases, however, the mobile phase may contain one or more different solvents to facilitate the separation. A simple chromatograph separation of vitamin C, caffeine, and sodium benzoate from a soft drink is shown in Fig. 8.17. The sample is normally injected into the flowing mobile phase that is forced under pressure through the column by means of a pump. It is important that the operation of the pump is smooth and does not give rise to pulses of flow of the mobile phase but rather ensures a uniform passage of solvent under constant pressure. The separation occurs by partitioning of the solutes (analytes) within the mobile phase and the stationary phase (packing material) of the HPLC column.

8.11 Detectors for HPLC

8.11.1 Photodiodes and UV–visible absorption detectors

UV–visible detectors are the most commonly used type of detector due to their sensitivity, wide linear response range and ability to monitor many different analytes as they elute from a HPLC column. Most UV–visible detectors employ a UV–visible light source in conjunction with one or more photodiodes and a flow cell through which the eleuent passes, as shown in Fig. 8.18.

Optical (photon) absorption detectors may be used in the wavelength ranges of ~200–900 nm provided that the eluent possesses chromophores that absorb in the UV, visible, or near-infrared ranges. Chromophores in this context can include unsaturated double bonds (including aromatic rings), bromine-, iodine-, or sulphur-containing moieties or carbonyl groupings. As with other UV–visible approaches, absorptivity will differ from one analyte compound to another and will be determined by the molar absorptivity, ε, of each respective analyte. The spectral band in instruments of this type are not as narrow as those that employ monochromatic radiation sources. However, instruments with variable wavelength capabilities by their very nature possess a far greater applicability for many different analyses since they can be tuned for differing λ_{max} corresponding to different analytes as they elute from HPLC columns.

Many different monochromatic UV–visible light sources including low-pressure mercury vapour cadmium and zinc lamps can be used for specific applications; however, the most widely used detectors employ a tungsten lamp in conjunction with a deuterium lamp to provide a continuous emission spectrum over a 200–900 nm wavelength range. A monochromatic filter or diffraction grating is then used so that the flow cell and hence the eluent are illuminated within a narrow wavelength range.

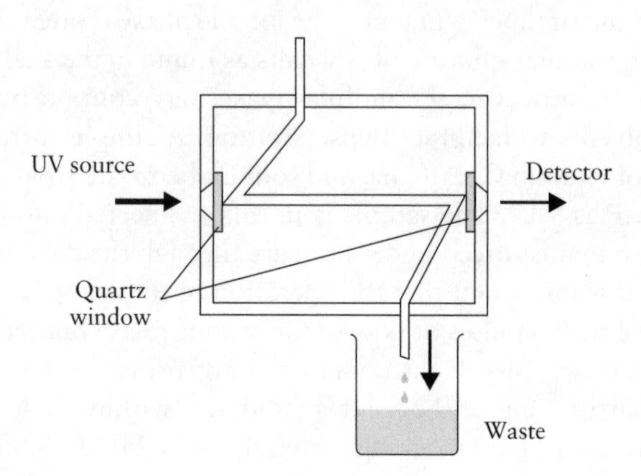

Figure 8.18 Schematic of a UV–visible detector for HPLC.

8.11.2 Fluorescence detectors

Fluorescence detectors can obviously only be used to monitor fluorescent compounds, yet may in some instances offer sensitivities 1000 times greater than those obtainable via UV–visible detection. Fluorescent detectors offer, moreover, a further level of selectivity over non-fluorescent compounds and this may prove extremely useful when we wish, for example, to determine trace compounds within complex mixture—even following an HPLC separation. Monochromatic light is used to irradiate (excite) the sample within a flow cell; fluorescent light is emitted in all directions but is typically monitored at a 90° angle to the incident, Fig. 8.19. As with all fluorescent techniques, the emitted light has a longer wavelength (lower energy) than the incident light. Fluorescence detectors typically use relatively large flow cells of 20 μl volume or greater, which further increases the sensitivity of the technique, particularly when monitoring very low concentrations of the eluted analyte.

Many instruments allow the selection of both the excitation (incident) and fluorescent (monitoring) wavelengths via the use of (diffraction prism or grating based) monochromators to offer greater flexibility. Excitation and fluorescent bandwidths are narrow and highly specific for individual compounds and should, therefore, be selected following either consultation with the literature or via empirical determination (Chapter 16).

The analyst should also be aware of the effect of fluorescent quenching by other compounds and/or ions in the sample. Fluorescent quenching (Chapter 7) is caused by molecular collisions between the analyte and any other compounds/ions. It follows that any factors leading to more molecular collisions (e.g. heating, ionic strength of the eluent, etc.) will lead to a greater degree of quenching.

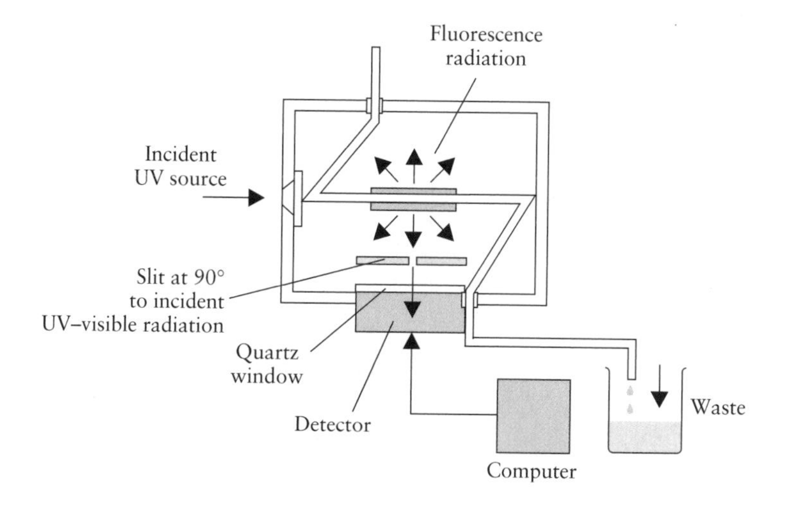

Figure 8.19 Schematic of a fluorescence detector for HPLC.

8.11.3 Electrochemical amperometric detectors

Electrochemical amperometric detectors generically offer high sensitivities together with very favourable selectivities towards differing analytes. The detectors are, moreover, robust, simple to use, as well as being inexpensive to maintain and replace. Amperometric electrochemical detectors monitor the current when an electroactive analyte is either reduced or oxidized at an electrode known as a ***working electrode***. Detectors of this type always contain two other electrodes. The working electrode is often formed from a precious inert metal such as platinum or gold although materials such as glassy carbon or even carbon pastes are sometimes used. One of the advantages of carbon and carbon paste electrodes is the relative ease by which new electrochemical surfaces may be obtained, although in practice electrodes of this type are more cumbersome to use than their metallic counterparts. The potential imposed (polarization) at the working electrode is set with respect to a reference electrode using a potentiostat (see Chapter 10). Current is prevented from flowing through the reference electrode (e.g. Ag/AgCl—see Chapter 10) since the solution–metallic interface of this electrode is designed to have an extremely high electrical impedance. Current is drawn from a third electrode known as the counter or auxiliary electrode, which is designed to be at least 5–10 times larger than the working electrode. In this way, the auxiliary electrode will not limit the rate of the electrochemical reaction occurring at the working electrode surface.

It is important to ensure that the mobile phase is sufficiently conductive to allow the flow of current across the solvent and so between the electrodes. If necessary, ionic electrolytes (e.g. phosphate buffer salts) can be added to increase the conductivity of the mobile phase.

Amperometric detectors are normally designed with the three electrodes suspended in a thin-layer flow cell to allow the continuous monitoring of analytes as they elute from the HPLC column, Fig. 8.20.

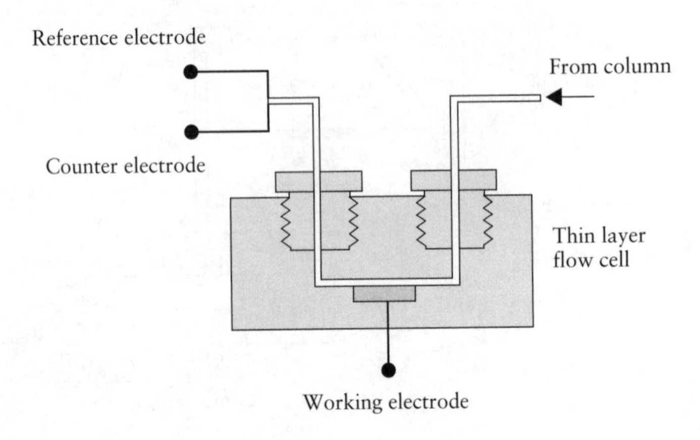

Figure 8.20 Amperometric thin-layer detector cell for HPLC.

8.11.4 Electrochemical conductivity detectors

Electrochemical conductivity detectors monitor the conductivity of the mobile phase as it elutes from the column and passes through a flow cell. This mode of detection may be used to monitor the presence of ions since these will modulate the conductivity of the mobile phase. The greater the concentration of ions, the greater will be the conductivity between the electrodes in the flow cell. Temperature correction is often needed since conductivity varies considerably with temperature. Mobile phases that contain high ionic strengths and, therefore, conductivities should also be avoided since small conductivity changes will be difficult to monitor in solvents that are already highly conductive.

8.11.5 Infrared detectors

Infrared detectors monitor the absorption of organic molecules as they elute from a HPLC column. One problem is that almost all organic compounds absorb infrared radiation to some extent, which includes the mobile phase if solvents such as acetonitrile, dichloromethane, or hexane are used. It is, therefore, important to ensure that the monitoring wavelength is chosen to minimize absorption of the mobile phase.

8.11.6 Differential refractive index detectors

Differential refractive index detectors monitor changes in the refractive index of the mobile phase due to the presence of a dissolved analyte as depicted in Fig. 8.21. As the mobile phase elutes from the HPLC column, its refractive index varies depending on the presence and concentration of each analye. The refractive index of the eluent is differentially monitored with respect to a portion of the mobile phase that has not been passed through the HPLC column. The eluent passes through an optically transparent flow cell as the pure mobile phase passes through another. An incident light beam passes through both cells and detectors monitor the

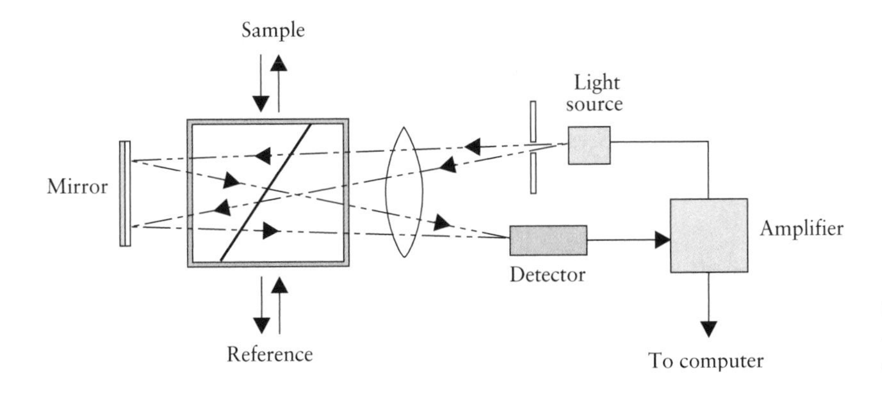

Figure 8.21 Schematic of a differential refractive index detector.

deflection of the beams and hence the refractive indices of the two solutions. By monitoring the refractive indices of both the mobile phase and the eluate, changes in the refractive index may be recorded in the form of a chromatograph.

Differential refractive index detectors offer many advantages since virtually all analytes can be reliably monitored via this approach and, moreover, the sensitivity of the technique is unaffected by slow flow rates. The technique is, however, highly sensitive to temperature fluctuations and, indeed, it is normally necessary to maintain the temperature of the column to within a few hundredths of a Kelvin, which is difficult. One of the principle disadvantages is that sensitivity levels cannot typically match those offered by many other detectors such the as infrared or electrochemically based approaches. It should also be realized that differential refractive detectors cannot be used in conjunction with HPLC protocols that involve solvent gradient elution or temperature ramping approaches since these will alter the refractive index of the mobile phase. It is not possible to differentially monitor the refractive index of the eluent with respect to a mobile phase of variable refractive index because the time taken for the mobile phase to pass through the column relative to the by-pass loop provided for the pure solvent is high.

8.11.7 The evaporative light scattering detector

Evaporative light scattering detectors (ELSDs) are sometimes used to monitor *non-volatile* compounds as they elute from HPLC columns. ELSDs typically offer greater sensitivities than those obtainable from refractive-index detectors, and typically offer similar sensitivities to the majority of organic compounds. The HPLC effluent in this type of detector is first passed through a nebulizer to form a fine solvent/analyte mist within a flow of either air or nitrogen. This mist then passes through a temperature regulated drift tube in which the solvent evaporates to produce a fine gaseous suspension of analyte particles. This cloud is directed through the path of a laser using a flow cell. The particles scatter the light and scattered light is normally monitored at 90° to the laser beam by means of a photo-diode tube.

Since the solvent is evaporated, the analyte(s) must have a significantly higher boiling point than that of the solvent within the mobile phase. In some instances this represents a drawback, although this can be exploited as a further approach for enhancing the selectivity of the analytical determination.

8.11.8 Mass spectrometers as detectors for HPLC

Mass spectrometry can be used as an extremely powerful and also selective form of detection for HPLC since the mass of each molecular species can be identified as it elutes from the HPLC column. The fundamental problem with coupling HPLC with mass spectrometry is that HPLC utilizes relatively large quantities of solvents in the mobile phase, while mass spectrometry requires that the sample be introduced into an evacuated chamber. Several different approaches have been employed but we shall only describe the most important.

One approach involves the use of a ***thermospray interface***. In this device, the HPLC effluent passes through a heated stainless steel capillary tube to evaporate much of the solvent to form an analyte/solvent aerosol. A salt such as ammonium acetate is also normally incorporated as a further solute into the mobile phase since this encourages ionization and so mass spectrometric identification of many polar analytes. Despite these limitations, thermospray-mass spectrometry HPLC can offer extremely high levels of sensitivity and minute detection limits.

Another approach involves only allowing a tiny fraction of the HPLC eluent to be introduced directly into the mass spectrometer.

8.12 Columns for HPLC

Columns for HPLC are normally packaged within stainless steel tubings to withstand the high pressures used. Typical internal bore diameters are 3–10 mm and normally have lengths of 10–30 cm. Particle diameters range from 3 to 10 μm so as to provide 40 000–100 000 (theorectical) plates per metre.

8.12.1 Types of packing material

Two main types of packing material, ***pellicular*** and ***porous***, are commonly used. Pellicular packing particles are formed from spherical, non-porous glass or polymer beads with diameters in the range of 30–40 μm. Thin porous layers of, for example, an ion exchange resin, silica, or alumina are deposited on the surface of the beads and these may, in turn, be used as a further support for a liquid stationary phase for use within partition HPLC. Pellicular packings are sometimes used within guard columns introduced before the main analytical column to filter out particulates that might otherwise be retained by the main column and so impede the HPLC separation. Analytical separations are normally performed with columns packed with porous particles of an ion-exchange resin, silica, or alumina of typical diameters 3–10 μm. It is important that the particle diameter is maintained as uniform as possible since this is one of the factors that crucially affects the reproducibility of the separational performance of packed columns.

8.12.2 Silica and chemically modified silica columns

Silica is widely used as a packing material and has excellent adsorbent properties. Silicon forms a three-dimensional oxygen-linked lattice loosely based on a tetrahedral-type structure. Oxygen atoms bridge neighbouring silicon atoms with Si–O–Si bonds with the structure being saturated with many terminal silanol (Si–OH) groupings that are both polar and very reactive; for this reason, silica is often used to prepare chemically modified stationary phases with functionalized properties.

Silica can be prepared in several ways depending on the size of the particle desired and whether or not the silica is to be chemically modified. One of the simplest approaches for chemical modification involves hydrolysis and subsequent condensation of silicon tetrachloride, sodium silicate, or tetralkoxysilane. This approach gives rise to irregularly shaped silica particles of varying size that require dehydrating and separating according to size. More regularly sized and spherically shaped silica particles can be formed via the partial hydrolysis of liquid polyethoxysiloxane and subsequent emulsification with an ethanol–water mixture.

The specific surface area of the column packing (measured in $m^2\ g^{-1}$) is inversely proportional to the pore width of the silica, which we determine by the reaction conditions that are tailored to produce packing materials with differing properties. Columns with smaller pore sizes (e.g. 5 nm in diameter) and larger specific surface areas give rise to longer retention times and may be used to separate analytes with very similar chemical properties that are otherwise difficult to separate. By contrast, silica packing materials with larger pore sizes (30 nm or more) may be used for the separation and analysis of macromolecules such as proteins.

Silicas can either act as acids or bases depending on how they are formed. Silicas with irregularly shaped particles generally behave as neutral to mildly basic materials whereas those with spherical particles are neutral to mildly acidic in nature. Acidic silicas should normally be used for the separation of acidic compounds, and in a similar manner basic silicas are normally chosen for the separation of basic compounds. Silicas are normally only stable within a pH range of 1–8 and should, therefore, ideally be used with mobile phases within these limits. The silanol groups on the surface of the silica may be chemically modified to produce stationary phases with a range of differing properties. We shall describe four major classes of chemically modified silica.

Esterified silicas

Silanol groupings of the silica can be esterified with alcohols to produce surface-modified materials with tail-like protrusions as depicted in Fig. 8.22. Esterified silicas are, however, prone to hydrolysis and, therefore, should not be used with highly polar solvents such as water or ethanol.

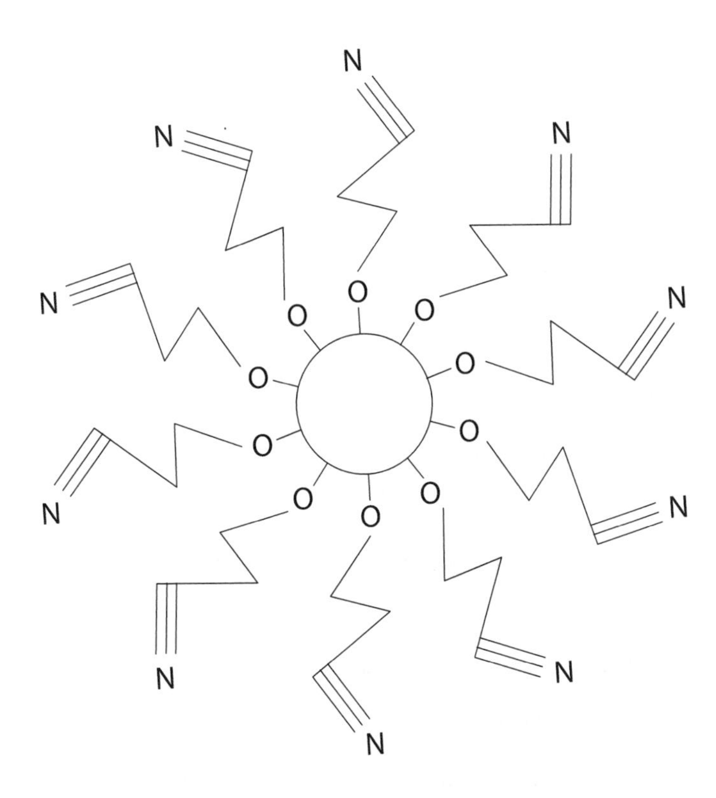

Figure 8.22 Esterified silicas.

Si–N functionalized silicas

Silicas can react with thionyl chloride, $SOCl_2$, to give chlorides that combine with amines ($R–NH_2$) to give Si–N linked derivatized silicas. The radical group may be varied and chosen at will. These support materials generally offer better hydrolytic stability than esterified silica supports.

Si–O–Si–C functionalized silicas

Silicas derivatized via reaction with mono- or dichlorosilane yield the most stable of all the chemically modified silica stationary support phases and possess Si–O–Si–C linkages. The most widely encountered example of this type of chemically modified silica is octadecylsilane in which $R = -(CH_2)_{17}CH_3$. This material is extremely non-polar and is most commonly used for reverse-phase applications. Unreacted silanol groups may be subsequently treated with trimethylchlorosilane via a process known as 'end-capping' to render them un-reactive.

Polysiloxane/silicone derivatized silicas

Silicone (polysiloxane) derivatized silicas are formed via reaction with chlorosilane (typically a dichlorosilane) to form a bonded polymer coating layer upon the silica support. These packing materials typically offer extremely favourable stability to pH due to the protection that the polymer coating offers to the silica backbone.

8.12.3 Styrene divinylbenzene

Styrene divinylbenzene may be used as a versatile stationary phase support and is formed via the co-polymerization of styrene and divinylbenzene. The degree of cross-linking and pore structure can be controlled by varying the ratio of the two monomers. Semi-rigid supports containing greater than 8% divinylbenzene can be used with pressures up to ~60 bar. Supports of this type may swell and shrink in differing solvents and, therefore, should be used exclusively with only one solvent. Rigid polystyrene–divinylbenzene co-polymeric supports, by contrast, do not swell or shrink even when exposed to differing solvents and are stable up to pressures of 3300–350 bar or more.

Polystyrene–divinylbenzene supports offer greater stability to variation in pH than silica can offer, and hence can be routinely used in a pH range of 1–13. Either ion exchange resins or octadecylsilane can be incorporated into the polymer matrix to provide functionalized stationary supports.

8.12.4 Alumina stationary phases

Alumina is naturally basic but may with suitable processing be rendered either neutral or acidic. Acidic alumina may be used as a weak anion exchanger while the basic form may be used as a weak cation exchanger. Alumina offers good pH stability and may be routinely used throughout a pH range of 2–12. Alumina is relatively unstable to elevated temperatures and, therefore, columns packed with this material should not be heated above 150°C.

8.12.5 Controlled pore glass

Controlled pore glass or CPG is formed by de-glassing borosilicates first dispersing and then removing B_2O_3 from within the borosilicate matrix. In the resultant material, as its name suggests, the size of the pores can be easily controlled in size. CPG is generally inert to extreme pressures and is chemically inert except towards strong bases, and may be used for partition, adsorption, ion exchange, size exclusion, and affinity forms of HPLC.

8.12.6 Hydroxyapatite

Hydroxyapatite is a crystalline form of calcium phosphate $Ca_{10}(PO_4)_6(OH)_2$ and may withstand pressures of 150 bar or more. Hydroxyapatite columns are used primarily for the separation and identification of mixtures containing macromolecules such as proteins.

8.12.7 Agarose

Agarose is an extremely pH stable cross-linked polysaccharide that may be derivatized for applications within, for example, affinity chromatography.

8.12.8 Porous graphite carbon

Porous graphite carbon provides extremely chemically stable, homogeneous, and non-polar surfaces and may be used with high-pressure–high-performance reversed-phase chromatography columns.

8.13 Zone electrophoresis

Electrophoresis separates substances by electrical migration so that ionized or at least highly polar substances move under the influence of an electric field. The rate of movement of a substance is a function of both its molecular weight and its charge due to the resistance to movement, that is, viscous drag offered by the medium through which it will have to travel.

There are many forms of electrophoresis; one of the most versatile and probably the most commonly encountered forms is known as ***zone electrophoresis***, in which solutes travel within a mobile phase across a stationary support. Analyses of this type, therefore, exploit both chromatographic and migratory approaches to achieve the required separation.

Support stationary phases are normally packaged or manufactured in some form of a block or plate, made of starch, polyacrylamide gel, polyurethane foam, or even paper.

Starch or polyacrylamide gel electrophoresis involves depositing a narrow band of the sample mixture across a midway line between the ends of the block of the material. The two ends of the block are connected via electrodes to a high-voltage potential source and upon polarization of the block, different components (ionized solutes) migrate towards either the anode or cathode depending on their respective polarities. Separated bands or lines correspond to solutes within the mixture and are typically visualized by staining the components following the separation. A ***densitometer*** may then be used to determine the intensities of coloured bands and in this way the relative concentrations of different components may be quantified within the mixture by reference to a pre-determined colour intensity calibration profile.

The electric field across the plate or block is expressed in terms of volts per centimetre and may range in values from 500 to 5000 V cm^{-1} or even greater. Larger molecular weight solutes (such as macromolecular proteins) often require larger electrical fields for separation than solutes of lower molecular weight.

It should be mentioned here that the pH of the mobile phase will, of course, affect the degree of ionization of solutes within the mixture and hence the rate and degree of separation. This is particularly relevant when analysing the amino acid content of mixtures, which are often separated within clinical biochemistry laboratories by zone electrophoresis. At a certain pH (the so-called isoelectric point), the net charge held by a particular amino acid will be zero since it will exist in the form of a zwitterions; under these conditions the amino acid will not move either towards the anode or the cathode.

Capillary zone electrophoresis

Capillary zone electrophoresis (CZE) is becoming widely used in biological laboratories since small sample volumes (in the order of 1 nL) of even complicated mixtures can be resolved and quantified with relative ease. A schematic of the instrumentation required is shown in Fig. 8.23. A fused silica capillary tube with a typical length of 1 m and 10–100 μm internal diameter is filled with the electrolyte of choice. A small volume of the mixture to be separated is then normally introduced to one end of the tube either via a simple immersion of the tubing within the mixture or via a hydrostatic or pneumatic uptake of the sample. Samples may in some circumstances be introduced by applying a small potential across the ends of the capillary to induce an uptake of solutes via electrical migration.

Opposing ends of the capillary tubing are then immersed in two separate reservoirs of electrolyte buffer in which platinum electrodes are immersed. Capillary tubing is often used containing ionizable silanol groupings that will be negatively charged within environments of pH 2 or greater. Cations (positively charged ions) adsorb along the internal wall of

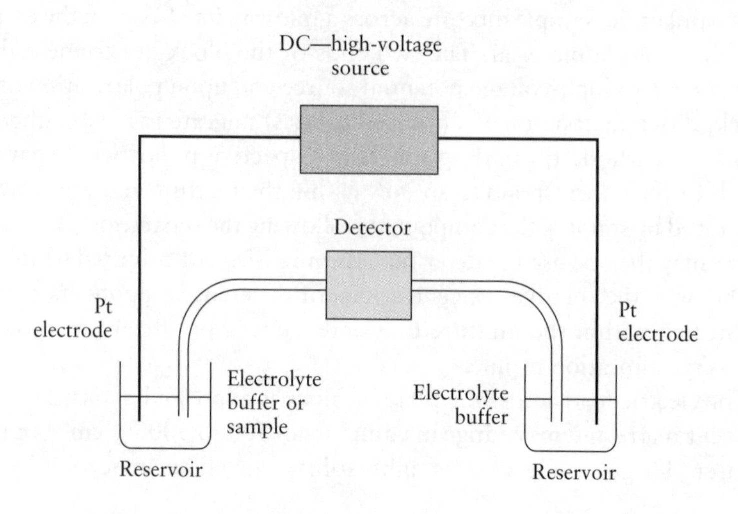

Figure 8.23 Schematic of capillary zone electrophoresis apparatus.

the capillary and form a double or counter-layer of charge. A potential of 1000–30 000 V or more is then applied across the capillary and this causes anions (negatively charged ions) to migrate towards the cathode (negative electrode). Cations at the surfaces of the capillary within the mobile phase move towards the cathode (negative electrode) under the influence of electrical migrations. Since these ions are solvated, they carry solvent molecules and all other solutes along in the same direction causing a unidirectional flow of fluid (***electroosmosis***) towards the cathode. Osmotic flow rates are often in the order of several hundred nanolitres per minute, although this is dependent on the pH of the electrolyte, the applied potential, and the buffer concentration.

The most positively charged (cationic) analytes with the lowest molecular weights will be detected first. Neutral molecules typically travel at a flow rate that is *slightly less* than that of the osmotic flow rate since they are not influenced directly by the electric field but only by the flow of solvent. The most negatively charged anions will travel at the slowest flow rate and will, therefore, be detected last since the electric field will retard their travel.

Exercises and problems

8.1. Why should you always use pencil to mark the starting line within paper chromatography?

8.2. What is the R_f value of a chromatogram and what are the largest and smallest R_f values possible?

8.3. What is meant by a stationary phase and a mobile phase?

8.4. What is the basis of electrophoretic separations?

8.5. What phenomenon gives rises to band broadening in chromatography and why does band broadening increase with longer columns?

8.6. Discuss the relative advantages of using paper and thin-layer chromatography.

8.7. Describe the principles of gas chromatography.

8.8. What samples can be separated by HPLC and not GLC.

8.9. Why might gas chromatography be coupled to mass spectrometry and how can this be achieved?

8.10. A solute has a distribution coefficient of 9.5 between water and hexane. Ten grams of the solute is dissolved in 50 cm^3 of water. Calculate the concentration of the solute remaining in the water sample, following extraction with one, two, and three extractions with 10-cm^3 aliquots of hexane.

8.11. How might the number of theoretical plates in a chromatographic column be determined?

8.12. If the width of a solute peak is 7.5 mm at one-half its height, what will be its base width by extrapolation?

8.13. If the base width of a chromatographic peak is equivalent to 2.4 s duration calculate the half-height peak width for this peak.

8.14. A chromatographic peak is found to have a retention time of 72 s. The base width of the peak is determined to be 6.5 s by intersection of the sides of the peak with the baseline. If the column is 1 m in length, calculate the HETP in terms of centimetres per plate.

8.15. A gas chromatogram of a mixture of *ortho* and *para* isomers of cresol gives peaks with integrated areas of 35.7 and 10.5, respectively. Assuming the detector responds equally to both isomers calculate the percentage of each isomer within the mixture.

8.16. If the retention time for a chromatographic peak, t_R, is 85 s and t_{mob} is 40 s, calculate the capacity factor, k'.

8.17. Calculate the capacity factor, k', for a chromatographic peak if its retention time, t_R, is 95 s and t_{mob} is 45 s.

8.18. If the HETP of a column is 0.01 cm per plate and the number of theoretical plates is 5000, calculate the length of the column.

8.19. Suggest approaches for helping to improve the resolution between different substances within a chromatographic separation.

8.20. Define what is meant by (a) the selectivity factor and (b) the capacity factor.

8.21. If two chromatographic peaks have capacity factors of 1.4 s^{-1} and 3.4 s^{-1}, calculate the selectivity factor.

Summary

1. Solvent extraction methods based on the mixing of two immiscible liquids are known as liquid–liquid extractions.

2. The distribution coefficient K_D is equal to the ratio of the concentration of a solute between two phases at equilibrium, that is, $K_D = [S]_{org}/[S]_{aq}$.

3. Multiple solvent extractions involving several small volumes are more efficient than one cumulative volume of solvent.

4. Solid-phase extractions may be used as an alternative separative approach and may involve the use of, for example, hydrophobic groups such as ^{18}C moieties bonded to solid supports.

5. Chromatography is a term used to describe separatory techniques that utilize a stationary and a mobile phase.

6. Elution chromatography involves the use of two liquid phases in which one liquid phase is immobilized onto a solid support.

7. The retention time, t_R, describes the time a solute is retained by a chromatographic column.

8. The capacity factor, k', may be used to compare relative rates of solute migration along columns:

$$k' = \frac{t_R - t_{mob}}{t_{mob}}$$

where t_{mob} is the spread of the mobile phase.

9. The selectivity factor, α, for two solutes is the ratio of the larger partition coefficient to smaller partition coefficient for two phases, that is,

$$\alpha = \frac{K'_l}{K'_s}$$

10. The efficiency of a chromatography column may be described in terms of the number of theoretical plates, N, or the height equivalent of a theoretical plate and these are summarized by the van Deemter equation, that is:

$$H = \frac{L}{N} = A + \frac{B}{\overline{\mu}} + C\overline{\mu}$$

where H is the height equivalent of a theoretical plate, L is the length of the column (normally in cm) and A, B, and C are constants relating to a particular system. $\overline{\mu}$ is the average linear velocity of the mobile phase.

11. Chromatographic band broadening can be caused by a number of effects such as eddy diffusion of solvent, flow distribution of the mobile phase, and the presence of areas of stagnant mobile phase within the column.

12. Band broadening increases with increasing column lengths.

13. The R_f of a paper or TLC chromatogram may be calculated according to:

$$R_f = \frac{\text{Distance travelled by centre of spot}}{\text{Distance travelled by solvent front}}$$

14. Gas chromatography utilizes a carrier gas as the mobile phase together with a solid stationary phase within a packed column.

15. Gas–liquid chromatography involves a liquid stationary phase being adsorbed onto a solid phase.

16. A number of differing detectors may be used with GC and GLC approaches including flame ionization, flame photometric, atomic emission, electron capture, and thermal conductivity approaches.

17. High-performance liquid chromatography is used for the separation of mixtures using a number of solvents and involves passing a mobile phase through a finely divided stationary phase under high pressure.

18. A number of different detectors may be used in conjunction with HPLC columns and include photodioides, fluorescent electrochemical, infrared differential refractive index, evaporative light scattering, and mass spectrometry based devices.

19. Many different packing materials may be chosen according to the mixture to be resolved but may be classified as being either pellicular- or particle-type packings. Pellicular packing materials are formed from spherical glass or non-porous polymer beads that may be coated.

20. Electrophoresis allows separating substances via electrical migration. In zone electrophoresis, solutes travel in a mobile phase across a stationary support.

21. Capillary zone electrophoresis involves the use of fused silica columns to allow the resolution of the most complicated mixtures.

Further reading

Jennings, W., Mittlefehldt, E., and Stemple, P. (1997). *Analytical gas chromatography*. Academic Press.

McNair, H. and Miller, J. M. (1997). *Basic gas chromatography*. Techniques in Analytical Chemistry Series, Wiley.

Meyer, V. (2004). *Practical high-performance liquid chromatography*. Wiley.

Miller, J. M. (2004). *Chromatography: concepts and contrasts* (2nd edn). Wiley.

Robards, K., Haddard, P. R., and Jackson, P. (1994). *Principles and practice of modern chromatography*. Academic Press.

Mass spectrometry 9

Skills and concepts

This chapter will help you to understand:

- The principle underpinning mass spectrometry for the separation of ions according to their m/z ratios.
- The main components that all mass spectrometers possess in terms of sample inlet systems, ionization chambers, and ion detectors.
- The operation of batch inlet, direct probe, and chromatographic inlet systems.
- The use of electron impact, fast atom bombardment, secondary ion, field ionization/desorption, thermospray, electrospray, inductively coupled plasma, and atmospheric pressure electron impact ionization sources for use in mass spectrometry.
- The operation of and main components of magnetic sector, double focusing, quadrupole, ion cyclotron resonance ion trap, and time-of-flight mass analysers.
- How electron multiplier, Faraday cup and scintillation detectors, and photographic films may be used for ion detection in mass spectrometry.
- What is meant by tandem mass spectrometry and why it is used for analytical purposes.
- How mass spectrometry may be coupled to other analytical techniques such as gas chromatography.

9.1 Introduction to mass spectrometry

Mass spectrometry is a widely used instrumental technique that relies on separating gaseous charged ions according to their mass-to-charge ratios. There are many forms of mass spectrometry and this chapter will give an introduction to the most widely practised forms. Mass spectrometry has no link or similarity to any of the other spectroscopic techniques discussed within this book (that exploit electromagnetic radiation) but rather historically became known as a form of 'spectroscopy' since early instruments recorded data photographically in the form of lines and in this way possessed some resemblance to optical line spectra. Almost all forms of mass spectrometry now give spectra in the form of electronic data and so this historical naming can be rather misleading.

Mass spectrometry is an extremely powerful and widely used analytical tool capable of providing both qualitative and quantitative information relating to: (a) the structure of inorganic and organic components within complicated mixtures; (b) the relative and absolute concentrations of components within mixtures; and (c) the isotopic composition and relative ratios of isotopes within unknown analyte samples.

Mass spectrometry is widely used in conjunction with other analytical techniques as described in other chapters of this book. Examples in this context include the use of mass spectrometers as detectors within high performance liquid chromatography (HPLC) and gas chromatography (GC) (see Chapter 8).

Atomic, ionic, and molecular weights in mass spectrometry are normally expressed in terms of ***atomic mass units*** (**AMUs**). One atomic mass unit is defined as one-twelfth the mass of a ${}^{12}_{6}C$ atom. It should be noted in this context that chemical molecular weights are *normally not* whole numbers since they take into account the relative natural ratios of isotopes that are found in samples of a compound.

There are many different types of mass spectrometers. The basic principles of mass spectrometry are common to each. Such a device may be thought of as involving: (a) a sample inlet system; (b) a means by which samples are ionized; (c) an accelerator of the ions by an electric field; (d) a dispersion of ions according to their mass-to-charge ratio; and (e) identification of ions together with appropriate signal processing and data output. In essence, instruments are classified and named according to how ions are dispersed and subsequently detected. The most commonly encountered forms of mass spectrometry employ double focus, magnetic sector, quadrupole, or time-of-flight type instrumentation, all of which are described in the following sections.

The instrumentation through which the ion beam passes (namely the ion accelerator, ion dispersion chamber and detector), must be evacuated down to a pressure of typically in the range of 10^{-4}–10^{-8} Torr, which is often achieved with an oil-diffusion pump.

A very small quantity of sample is first introduced through a sample inlet system. Inlet systems normally contain a nebulizer or atomizer together with a heater in order to vaporize the sample prior to ionization. Samples are then ionized by bombarding them with electrons, photons, molecules, or ions. The inlet system and ionization sources may be combined within one unit—or may be fabricated as separate components of the mass spectrometer; in either case, they serve to produce an ionized stream of the analyte sample, which may then be accelerated and passed into the mass analyser. The stream of ions that is produced may either consist of positively or negatively charged ions, although the former is more common. Since the ions carry a charge they may be accelerated and focused into a beam by an electric field and then passed into the mass analyser. Each of the ions possesses a characteristic momentum and charge; it should be noted that the momentum of an ion is determined by both the mass of the ion and how the ion is accelerated by the electric field, which itself depends on the charge it carries. In this way, ions may be dispersed or physically segregated within the mass analyser according to their mass-to-charge ratios. Ions are finally detected and characterized according to their mass-to-charge ratios and the information recorded in the form of a '***mass spectrum***'.

In the following sections, we shall first describe each component of a mass spectrometer in detail before discussing the possible uses and applications of mass spectrometry as an analytical tool.

9.2 Sample inlet systems

A sample inlet system must be designed to allow the introduction of a sample into the mass spectrometer with as a little loss in the quality of the vacuum as possible. The majority of mass spectrometers are equipped with two or more sample inlet systems to allow for the introduction and handling of gaseous, liquid, or solid samples. The three most commonly employed systems are the batch inlet, direct probe, and chromatographic type inlet systems, each of which we shall consider in turn.

9.2.1 Batch inlet systems

Batch inlet systems are the simplest type of inlet systems available and operate by allowing samples to leak through a microporous metallic or glass diaphragm into the ionization chamber; Fig. 9.1. The reservoir chamber is often lined with a vitreous coating to prevent sorption of the sample and subsequent contamination of future samples. Liquid samples (normally with volumes of the order of microlitres) are first injected into a reservoir chamber evacuated to a pressure of around 10^{-4}–10^{-5} Torr to allow evaporation of the sample prior to its introduction to the ionization chamber. The reservoir chamber is typically heated to a maximum temperature of around 500°C to facilitate sample vaporization.

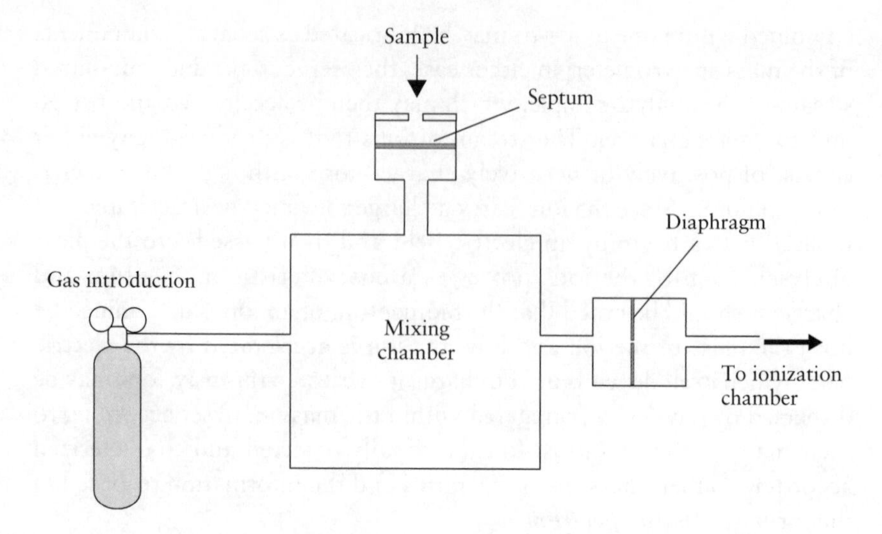

Figure 9.1 Batch inlet system.

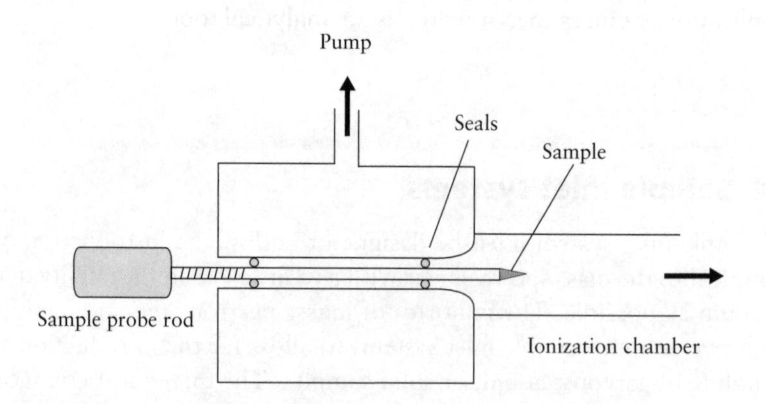

Figure 9.2 Direct probe inlet system.

9.2.2 Direct probe inlet systems

Extremely small samples (often down to a few nanograms) may often be introduced directly into the ionization chamber supported upon a sample probe; in this arrangement, the ionization chamber is normally designed such that it is isolated from the main instrument by means of a vacuum lock; Fig. 9.2. In this way, the evacuation of the system is maintained with minimal re-pumping between analyses. Truly minute quantities of sample may, moreover, be analysed in situations where the amounts of sample are limited. Solid and liquid samples are typically introduced into a probe inlet system via a glass or aluminium capillary, sample cap, or on a wire support, and positioned a few millimetres from the ionization source. Vaporization of the source is achieved by evacuation of the ionization chamber and this may also be aided by heating of the chamber. Thermally unstable samples

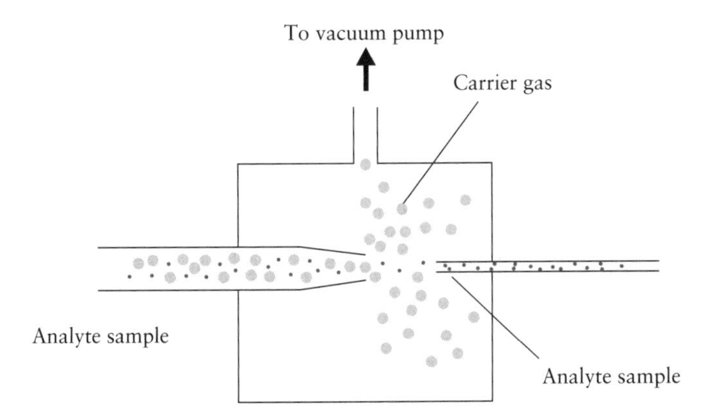

Figure 9.3 Jet separator inlet systems.

may often be vaporized and so analysed by simply evacuating the ionization chamber to a sufficiently low pressure.

9.2.3 Chromatographic inlet systems

Mass spectrometers are often used as detectors in gas–liquid chromatography (Chapter 8) to quantify analytes as they elute from the column. Mass spectrometry is an extremely powerful technique in this context since analytes may be both identified (from their characteristic molecular/ionic masses) as well as being quantified.

It should be remembered that the analyte elutes from the column within a stream of gas acting as the chromatographic mobile phase. This gas cannot be passed directly into the mass spectrometer. In order to maintain the requisite level of evacuation within the instrument the carrier gas (normally hydrogen or helium) must be removed before the stream of analyte passes into the ionization chamber. In practice, this is normally achieved by means of a jet separator (Fig. 9.3) designed so as to remove most of the carrier gas. The less dense carrier gas is deflected by the vacuum, while the denser analyte molecules with their greater momentum carry on towards the output of the jet separator and so to the ionization chamber.

9.3 The ionization chamber

As we saw earlier, the sample must either originally be in the gaseous state or be vaporized prior to ionization. There are several methods by which samples may be analysed, which we shall now consider in turn.

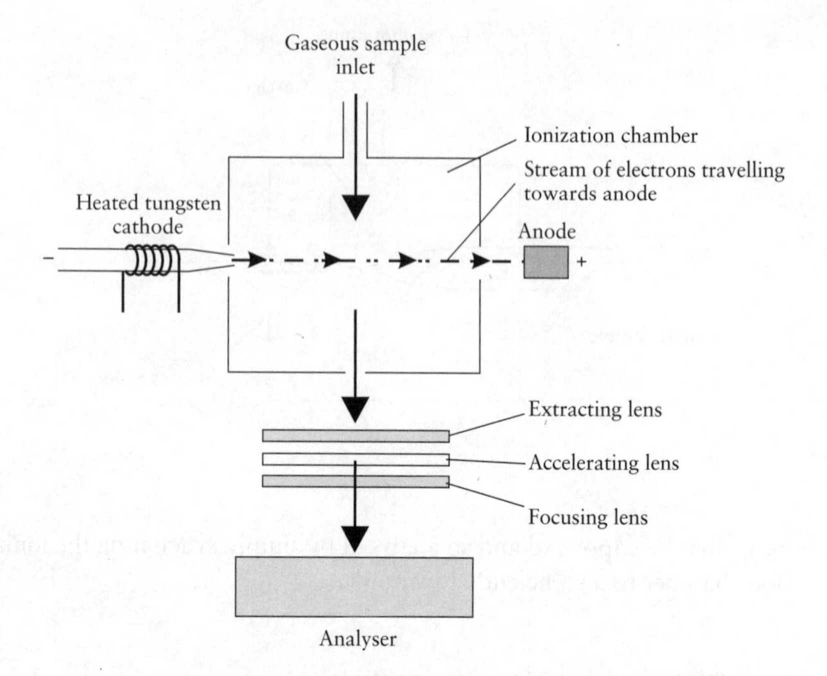

Figure 9.4 Electron impact ionization source.

9.3.1 Electron impact ionization sources

Electron impact ionization sources are amongst the most widely used in mass spectrometry. The gaseous sample stream within this arrangement passes directly across a path of electrons being accelerated towards an anode; Fig. 9.4. The electron beam is normally generated by thermionic emission from a heated tungsten electrode which acts as a cathode. The gaseous sample is introduced at 90° to the stream of electrons. The electrons cause ionization of the gaseous sample. The majority of the ions that are produced are positively charged (cations) which may be accelerated towards cathodes polarized at a given potential, V. A series of extracting and focusing electrodes are provided as lenses to direct the ions formed by collision with the stream of electrons onto the analyser. If we assume that an ion has a charge of one, then the kinetic energy E_{kin} imparted will be equal to:

$$E_{kin} = eV = \frac{1}{2}mv^2 \tag{9.1}$$

where m is the mass of the ion and v its velocity and e is the charge on an electron.

It is important that a very *similar* kinetic energy is imparted to all the ions; in practice, we achieve this by careful design of the ionization chamber so that ionization occurs as close as possible to the accelerating electrode assembly.

9.3.2 Fast atom bombardment ionization sources

Fast atom bombardment (***FAB***) mass spectrometry is a technique in which samples are ionized by bombardment with a beam of highly energetic (several kilovolts) atoms of, for example, xenon or argon. Samples are normally dissolved in a solution of glycerol or another non-volatile solvent since this helps decrease the lattice energy that must be overcome if the ions are to be desorbed and liberated from a sample.

This beam of fast atoms is normally produced via the electronic acceleration of xenon or argon *ions* from a separate electron impact ion source into a chamber containing xenon or argon atoms at a pressure of around 10^{-5} Torr. The speeding ions undergo electronic transfer reactions with the argon or xenon atoms within the chamber and in this way, are neutralized to form a beam of fast *atoms*. A schematic of an FAB gun is shown in Fig. 9.5. Upon leaving the FAB gun the beam of fast atoms then ionizes the sample described above. FAB ionization techniques are particularly suitable for analysing higher molecular weight samples (e.g. for the analysis of samples with molecular weights in the range of 3000–10 000 Da), thermally unstable samples and many biological samples.

9.3.3 Secondary ion mass spectrometry

Secondary ion mass spectrometry (SIMS) may be thought of as a derivative of FAB mass spectrometry. SIMS is normally used for the analysis of solid surfaces, whereas ***liquid secondary ion mass spectrometry (LSIMS)***,

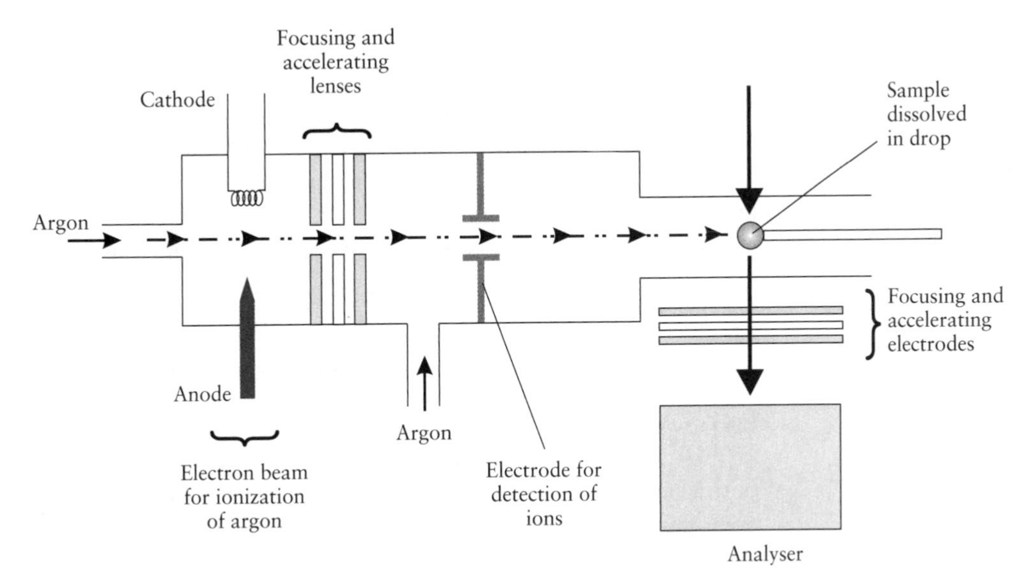

Figure 9.5 Fast atom bombardment source.

as the name suggests, may be used for the analysis of samples that comprise a non-volatile solvent in which the analyte is dissolved.

Solid surfaces are bombarded within SIMS by high energy (5–20 kV) Ar^+, Cs^+, N^{2+}, or O^{2+} ions formed within an electron impact ion source or gun. These primary ions give rise to the desorption of atoms, a few of which gain sufficient energy to become ionized, and these are known as the ***secondary ions***. These secondary ions are then electronically accelerated and pass into the mass analyser for determination.

The ability of SIMS to study localized areas of surfaces is one of the principal advantages of this approach. Indeed, the primary ion beam may be focused onto an area of no more than 0.5 mm^2 in conventional SIMS instruments. Depth profiling down to 100 Å or more is often also possible by varying the energy of the primary ion beam and so the depth to which primary ions may penetrate the sample.

Ion micro-probe SIMS allows the primary ion beam to be focused on surface areas of 1–2 μm^2 to allow truly micro-analyses to be performed. A microscope is normally used in conjunction with SIMS to allow micro-positioning of the ion beam to the desired area. The primary ion source is sometimes passed through a mass filtering device such as a quadrupole (see Section 9.4) to allow ions of only one energy to impact the sample surface, which offers a further approach for selectively ionizing analytes within complex surfaces.

SIMS is performed on samples dissolved within a non-volatile solvent such as glycerol. The sample is bombarded with highly energetic heavy ions of, for example, Cs^+ that possess energies of 30 keV or more.

9.3.4 Field ionization/desorption sources

Field ionization chambers produce electrostatic gradients of the order of 10^3 V cm^{-1} close to the tip of a sharp metallic (or carbon) electrode. The electrodes is shaped in the form of a blade or needle-like assembly which facilitates the focusing of the electrical field gradient towards the tip of the electrode. Samples can be ionized by removing electrons from gaseous molecules.

9.3.5 Laser desorption ionization sources

Laser desorption operates by focusing a pulsed laser capable of delivering 10^6–10^8 W onto a localized area of no more than 10^{-3} cm^2 of a sample to both evaporate (desorb) and ionize the sample. Since very small surface areas can be analysed, laser desorption is particularly applicable for the study of surfaces with varying composition (e.g. minerals), or even, for example, for microbiological structures (e.g. within cellular organelles). The wavelength of the laser may moreover often be tuned so as to allow for selective ionization and thus quantification of particular compounds within complicated samples.

Matrix-assisted laser desorption ionization (MALDI) is a technique in which the sample is mixed within a solvent containing organic molecules (the matrix) that strongly absorb the laser radiation. The radiation that is absorbed by the matrix first causes evaporation of the solvent and second photo-excites the sample, which, in turn, leads to the ionization of the sample. This approach negates the need for selecting laser frequencies to match differing analytes since it is the matrix and not the analyte that absorbs the laser radiation.

9.3.6 Spark ionization sources

Spark ionization sources operate by passing a radio-frequency (typically 80–100 kHz) high voltage (kV) spark between a pair of electrodes, one of which is either coated with the sample, or alternatively, is designed to hold the sample, in the form of an orifice or a thimble-like cup.

9.3.7 Surface thermal ionization sources

Surface thermal ionization operates by directly heating a tungsten filament that is designed to be coated with the analyte. This approach is particularly useful for ionizing non-volatile samples.

9.3.8 Plasma desorption ionization sources

In ***plasma desorption ionization (PDI)*** approaches, a sample is first adsorbed on a small aluminized nylon foil before being bombarded with the highly energetic fission fragments of a radionuclide such as ^{252}Cf to induce the desorption and ionization of analyte ions. In practice, PDI is not so widely used following the advent of alternative techniques such as MALDI, although plasma desorption may still prove useful for the ionization of analytes with relative molecular masses of 10 000 or more.

9.3.9 Thermospray ionization sources

Thermospray (TSP) ionization operates by passing a sample of analyte solution through a thin steel capillary tube at very high (supersonic) velocities into a vacuum chamber to produce a fine spray of analyte ions. The steel capillary tube must be heated to prevent freezing of the droplets as they leave the capillary tube and enter the vacuum chamber. Ions are focused, extracted, and accelerated from the vacuum chamber and into the mass analyser by applying an electric field.

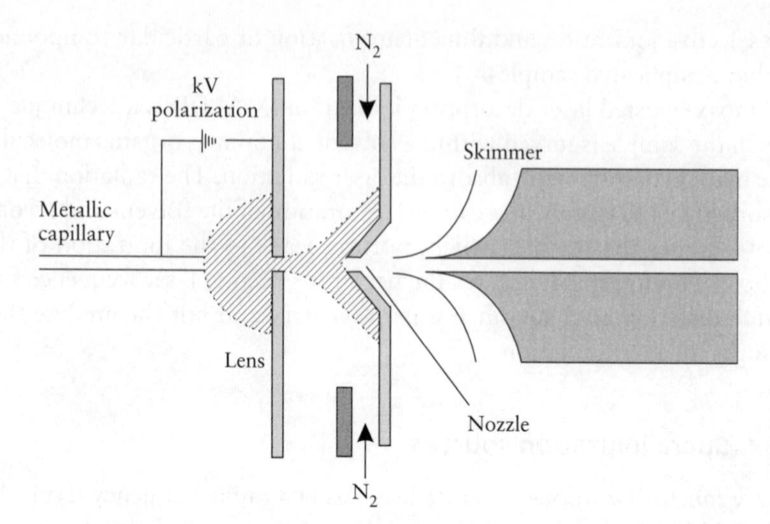

Figure 9.6 Electrospray ionization source.

9.3.10 Electrospray ionization sources

Electrospray ionization sources are often abbreviated to either ESI or EI within the literature. A schematic of an electrospray ionization source is shown in Fig. 9.6. A solution analyte is first passed through a thin metallic capillary tube within electrospray ionization sources to produce a fine spray of sample droplets. A potential of between 3 and 6 kV is then applied between the tip of the capillary tubing and a counter electrode positioned a few millimetres away, so that the droplets pass through a large electric field and are ionized. Ions are often produced with multiple charges if more than one ionizable moiety is available on a molecule. This occurs most frequently with higher weight molecular compounds, and therefore is often of relevance to the analysis of biological samples containing high molecular weight compounds.

9.3.11 Inductively coupled plasma ion sources in mass spectrometry

Inductively coupled plasma (ICP) sources are used to ionize samples prior to analysis in a similar manner to atomic absorption spectroscopy (Chapter 7). The ICP source consists of a torch that contains a plasma stream of an inert gas (usually argon) that is maintained via an induction radio-frequency coil, which concentrically surrounds it. The sample is introduced into the flowing gas stream and the resultant plasma passes through the region surrounded by the induction coil in the form of a spray. The temperature of the plasma is extremely high, being maintained at temperatures in excess of 10 000 K.

9.3.12 Atmospheric pressure electron impact ionization sources

Standard electron impact ionization sources operate under high vacuum conditions since the ions must enter an evacuated mass analyser. Ionization via an electron impact approach may, however, be performed under normal atmospheric conditions with an efficiency of 10^3–10^4 times greater than that obtainable under high vacuum conditions. One problem is that ions must leave the ion source (at atmospheric pressure) and enter the mass analyser (under vacuum)—and so the ion source must in some way be coupled to the ion mass analyser. In practice, this is often achieved by allowing ions to leak through a diaphragm (with an opening of no more than 10 μm diameter) that separates the ion source from the rest of the mass spectrometer's instrumentation. A heated section of tubing is often used to connect the ion source and the mass analyser since ionized samples that are not heated may freeze as they pass through the diaphragm at high velocity into the evacuated chamber of the mass analyser.

9.4 Mass analysers

Mass analysers disperse ionized samples according to differences in their mass-to-charge charge ratios as they emerge from the ionization source or chamber. It should be remembered that ions are drawn towards and focused through oppositely charged polarized electrode slits to impart kinetic energy and, in this way, ions are accelerated into the mass analyser. Every effort is taken in the design of the ionization chamber to ensure that the same kinetic energy is imparted to all of the ions. The ions possess differing masses, however, and will therefore enter the mass analyser with a range of velocities. Ions with the smallest masses will travel with the greatest velocities and ions with the greatest masses will travel with the slowest velocities. There are several different types of mass analyser that are commonly used within modern commercial instruments each of which we shall consider in turn.

9.4.1 Magnetic sector analysers

A schematic of a ***magnetic sector analyser*** is shown in Fig. 9.7. Magnetic sector analysers incorporate a metallic tube through which the gaseous ions are accelerated towards a detector positioned behind an exit slit. The tube is shaped with a 60, 90, or 180° curve. Ions are deflected by a magnet, (normally an electromagnet), so that they travel around the bend and on towards the exit slit and ion detector. The angle through which the ions

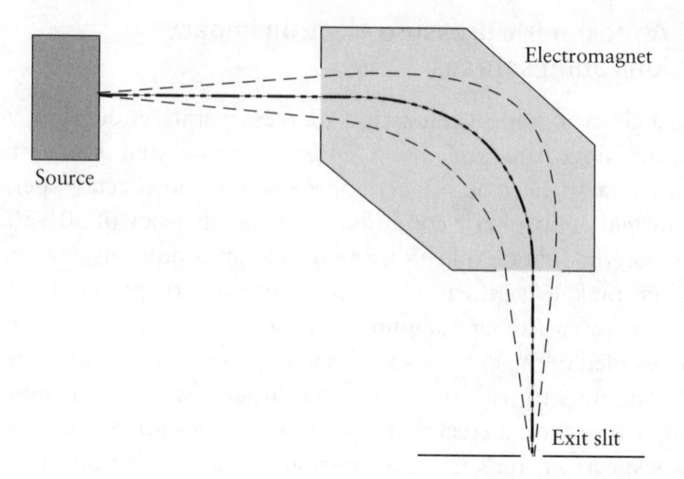

Figure 9.7 Magnetic sector analyser.

are deflected, and therefore the path that the ions take, depends on both the charge of the ion and its mass. Ions with similar mass-to-charge ratios will travel along the same trajectory and ions with different mass-to-charge ratios will travel along different trajectories, In this way, ions can be separated according to ionic mass-to-charge ratios. Ions initially may enter the mass analyser chamber with slightly differing trajectories. However, the magnetic sector focuses the ions of similar mass-to-charge ratios along common trajectories. Instruments of this type are sometimes known as single focusing mass spectrometers as opposed to the double focusing instruments, which we describe in the next section.

The exit slit leading from the mass analyser to the ion detector is designed so as to be extremely narrow to only let ions within a very narrow limits of mass-to-charge ratios to pass through. The trajectory of the ions may be altered by either: (i) altering the strength of the magnetic field through which the ions pass or (ii) modulating the accelerating potential of ion accelerating electrodes. These two approaches allow ions of differing mass-to-charge ratios may be focused towards the exit slit. For any one set of conditions, the ion detector will only detect ions with a certain mass-to-charge ratio since only ions with a certain mass-to-charge ratio will be able to pass through the exit slit of the ion analyser and through to the ion detector.

By varying either the strength of the electromagnet or the velocity of the ions (via the polarized accelerating electrode), ions of all charge-to-mass ratios may be focused to pass through the mass analyser exit slit and thus be detected.

The resolution of magnetic sector mass spectrometers is, in practice, limited by the natural distribution of kinetic energies of ions emerging from the ion source. While all the ions receive the same kinetic energy as they are accelerated by the electric field, they will initially possess differing kinetic energies and so the energy of any ion equates to:

$$E_{kin} = E_{int} + eV \quad (9.2)$$

where E_{kin} is the total kinetic energy of an ion, e the charge of an electron, and V the potential through which the ion is accelerated. It is this natural distribution of ionic energies within the mass analyser that gives rise to a small spatial distribution or spread of ions around the exit slit. This spread typically limits the resolution, R, of instruments of this type to relative molecular weights of approximately 2000 or less.

9.4.2 Double focusing mass analysers

Double focusing mass spectrometers are designed to overcome the shortcomings of magnetic sector instruments by the inclusion of two focusing devices for the ions within the mass analyser. Electromagnets are in such cases firstly used to focus ions with similar mass-to-charge ratios but with narrow deviations in a direction along a common trajectory. A device known as an ***electrostatic analyser*** (**or ESA**) that consists of a pair of smooth curved plates across which a dc potential is applied, is also used to correct for the natural spread of kinetic energies imparted to ions occurring within the ion source.

The ESA is positioned in between the ion source and the magnetic sector so that the ions emerging from the ion source pass into the gap between the plates. The dc potential applied across the plates then limits the spread of kinetic energies of ions which are allowed to proceed onto the magnetic sector. Ions with kinetic energies above a given cut-off value strike the top plate and thereby removed. Ions with energies below a cut-off value strike the bottom plate and so only ions within a very narrow band of kinetic energies are permitted to proceed to the magnetic sector. Ions are again focused between polarized electric plates so as to correct for deviations in directional trajectories of ions with similar mass-to-charge ratios. Instruments employing a double focusing approach in this manner are often capable of resolutions of relative molecular mass, R, of 10^5 or greater and this represents an improvement of some two orders of magnitude greater than those obtainable with single focusing mass spectrometers.

See Section 9.5.1 for a description of the resolution of mass analysers.

Magnetic sector mass analysers of double focusing instruments operate in a similar manner to those found in single focusing magnetic sector mass spectrometers. There are a number of geometries used within different magnetic sectors and these configurations are normally known by the names of the people who first described these arrangements such as the 'Mattauch–Herzog' (Fig. 9.8) and 'Nier–Johnson' (Fig. 9.9) designs of double focusing mass spectrometers.

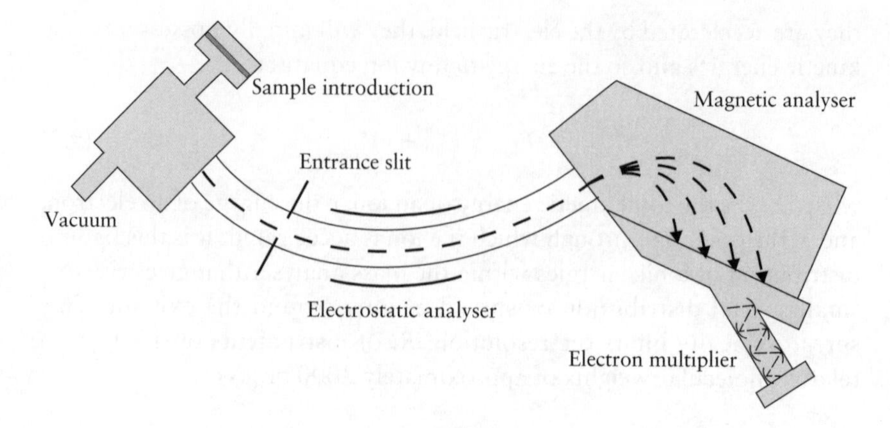

Figure 9.8 Mattauch–Herzog double focusing mass spectrometer.

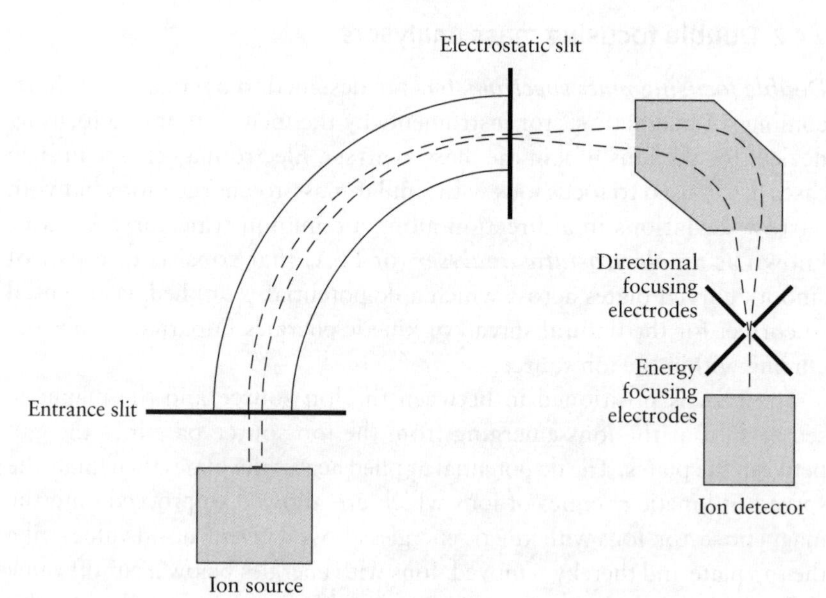

Figure 9.9 Nier–Johnson double-focus mass spectrometer.

9.4.3 Quadrupole mass analysers

Quadrupole mass analysers offer very fast analyses in the order of 100 ns time scales or less. In practice, this means that analytes may be analysed in real-time as they elute from a chromatographic column and, for this reason, quadrupole mass spectrometers are often employed as the detectors within gas chromatography–mass spectrometry (GC–MS) coupled systems.

A quadrupole is a term given to four metals rods or electrodes found at the heart of machines of this type, a schematic of which is shown in Fig. 9.10. The rods are typically 10–15 cm in length and 5–6 mm in diameter, although many configurations are commercially available. Typically ions are first accelerated through a potential of 10–15 kV and onto the space between the four rods. An ion detector is positioned to

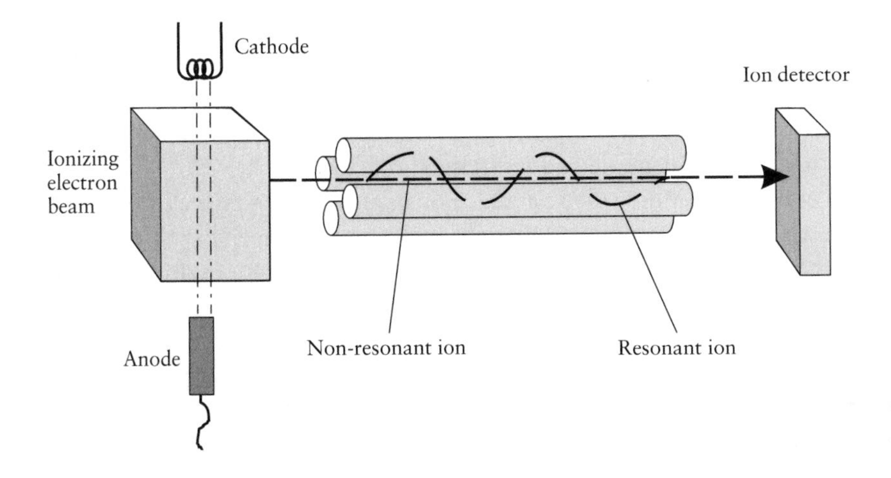

Figure 9.10 Schematic of a quadrupole mass spectrometer.

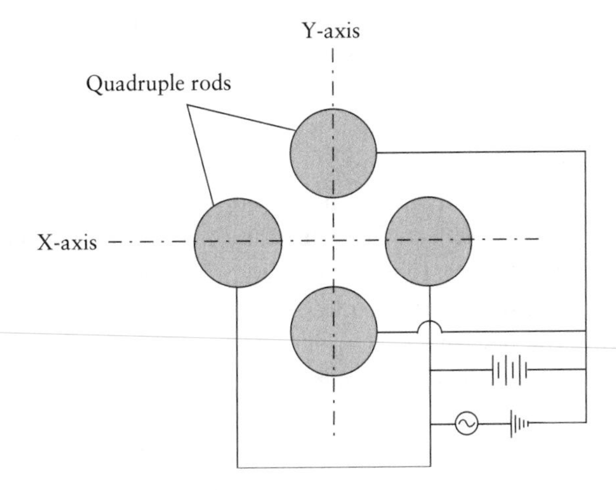

Figure 9.11 Geometry and polarization of rods within a quadrupole mass spectrometer.

permit monitoring of any ions that emerge from the other end of the quadrupole. The four metal rods are polarized to only allow ions of a particular mass-to-charge ratio to pass through the space between the rods. Quadrupoles are often referred to in this context as strictly being ion filters as opposed to ion analysers.

Each rod is electrically connected to its opposite neighbour; the four rods of the quadrupole are therefore connected as two pairs of rods, Fig. 9.11. A dc potential is applied across the two pairs of rods so that one pair acts as anodes and the other pair as cathodes. Two low amplitude–high frequency alternating potentials are applied to the two pairs of rods such that the ac potentials across the anodes and cathodes are 180° out of phase with each other.

For ions to be detected, they must travel through the space enclosed by the quadrupole and onto the ion detector. If the ions are attracted to and collide with any of the four electrodes, their charge is lost to ground and the ions fail to pass through the quadrupole. The combined ac and dc potentials allow the two anodes within the quadrupole to act as a low pass mass filter and the two cathodes as a high pass mass filter. It follows that ions of only very narrow mass-to-charge ratio manage to pass through the quadrupole and on to the ion detector.

Let us consider how the ac and dc potentials applied to the electrode rods affect the passage of ions through the quadrupole. Remember that the ions are predominately positively charged, that is, they are cations. In addition to the ac potential, a dc potential is applied across the two pairs of electrodes.

The anodes are polarized positively with respect to the cathodically polarized rods and these are consequently polarized with a negative potential with respect to the anode rods. In the absence of the ac potential, the ion beam consisting predominately of cations will therefore be attracted to the cathodes and repelled away from the anodes. It follows that the alternating component of the applied potential, will tend to focus ions towards the centre of the space between the four electrodes during positive half cycles whereas during the negative half of the cycle the potential will tend to diverge the ion beam towards the rod electrodes. It should be remembered that ions of smaller masses are easier to deflect than ions with greater masses—and it follows therefore that ions with smaller masses will be deflected to a greater extent by the ac potential than those with greater masses. Ions will naturally be repelled from the anodes, however, ions below a certain mass may be forced to collide with the anodes during the negative excursions of the ac potential and so will be lost from the ion beam which reaches the ion detector. In this way, the anodes act as a filter for ions below a particular molecular weight and are thus sometimes refereed to as a form of ***high-pass filter***.

We now consider what effect the ac and dc potentials applied to the cathodes, the (negative pair of electrodes), have on the passage of ions through the quadrupole. We should remember that the pair of electrodes acting as cathodes within the quadrupole are positioned in a plane 90° relative to the anodes. The ions (cations) within the ion beam will, in the absence of the ac potential, be attracted towards the negative cathodes. Any ions that collide with the electrodes will again be neutralized and so lost from the ion beam. For ions of sufficiently low mass, this migratory movement towards the cathodes may be offset by the effect of the ac potential during the positive half cycle of the ac potential waveform. More massive ions will strike the cathodes and in turn are eliminated. In this way, the cathodes act as a form of ***low-pass*** filter.

The thresholds for the anodic high- and low-pass cathodic filters may be regulated by modulation of applied ac and dc potentials applied across the rods. The anodes and cathodes are normally regulated so that only ions with a very narrow band of mass-to-charge ratios can pass through the quadrupole and so onto the ion detector.

Mass spectra are obtained by linearly ramping the dc potentials while simultaneously increasing the ac potentials applied to the two pairs of electrodes. In this way, ions of smaller mass-to-charge ratios are first selected to pass through the quadrupole followed by ions of ever increasing mass-to-charge ratios.

9.4.4 Ion trap analysers

An ion trap, as its name suggests, is a device in which ions may be formed and then stored for some period of time. There are a number of differing ion traps that may be used as mass analysers within mass spectrometry. We shall first consider a simple ion trap and then go on to describe an ion cyclotron resonance (ICR) approach.

The simplest instruments possess a doughnut-shaped ring electrode, together with a pair of end-cap electrodes; Fig. 9.12. A radio-frequency voltage is applied to the ring electrode and this may be altered to vary the radius of orbit of ions with differing *m/z* ratios. As the voltage is swept, ions of differing *m/z* ratios become stabilized and may leave the ring electrode cavity via openings in the end-caps to make contact with a detector. Simple ion trap detectors of this type offer advantages in terms of simplicity and cost with respect to more complicated approaches such as quadrupole-based instrumentation.

In ion cyclotron-based instruments ions are normally generated via an electron impact ion source and then injected into an enclosure or ion trap known as a ***cyclotron***; Fig. 9.13. Strong electromagnets (often with

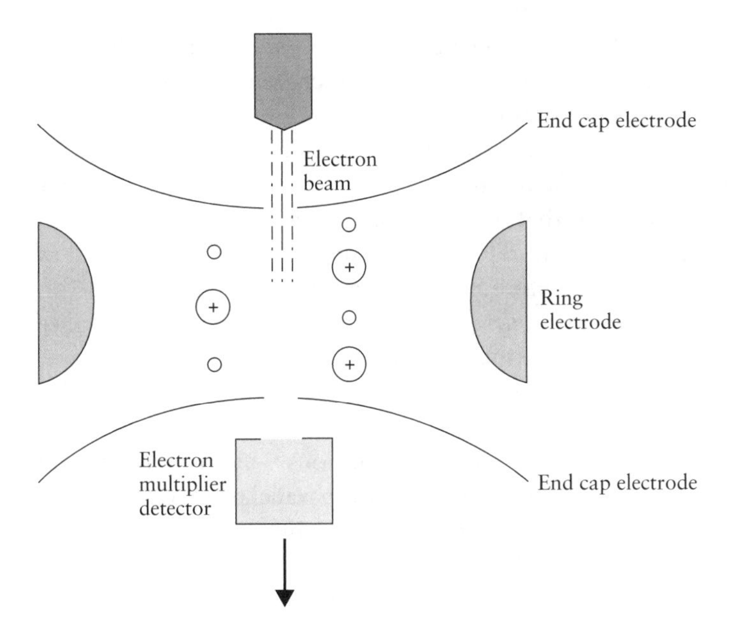

Figure 9.12 Simple ion trap analyser.

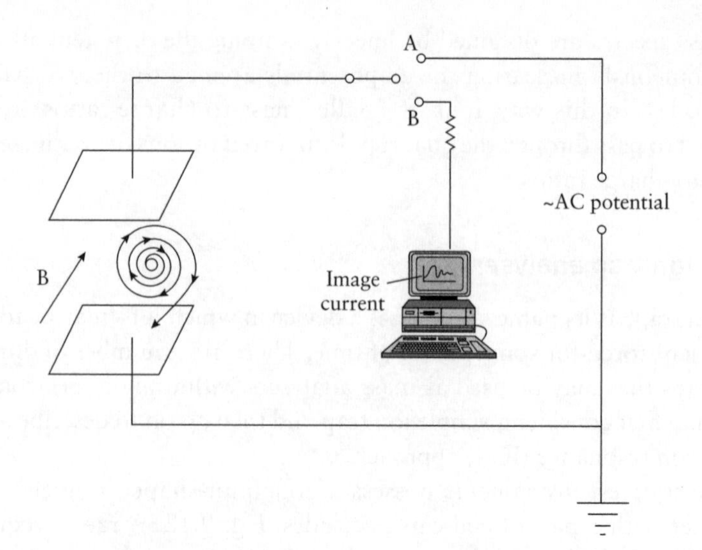

Figure 9.13 Schematic of an ion cyclotron resonance (ICR) trap analyser. Spiral path of ion shown occurs when switch is moved briefly to position A.

strengths of up to 1.5 T or more) cause the ions to rotate in a circular path in a plane of motion that is perpendicular to the direction of the field. For a given magnetic field strength, B, the angular motion of an ion, ω_c, may be shown to be dependent upon its charge, z, and inversely dependent upon its mass, m, Eqn (9.3):

$$\omega_c = \frac{zeB}{m} \tag{9.3}$$

The angular frequency, ω_c, of an ion under a particular set of conditions is also sometimes referred to as its cyclotron frequency. Equation (9.3) predicts that the greater the mass of an ion, the smaller will be its angular motion, ω_c, and, conversely, the greater charge the ion carries the greater will be its angular motion.

An ion trapped and travelling in a circular motion may absorb energy and so be accelerated by an ac electric field, provided that the frequency of the field is close to that of the cyclotron frequency. If an ac electric field is applied across the ion trap enclosure, then ions with a particular cyclotron frequency (and therefore mass-to-charge ratio) will be accelerated. Ions will continue to be accelerated provided the electric field is applied and so the ions are accelerated the radius of their orbital motion will increase. The electric field may be applied or terminated by moving the switch between positions A or B.

The circular motion of the resonant ions which are accelerated by the electric field gives rise to a current in the parallel plates either side of the ion trap enclosure upon termination of the frequency sweep signal; Fig. 9.13. This current (which is known as the ***image current***) decays with time and typically may be monitored over a few tenths of a second to a few seconds following application of the ac electric field. The current will be proportional to the number of resonant ions and, in this way, ions with a particular mass-to-charge ratio may be quantified.

Mass spectra could be obtained by sequentially accelerating ions of differing mass-to-charge ratios by the application of ac electric fields of differing frequencies although this approach would be excessively tedious and time consuming. In practice, the mass spectra from ion cyclotron analysers are obtained via Fourier transform signal processing techniques following the acceleration of ions via an ac electric field that increases in frequency linearly with time. ICR trap analysers now form the heart of most commercial Fourier transform mass spectrometers.

Fourier transform analysis is a mathematical operation that may be performed by a computer to allow de-convolution and analysis of complicated data sets from mass spectrometry, for example.

9.4.5 Time-of-flight mass analysers

Time-of-flight mass spectrometers operate by sequentially monitoring when and how many ions of differing mass-to-charge ratios reach a detector, following the injection of a single burst of ions from an ion source. A schematic of a simple time-of-flight mass spectrometer is shown in Fig. 9.14.

Bursts of positive ions (cations) are produced within the ion source by pulsed electron, photon, or secondary ion impact approaches. These ions are then rapidly accelerated into a drift tube towards the detector via an electric field pulse of between 10^3 and 10^4 V. Since the ions possessing the same charge will each receive the same kinetic energy they will travel through the drift tube at differing velocities and so reach the detector at different times. The time period over which the burst of ions reaches the detector is extremely small and is typically in the order of a few microseconds or less. The ion detector then monitors and quantifies the number of ions as they emerge from the drift tube and reach the detector. This form of mass spectrometer is hence called a *time-of-flight* instrument since ions are dispersed and monitored via the differing times taken for ions to travel the length of the drift tube.

Since all the ions from a single burst of ions reach the detector over such a short period of time, extremely fast electronics and signal processing

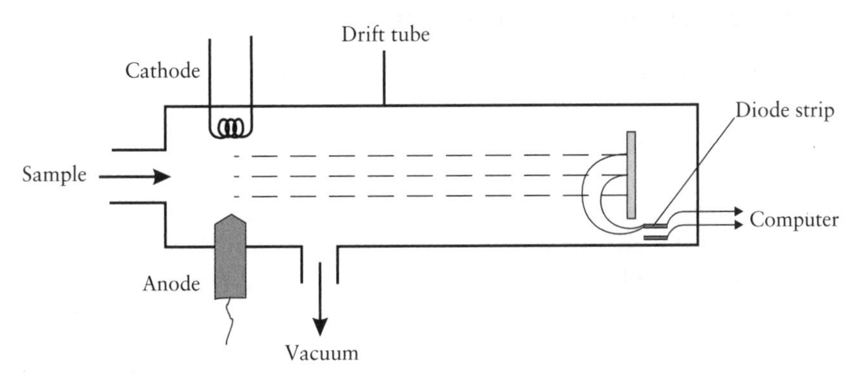

Figure 9.14 Time of flight mass spectrometer.

capabilities are needed to sequentially detect ions of differing mass-to-charge ratios. Resolutions of around 500–1000 atomic mass numbers are typically offered by instruments of this type and while this cannot match the performance of some magnetic sector and Fourier transform instruments—is often offset by the relative simplicity and robustness of time-of-flight instrumentation when a choice is made for purchasing a mass spectrometer.

9.5 Ion analysers

9.5.1 The resolution of mass analysers

There are several different ways in which the resolution of a mass spectrometer can be described. For the purpose of this book we will consider the two most widely adopted approaches.

In the first of these models, the resolution, R, is equal to the ratio of $M/\Delta M$, where ΔM is the difference in mass numbers that will give a valley of 10% between peaks of M and $M + \Delta M$, when the two peaks are of equal height; Fig. 9.15. If ΔM is the smallest mass difference whereby two peaks may be resolved, then the *resolution, R*, may be calculated according to Eqn (9.4), that is,

$$R = \frac{M}{\Delta M} \tag{9.4}$$

The resolution of individual or isolated peaks may also be defined in terms of δm, which is taken to be the Full peak Width at Half the peak Maximum, often abbreviated to FWHM; Fig. 9.16.

The resolution of instruments is often not uniform across differing masses and generally becomes poorer at higher masses. In practice, most instruments are able to resolve peak with differences in mass of less than one mass unit, and this means that ions of similar ionic formula but containing differing isotopes may be distinguished.

9.5.2 Electron multiplier detectors

Electron multiplier detectors are probably the most widely used of ion detectors within mass spectrometry, since they are both extremely rugged and may be used in conjunction with most forms of mass spectrometry.

The heart of an electron multiplier is a ***dynode***, which contains electrode surfaces coated with Cu and Be that gives rise to bursts of electrons when struck by energetic ions. Different designs of dynode are used within electron multiplier detectors although each are designed so that the initial bursts of electrons strike other dynodes or different parts of the same dynode so as to give rise to further bursts of electrons which in turn

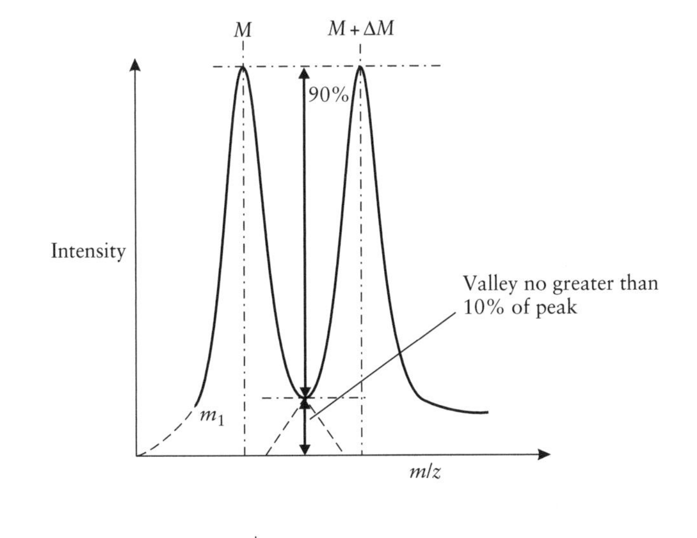

Figure 9.15 Resolution of peaks within mass spectrometer via 10% trough approach.

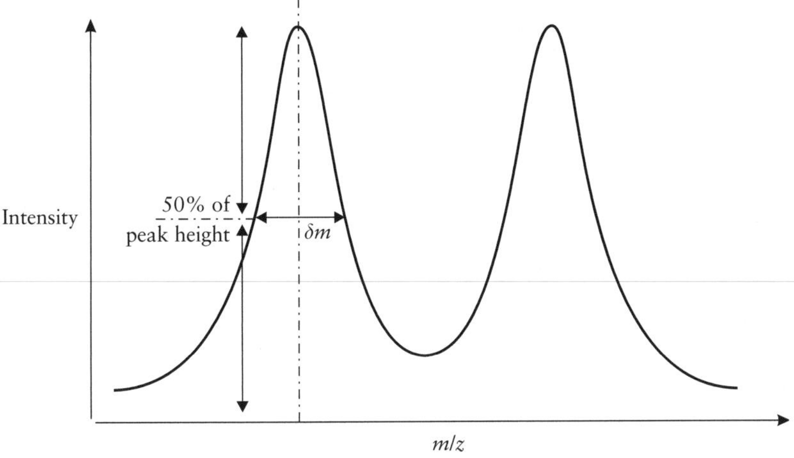

Figure 9.16 Resolution of peaks within mass spectrometry via full peak width at half peak maximum approach.

do the same—and this gives rise to a cascade process. Current gains in the order of 10^7 or more are often attainable. Electron multiplier ion detectors operate in a very similar manner to photo-multiplier tubes as used in UV–visible spectrometry. It should be noted that the ion detector is contained within the evacuated environment of the mass spectrometer and does not therefore require encasing within a glass envelope.

Electron multiplier detectors contain a number of dynodes (often up to 20 or more) which work together in the form of an array; Fig. 9.17. Each electrode plate or dynode is polarized at a greater potential than the previous one permitting electrons to be accelerated progressively towards each dynode and this, in turn, facilitates the impact of electrons at the dynode surface and so the release of further electrons.

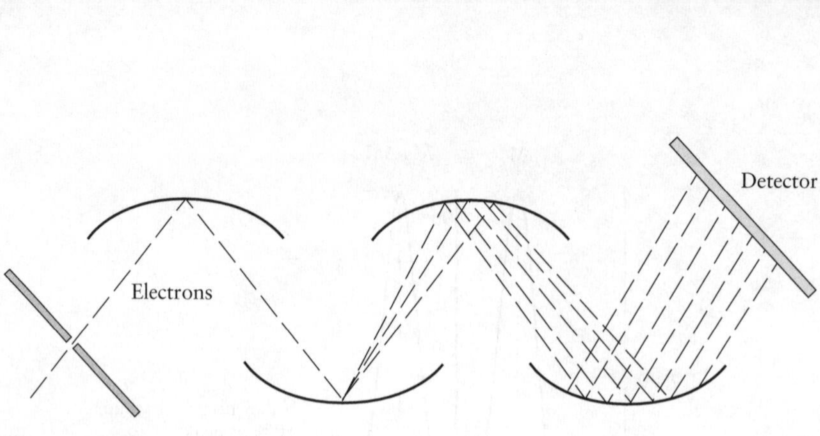

Detector

Electrons

Detector strip

Figure 9.17 Dynode electron multiplier (Dynodes are polarized at successively higher potentials).

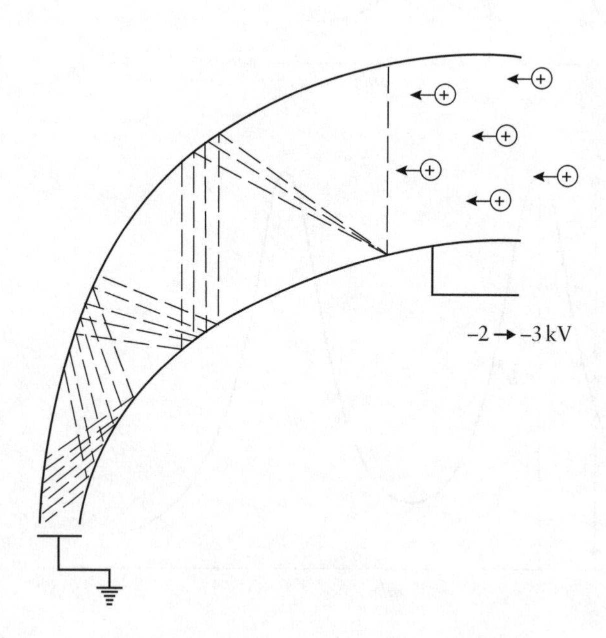

Figure 9.18 Continuous dynode electron multiplier.

Continuous dynode electron multiplier

Continuous dynode electron multipliers utilize a single dynode in the form of a large curved glass horn doped with lead; Fig. 9.18. Detectors of this type are again immersed in the evacuated chamber of the mass spectrometer and so do not require any protective encapsulation. The entire internal surface of the horn is poised at a potential of around 2 kV relative to ground. Ions entering the horn impact upon some part of the internal surface due to its curved shape and so give rise to a burst of electrons from the surface. These, in turn, hit the opposite face of the horn and give rise to further bursts of electrons as a cascade which progresses to the base of the dynode. Currents may be recorded and used to quantify the number of ions originally entering the detector.

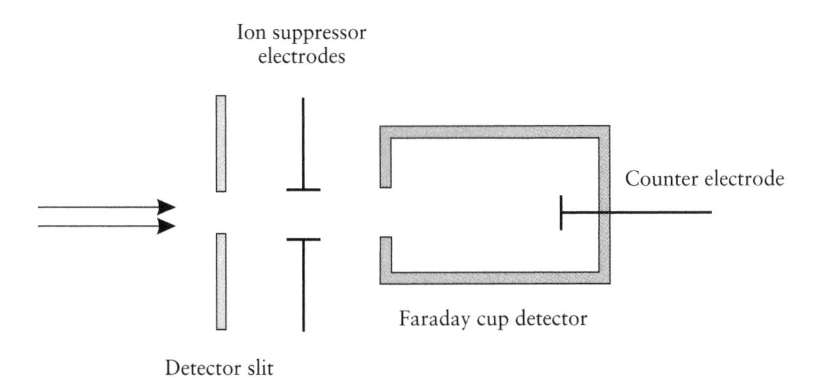

Figure 9.19 Faraday cup detector.

9.5.3 Faraday cup detectors

A Faraday cage is a metal shield that is grounded.

In this form of ion detector, a Faraday cage in the shape of a thimble or cup surrounds a collector electrode; Fig. 9.19. Ions from the mass analyser first pass through an ion entrance slit and thus into the ***Faraday cup detector*** assembly. The Faraday cup entrance slit is clearly aligned to allow all ions to pass directly though its entrance and onto the collector electrode. The Faraday cage and the collector electrode are electrically connected. Ions striking any part of this assembly can give rise to a signal and/or the emission of secondary electrons, which, in turn, may enhance the signal if these strike a further part of the Faraday cup/collector electrode surface. The collector electrode itself is aligned at an angle to the incoming ion beam so that ions reflected from the electrode strike the internal wall of the electrode and in this way the signal obtainable from the incoming ions may be maximized.

The collector electrode and Faraday cup are poised at a high potential relative to ground via a high resistance load resistor. Ions and/or secondary electron collisions striking the collector electrode and Faraday cup modulate the flow of current. Fluctuations in the current may be recorded and so used to quantify ions as they emerge from the ion analyser.

The sensitivity offered by Faraday cups detectors may in certain circumstances only be 0.1% of that offered by the best electron impact detectors since, as we have seen, little scope for internal signal amplification is possible within the simple cup assembly. Faraday cup ion detectors are, however, amongst the simplest, most rugged, and least expensive types of detector for mass spectrometry, and for this reason, are still widely used within routine analyses where sensitivity is less important.

9.5.4 Scintillation detectors

Scintillation ion detectors operate by recording ions via the emission of visible light following collision of the ions (or secondary electrons) with a phosphor-coated surface. There are a number of differing configurations of scintillation ion detector which may be encountered. In one arrangement, the detector takes the form of a photo-multiplier tube with a thin aluminium window coated with a phosphor. Ions striking the widow give rise to scintillations that may be counted and so used to quantify the ions as they leave the mass analyser. Another form of ion scintillation detector consists of a cathode to attract the ions as they emerge from the mass analyser. Secondary electrons strike a phosphor-coated surface and the scintillations of emitted light are again quantified.

9.5.5 Photographic detection of ions in mass spectrometry

Photographic plates or films are very rarely used these days for recording mass spectra although this approach may in certain circumstances still offer some advantages. Ions directly expose a photographic film and therefore can only be used in conjunction with mass analysers that *spatially disperse* ions according to their mass-to-charge ratio. In practice, photographic films are almost always used in conjunction with spark ion sources and certain designs of magnetic sector based instrumentation. Silver bromide based photographic films are most commonly employed as these are more readily exposed by energetic ions than other types of film. Different areas of exposure on the photographic film correspond to ions of characteristic mass-to-charge ratios, and can be quantified by means of a densiometer to determine the relative levels of exposure corresponding to different ions. The film essentially integrates the exposure over some period of time as ions continue to strike are recorded on a film. Excellent sensitivities can be obtained by using a photographic-based ion detection approach in situations where extremely low concentrations of analytes must be quantified.

A densitometer is an instrument that allows the intensity of a photographic film exposure to be quantified.

9.6 Mass spectrometry coupled with other analytical techniques and instrumentation

Mass spectrometry allows us to determine often extremely small quantities of a sample and is therefore one of the most sensitive techniques available to the analytical chemist. Two or more closely related ions may be produced from molecules of the same analyte if more than one ionizable group is present on the molecule, or indeed, if the analyte is prone to fragmentation. It follows that mass spectra are quite commonly difficult to interpret and this becomes progressively more of an issue if a number of different molecular species are present within the sample. One approach for simplifying the situation involves coupling the mass spectrometer with another instrumental technique to remove many of the unwanted components of the mixture prior to analysis. Two examples of this approach involve either using gas chromatography or high performance liquid chromatography in conjunction with mass spectrometry and these are known as ***gas chromotography–mass spectrometry*** (GC–MS) or ***high performance liquid chromatography–mass spectrometry*** (HPLC–MS), respectively. Another approach involves coupling two mass spectrometers in situations where larger ions may undergo fragmentation reactions; in this approach one mass spectrometer is used to isolate the ionized analyte of interest while the second analyses the fragmentation products following decomposition of the primary ion.

9.6.1 Gas chromatography–mass spectrometry

Mass spectrometry may be used as the detection system as analytes are separated and eluted from a gas chromatograph (Chapter 8). The basic problem encountered with coupling gas chromatography with mass spectrometry is that large volumes of a carrier gas can obviously not be introduced into the evacuated environment of the mass analyser with a mass spectrometer.

There are several ways in which the output from the gas chromatograph may be introduced to the mass spectrometer. The simplest approach for coupling the two techniques employs a molecular leak diaphragm with apertures in the order of micrometres in diameter to allow only tiny quantities of sample to pass in to the mass analyser and to prevent compromise of the evacuated conditions that are required. More elaborate coupling devices exist, but we shall not consider these further since the task these perform is in essence the same in all instances. One factor that should be considered, however, is the concentration of the analyte within the carrier gas as it elutes from the gas chromatograph. It is clearly beneficial to introduce as little carrier gas into the mass spectrometer as possible since this compromises the vacuum of the mass spectrometer; conversely, sufficient analyte must be introduced on the other hand to facilitate its analysis. It follows that the greater the concentration of the analyte within the carrier gas stream, the smaller the volume that needs to be introduced to allow for the mass spectrometer to allow an analysis to be performed. It should be remembered that gas chromatography columns and experimental conditions that enhance the resolution of the chromatograms have the effect of narrowing chromatographic peaks and so concentrating the analyte in the carrier gas as it elutes from the column.

9.6.2 High performance liquid chromatography–mass spectrometry

The coupling of HPLC with mass spectrometry is even more problematic than GC–MS, since gaseous ions must be introduced into the mass analyser of a mass spectrometer, whereas HPLC uses a liquid mobile solvent phase in which the analyte is dissolved.

Many approaches are based upon the selective evaporation and thus removal of the solvent prior to the introduction of the sample into the mass spectrometer. Various types of interface are based on this principle and include ***particle beam*** (**PB**) or ***moving belt coupling*** (**MBC**) interfaces although the treatment and description of these devices is beyond the scope of this book.

Other approaches may involve reducing the flow of the eluent from the HPLC column to an acceptably low level (e.g. via splitter valves) so that ***direct liquid introduction*** (**DLI**) of the sample to the mass spectrometer is permissible.

Lastly, there are also a number of techniques that are based upon the direct ionization of the sample from HPLC eleuent and then only introducing the ionized analyte to the mass spectrometer. Most of these approaches use sample ionization techniques that we have already considered such as ***thermospray*** (**TSP**), ***spark ionization***, and ***electrospray*** (**ES**).

9.7 Identification of mass spectra and differences in spectra obtained using different instrumentation

The identification of molecules or ions requires assignment of mass numbers (or more commonly) peaks with specific *m*/*z* ratios is often achieved using a marker such as mercury vapour or perfluoro kerosene (PFK). Mercury vapour gives a characteristic fingerprint spectrum with *m*/*z* values ranging from 198 to 204; another commonly used marker is PFK, which gives a more complicated spectrum with characteristic *m*/*z* peaks at 69 (CF_3), 93 (C_3F_3), 124 (C_4F_4), and 131 (C_3F_5).

Mass spectrometry most readily lends itself to qualitative type analyses—although quantification may be achieved by coupled mass spectrometry approaches such as GC–MS, which are considered in Section 9.6.

Almost all compounds for analysis will fragment upon ionization. The nature of mass spectra depends upon the ionization method used, since this affects how much fragmentation occurs. The greater the degree of fragmentation the more complex will be the final mass spectra.

The relative abundance of the fragments are often plotted against the *m*/*z* ratios as different heights on the *Y*-axis of the spectrum. This type of spectrum is sometimes known as a stick diagram format. The molecular ion is normally the fragment with the highest *m/z* ratio, although it is rarely the most abundant and indeed in some cases may not be detected at all, preventing direct determination of the molecular weight of the parent analyte.

It is clear that the fragment of a compound depends on its structure and it is this property that allows structural elucidation of unknown species, since particular moieties or groupings are associated with specific fragmentation particles.

Figures 9.20(a) and (b) show a comparison of the mass spectrum for 1-decanol following ionization by: (a) chemical; and (b) electron impact approaches. Chemical ionization sources typically cause less fragmentation than electron impact ionization approaches and so normally yield simpler spectra.

The most prominent first peak observable in the chemical ionization mass spectrum is that of the $(M-1)^+$ ion, an ion that has lost an OH^- group—and a number of other smaller peaks each differing by 14 mass units corresponding to the loss of subsequent CH_3 moieties.

By contrast, the electron impact spectrum shows evidence for extensive fragmentation. A grouping of peaks centred around *m/z* value of 41 is

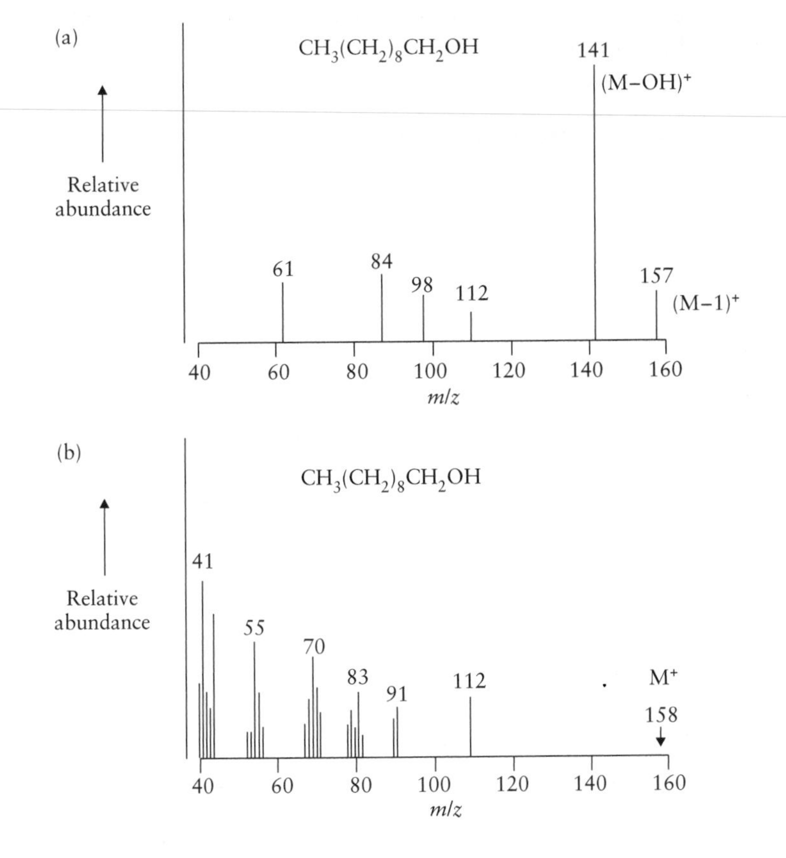

Figure 9.20 Mass spectrum for 1-decanol (a) via chemical ionization, and (b) via electron impact approaches.

observed corresponding to $C_3H_5^+$ with further clusters of peaks being observed at *m/z* values of 55, 70, 83, 91, and 112, each of which corresponds to the addition of further CH_2 groups.

EXAMPLE 9.1 A compound has a molecular ion peak at 130 mass units. What would be the mass for a peak corresponding to an ion having undergone the loss of a methyl group CH_3?

Method

Calculate mass of group to be lost and subtract from mass of molecular ion.

$$\text{Mass of } CH_3 = 12 + (1 \times 3) = 15$$

Therefore, mass corresponding to new fragmentation peak will be $130 - 15 = 115$.

EXAMPLE 9.2 Identify the main peaks for a mass spectrum of ethylamine as below:

Method

Calculate mass values for the moieties within the molecule and compare with the structure of the molecule.

Identification of peaks

Ethylamine: $CH_3CH_2NH_2$ (Molecular weight: 45)

45: $M^{\cdot +}$

44: $M^{\cdot +} - H$

30: $M^{\cdot +} - CH_3$

28: $CHNH^+$

15: CH_3^+

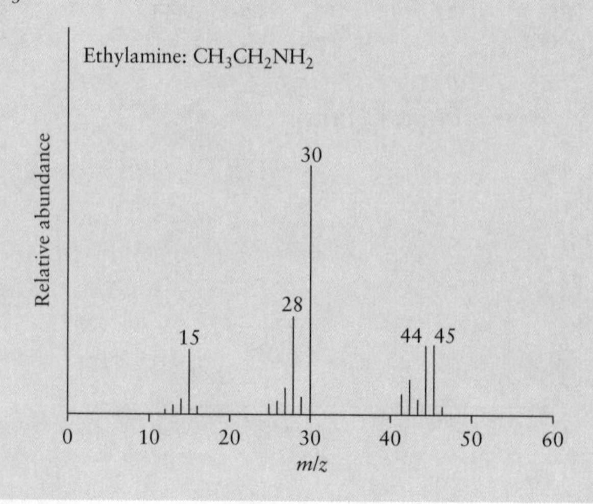

9.8 Tandem mass spectrometry

Tandem mass spectrometry (MS/MS) is a technique in which two mass spectrometers are used in conjunction with each other. The first mass spectrometer isolates a particular analyte ion, which then fragments (e.g. as in Eqn (9.5)) into a number of smaller ions and/or neutral products, which are detected by the second instrument.

$$m_p^+ \rightarrow m_d^+ + m_n \tag{9.5}$$

A third or even fourth mass spectrometer may be used if further fragmentation or dissociation reactions occur within the ionic sample. Similar or different types-of-mass spectrometer may be coupled together for differing applications; for example, time of flight instruments are often coupled with magnetic sector instruments so that ionic products within a time-of-flight instrument are subsequently detected by a magnetic sector mass spectrometer.

Mass spectrometers are often coupled via a ***collision cell***. In this approach, the first mass spectrometer selects a precursor or parent ion, which is then passed onto a collision cell; Fig. 9.21. Fragments or products produced via ***collision induced fragmentation*** (CIF) or ***collision activated reactions*** (CAR) are then detected by the second mass spectrometer. Coupled mass spectrometry is often used for mechanistic studies of fragmentation reactions as well as for purely analytical purposes.

The analytical applications for coupled mass spectrometry are numerous, although it should be borne in mind that no analyst will employ two instruments when one will suffice; it therefore follows that analytical applications for coupled mass spectrometry normally involve the determination of trace compounds within complicated mixtures that would otherwise defy analysis. The determination of both ethical and illicit

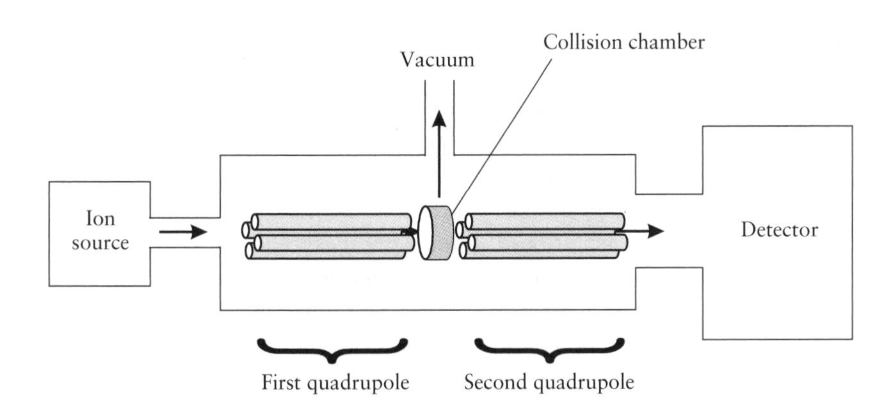

Figure 9.21 Schematic of a tandem quadrupole.

drugs within biological fluids (i.e. patient samples) is an established application for coupled mass spectrometry. Coupled mass spectrometry can be also used for the determination and/or verification of the elemental stoichiometry of analytes in mixtures that contain a number of compounds with closely related structures.

9.9 General applications of mass spectrometry

Until the mid-1970s, molecular weight determinations could only be performed by chromatographic, ultracentrifugation, and/or electrophoretic approaches. Each of these approaches were subject, however, to a number of inaccuracies and uncertainties relating to how molecular weights were calculated and this meant that the only true determination of a molecular weight was via a theoretical calculation following confirmation of the structure of an ion or molecule. The determination of molecular weights for *individual* ions or molecules within *mixtures* was not possible, however, until the advent of mass spectrometry. The applications of mass spectrometry are today extremely numerous and span the full range of environmental, industrial, and medical sectors. The majority of applications involve first confirming (by determining the molecular weight and/or via fragmentation reactions) the presence of trace analytes (e.g. a toxin) in a complex mixture. Sample mixtures in this context can range from river water samples through to foodstuffs. Many applications as we have already seen necessitate prior separation, since mass spectra of many mixtures would yield a myriad of peaks, all corresponding to different compounds. Particular consideration will be given, in the next section, to the biological applications of mass spectrometry since these now represent some of the largest application areas for mass spectrometry.

9.10 Biological applications of mass spectrometry

The determination of molecular weights from individual proteins, peptides, oligonucleotides, oligosaccharides, and lipids within biological fluids were amongst the first widespread applications for mass spectrometry. Mass spectrometry has become one of the most powerful analytical tools within the life sciences, and is finding widespread application in the rapidly expanding areas of proteomics and genomics (Section 14.10). The molecular weight of biological molecules may often be determined with accuracies of 0.01% or better, and this will often allow very small structural changes, such as single amino acid groups, to be identified. Mass spectrometry is now widely used for the determination of biomolecules and although a separate chapter is included covering Bioanalytical Chemistry, we shall in the next section briefly look at some of the most useful and widely encountered examples of mass spectrometry and its uses in biotechnology and medicine.

9.10.1 The analysis of peptides, polypeptides, and proteins

Proteins and polypeptides comprise long chains of amino acids. Mass spectrometry and, in particular, tandem mass spectrometry may be used to: (a) determine the sequence of the amino acids within a protein; (b) calculate the total molecular weight of a protein; and (c) quantify the amount of a particular protein within a mixture. Samples are normally ionized via FAB, ESI, and MALDI approaches, since these most readily give rise to fragmentation reactions that facilitate structural determinations. Individual proteins possess characteristic molecular weights and in many cases the identification of specific proteins may be easily achieved from a simple mass spectrum. Fragmentation reactions, however, often help to quantify the relative composition of a mixture in more complex samples that contain different compounds of very similar molecular weight. Mass spectrometry is now routinely used for amino acid sequencing, the determination of protein structures, and, as we see in the next section, oligonucleotide sequencing. Protolytic enzymes are sometimes first used to break larger proteins down into smaller polypeptide chains to facilitate their identification.

9.10.2 Peptide sequencing

Mass spectrometry peptide sequencing is performed by tandem mass spectrometry (Section 9.8). Peptide sequencing is now of immense importance to the field of proteomics (Section 14.10). Peptides tend to fragment in a predictable manner to give ions that may be easily identified and it is this information that may be used for peptide sequencing. A protein digest can normally be analysed without prior separation adding to power of the technique. In many cases a ***sequence tag*** of 4–5 amino acids from one peptide will provide sufficient information to identify a protein from a protein database.

Peptides fragment along the peptide backbone and this may be also be accompanied by some side chain fragmentation; it is the combination of this information that can allow the position of individual amino acids to be identified. The amount of sequence information obtainable varies from one peptide to another, although peptides of molecular weight 2500 tend to give the most useful information. In some instances the entire protein sequence can be verified, in others only part of the sequence can be identified.

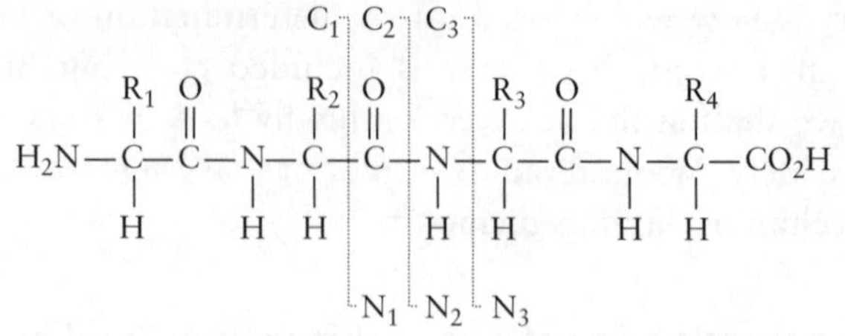

$$\mathrm{H_2N{-}C(R_1)(H){-}C({=}O){-}N(H){-}C(R_2)(H){-}C({=}O){-}N(H){-}C(R_3)(H){-}C({=}O){-}N(H){-}C(R_4)(H){-}CO_2H}$$

Figure 9.22 Fragmentation ions that may be formed from backbone cleavage of linear peptides.

There are three types of bond that can give rise to fragmentation within an amino acid backbone and so the peptide backbone. The bonds capable of fragmenting are NH–CH, CH–O and CO–NH, as shown in Fig. 9.22. Upon cleavage of one of these bonds two fragments are formed–one of these will be neutral, the other charged; which of the two fragments carries the charge depends on the chemistry and relative proton affinity of the two fragments. It is important to remember, however, that it is only the charged fragment that can be monitored by mass spectrometry.

Since three bonds may be cleaved to give rise to fragmentation, and any of the fragments may carry the charge following fragmentation, it follows that a total of six possible fragmentation ions may be formed for each amino acid residue. The sites of possible bond cleavage are depicted in Fig. 9.22 along with the possible fragmentation ions that may be formed. These are denoted in this figure as N1–N3, which retain the charge on the N-terminal fragment, and C1–C3, which carry the charge on the C-terminal fragment.

The degree of side chain fragmentation that occurs depends on the type of analyser used. Magnetic sector instruments, for example, which promote high-energy collisions, cause varied and multiple chain cleavage and so give rise to many differing fragments. Conversely, quadrupole–quadrupole and quadrupole–time-of-flight mass spectrometers, which cause lower energy collisions, give rise to the fewest side chain fragments.

It is beyond the scope of this book to consider fully the details of side chain cleavages although the fragments they form can be extremely useful in protein sequencing for proteomic applications (see Section 14.10).

The final point that should be considered is the formation of immonium ions ($H_2N^+{=}CHR$) from individual amino acid residues. Immonium ions are very useful for identifying amino acid residues within a peptide although their identification provides no direct information relating to their position within a peptide sequence.

9.10.3 The analysis of oligonucleotides

Oligonucleotides are long-chain linear nucleotide polymers found in ribose nucleic acid (RNA). Mass spectrometry is often used to determine the relative sequence of the four principal nucleosides (adenosine, cytidine, guanosine, and uridine) in RNA. Total molecular weight determinations are often performed via electron impact (EI), FAB or ESI approaches in conjunction with a simple magnetic sector or quadrupole based machine. More complex analyses in which the sequences of the individual nucleotides must be determined are often performed prior to enzymatic digestion of the oligonucleotide, followed by tandem mass spectrometry of the smaller chain nucleotide chains. Modifications of the individual nucleotides are sometimes possible within both t-RNA and r-RNA with mass spectrometry sometimes being used to help identify any covalent rearrangement and/or structural modifications that may have occurred in RNA samples.

9.10.4 The analysis of oligosaccharides

Oligosaccharides are formed from the glycosidic linkage of multiple monosaccharides such as glucose, fructose, mannose, and galactose. Determining the structure of longer chain polysaccharides or oligosaccharides is rather more complicated than with proteins, since branch chains are often formed. Mass spectrometry helps determine the exact sequence of components but often yields insufficient information to predict the exact structure of particular oligosaccharides. Further complications are often encountered since monosaccharide sub-units are isomeric in nature and this gives rise to many further possible structural permutations.

FAB–tandem mass spectrometry is the most widely used approach for analysing oligosaccharides via mass spectrometry, although ESI techniques are also used in some instances. Both of these approaches yield fragmentation ions following the cleavage of glycosidic bonds. In some instances, spectra allow us to predict both the total molecular weight of an oligosaccharide and its structure, although this often requires very careful interpretation.

9.10.5 The analysis of lipids

Lipids represent a wide-ranging class of compounds although they are all soluble in aprotic solvents. Lipids serve many biological functions and are found almost ubiquitously in nature. The determination and quantification of lipids finds many applications ranging from food manufacture to fundamental research.

Fatty acids must normally be separated from bulk samples and purified via HPLC or GC prior to a mass spectroscopic determination. Mass spectrometry may often be used to help determine the structure of fatty acids via a number of different regimes although the most popular approaches involve ionization via fast atom bombardment in conjunction with tandem mass spectrometry. Information is often provided via fragmentation reactions. This information often helps determine the structures and positions of side chains—which, in turn, helps elucidate the structure of more complex molecules.

Alkylglycerols are formed by the esterification of glycerol with fatty acids and again these have great biological significance. One, two, or three of the glycerol alcohol moieties may be esterified and this gives rise to mono-, di-, or tri-glycerides respectively. Mass spectroscopic analyses are again performed by first extracting and chromatographically purifying the alkylglycerols from a sample; alkylglycerols may then be treated with lipases to cleave the ester linkages to release the constituent fatty acids and glycerol—which may be analysed by mass spectrometry as before.

Bile salts are a family of compounds formed from cholesterol. Mass spectrometry is often routinely used for quantifying various bile salts within either urine or serum to facilitate the diagnosis and treatment of various metabolic disorders. Analyses of both free and conjugated bile salts are typically performed via FAB ionization of samples in conjunction with tandem mass spectrometry. Fragmentation of parent molecular ions again often allows total structural determinations to be performed in addition to obtaining the molecular weights of individual compounds.

Phospholipids and glycophospholipids have great biological significance since they form the basis of the bilayer unit membrane structure for both cellular membranes and the intra-cellular endoplasmic recticulum. Phospholipids and glycophospholipids are again most commonly analysed using FAB ionization approaches in conjunction with tandem mass spectrometry so that fragmentation reactions may be exploited to facilitate the identification of individual side groups and moieties.

Exercise and problems

9.1. Explain why mass spectrometers separate ions according to their m/z ratios.

9.2. Why do double focusing mass spectrometers allow higher resolutions and narrower peaks to be obtained?

9.3. How do spectra from electron impact and chemical ionization sources differ?

9.4. Calculate the kinetic energy that a singly charged ion will acquire if accelerated through a potential of 1.2×10^3 V.

9.5. Calculate the energy in J/mol that electrons acquire as they are accelerated through a potential of 100 V.

9.6. Why are spark source mass spectrometers normally designed to be double focusing?

9.7. A molecular ion has a mass equivalent to 59.97; suggest the identity of the parent molecule.

9.8. A compound shows a peak within a mass spectrum at m/z 145. Where might you expect a peak for an ion formed via the loss of CH_3?

9.9. Identify the main peaks within a simplified mass spectrum of 1-butene as below:

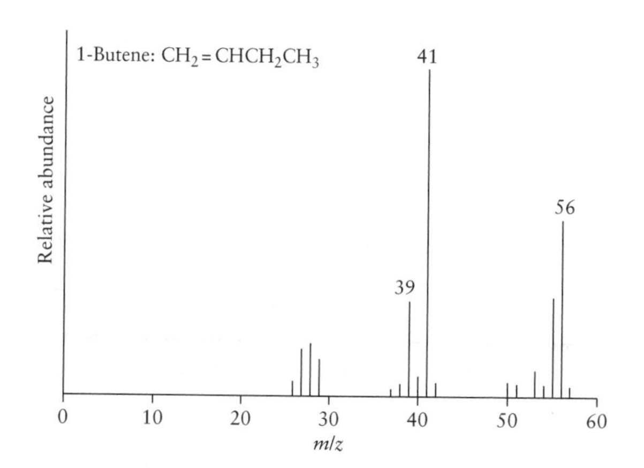

9.10. Identify the main peaks within a simplified mass spectrum of cyclohexanol as below:

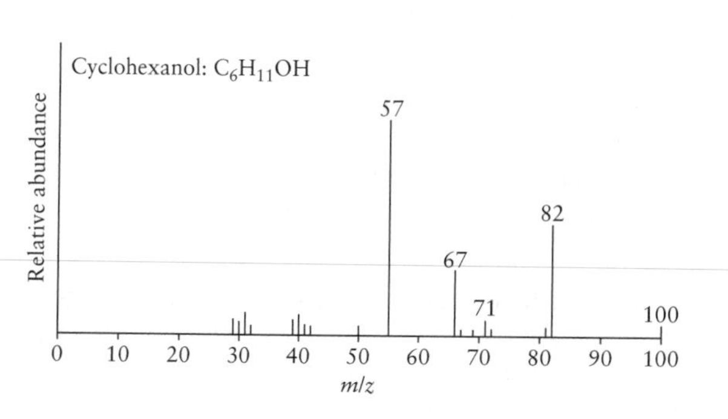

9.11. Identify the main peaks within a simplified mass spectrum of methylene chloride as below:

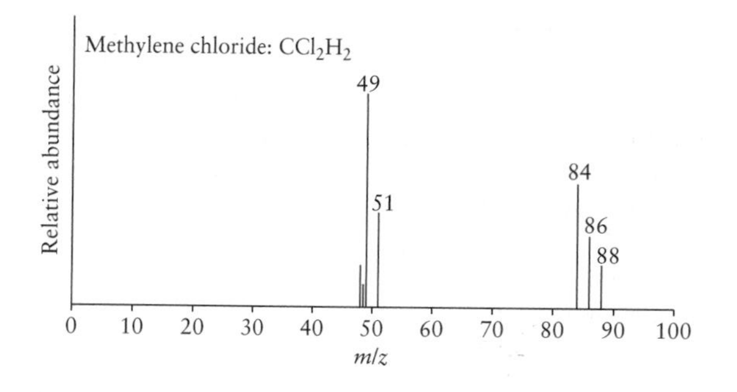

9.12. Identify the main peaks within a simplified mass spectrum of ethylbenzene as below:

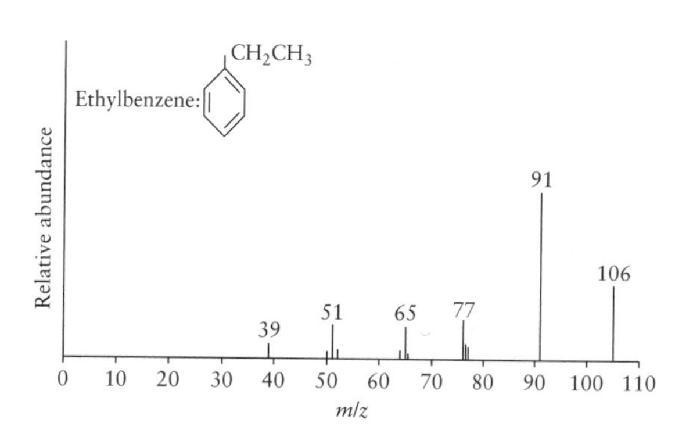

9.13. Identify the main peaks within a simplified mass spectrum of cyclohexane as below:

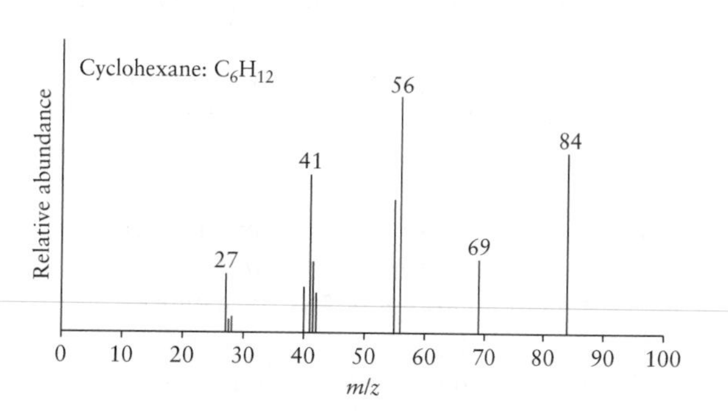

9.14. Compare and contrast the advantages and disadvantages associated with quadrupole and magnetic sector analysers.

9.15. How might mass spectrometry be used in the sequencing of polypeptides and proteins?

9.16. How might oligosaccharides be analysed by mass spectrometry?

9.17. Discuss the use of mass spectrometry in proteomics.

Summary

1. Mass spectrometry is an instrumental technique relying on separating gaseous charged ions according to their mass-to-charge ratios.

2. Sample inlet systems are designed to introduce a sample into the mass spectrometer with as little a loss in the quality of vacuum as possible.

3. The three most commonly used inlet systems used in mass spectrometry are: (i) batch inlet; (ii) direct probe; and (iii) chromatographic type inlet systems.

4. There are a number of different ionization chambers routinely being used and these include: electron impact, fast atom bombardment, secondary ion mass spectrometer (SIMS), field ionization/desorption, laser desorption, spark ionization, surface thermal, plasma desorption, thermospray, inductively coupled plasma, and atmospheric pressure electron impact ionization sources.

5. Mass analysers disperse ionized sampling according to differences in their mass-to-charge (m/z) ratios as they emerge from the ionization source or chamber.

6. There are a number of differing mass analysers routinely used and these include:

- o magnetic sector analysers,
- o double focusing analysers,
- o quadrupole mass analysers,
- o ion cyclotron resonance ion trap and Fourier transform mass analysers, and
- o time-of-flight mass analysers.

7. Magnetic sector analysers incorporate a tube with a 60, 90, or 180° curve through which ions are accelerated towards a detector; ions of different m/z ratios follow different trajectories and this is the basis of m/z separations.

8. Double focusing mass analysers as their name suggests use two ion-focusing devices. An electrostatic analyser (ESA) corrects for the Boltzmann distribution of kinetic energies of ions as well as a pair of smooth curved plates across which a dc potential is applied for the deflection of the ion beam.

9. Two widely used designs of double focussing mass analysers are known as Mattauch–Herzog and Nier–Johnson designs.

10. Quadrupole mass spectrometers contain four rods: each rod is electrically connected to its neighbour. A dc potential is applied between the two pairs of rods. An ac potential is superimposed on these and potentials are used to separate ions according to their m/z ratios. Ions of individual m/z ratios are sequentially allowed to pass through the space enclosed by the four rods and so onto the mass analyser.

11. Ion cyclotrons separate ions by injection of ions into an ion trap. Ions are accelerated by an applied potential in a circular path that is perpendicular to the direction of an applied magnetic field.

12. Mass spectra are recorded by sequentially accelerating ions of different m/z ratios by the application of ac electric fields that increases with time and data processing using a Fourier transform approach.

13. Time-of-flight mass analysers operate by sequentially detecting ions as they travel at differing velocities through a drift tube towards a detector. Although the resolution of mass spectra obtained via time-of-flight approaches cannot rival the performance of other approaches, the instrumentation is relatively inexpensive and robust.

14. The resolution of mass spectra peaks can be defined in a number of ways. In one model, peaks are said to be resolved if the valley between two peaks is equal to or less than 10% of the smaller peak intensity, that is, $R = M/\Delta M$, where ΔM is the smallest mass difference whereby two peaks of masses M and $M + \Delta M$ may be resolved. Another model defines resolution of peaks in terms of δ, which is taken to be the full width at half peak maximum (FWHM).

15. Electron impact detectors are widely used as ion detectors within mass spectrometry and may consist of electron multiplier detectors or continuous dynode electron multipliers.

16. Other ion detectors include: Faraday cup detectors, scintillation detectors, and photographic film detectors.

17. Mass spectrometry is often used in conjunction with other analytical techniques with examples including gas chromatography–mass spectrometry, HPLC–mass spectrometry, and tandem mass spectrometry (i.e. two mass spectrometers).

18. Mass spectrometry is widely used in many areas of analytical chemistry as well as finding many applications in life sciences.

Further reading

Constantin, E., Schell, A., and Thompson, M. (1990). *Mass spectrometry*. Ellis Horwood Series in Analytical Chemistry, Ellis Horwood.

Downard, K. (2004). *Mass spectrometry: a foundation course*. Royal Society of Chemistry.

Hamdam, M. (2005). *Mass spectrometry for proteomics*. Wiley.

Rose, M. E. and Johnstone, R. E. W. (1996). *Mass spectrometry for chemists and biochemists*. Cambridge University Press.

Siuzdak, G. (1996). *Mass spectrometry for biotechnology*. Academic Press.

Electro-analytical techniques 10

Skills and concepts

This chapter will help you to understand:

- The function of each of the main components of an electrochemical cell.
- How reference electrodes are used within electrochemical cells.
- How to quote potentials with respect to different reference electrodes.
- The features of a normal hydrogen electrode, a silver/silver chloride electrode and a standard calomel electrode.
- The operation and use of a simple pH electrode and a number of ion selective electrodes other than for pH.
- The logarithmic response behaviour of ion selective electrodes with respect to concentration.
- How linear sweep and cyclic voltammetry may be used for analytical purposes.
- The operation of polarography for the determination of metal ions in aqueous solutions.
- The use of differential pulse polarography and the advantages this approach may offer.
- How adsorptive stripping voltammetry may be used for electrochemical analyses with enhanced sensitivity.
- The advantages that micro-electrodes offer in comparison to using planar electrodes in terms of the hemispherical diffusional mass transport profiles they experience.
- How organic polarography may be used for practical analytical determinations and how micro-electrodes can be used for the electrochemical determinations of some organic compounds.
- The principle of electrochemical titrations including the use of pH and other ion selective electrodes.

- The operation of an oxygen electrode.
- The operation and use of electrochemical sensors including potentiometric, amperometric, and conductimetric based sensors for analytical purposes.

10.1 An introduction to electrochemical cells

Electrochemical analyses involve the use of ***electrodes*** that are housed within ***electrochemical cells***. A schematic of a simple electrochemical cell is shown in Fig. 10.1.

All electrochemical cells contain at least two electrodes to complete a circuit although in many cases three electrode arrangements are used.

If a two-electrode arrangement is used, the cell will contain a working or sensing electrode and a combined ***reference*** and ***secondary electrode***. We shall consider the importance of reference electrodes in greater detail in Section 10.3.

If a three-electrode arrangement is used, the cell will contain a working electrode, a reference electrode, and a secondary electrode. Secondary electrodes are sometimes known as ***auxiliary*** or ***counter electrodes***. The secondary electrode should ideally be designed to be at least ten times the size of the working electrode since this ensures that the secondary electrode will not limit the electrochemistry at the working electrode.

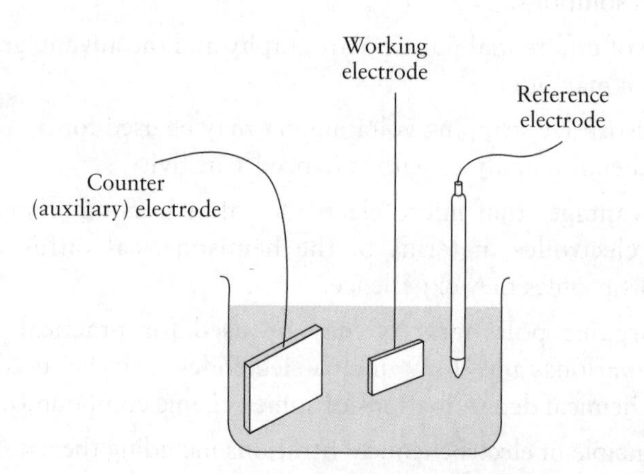

Figure 10.1 Electrochemical cell.

10.2 The Nernst equation and electrochemical cells

We have already seen that the all electrochemical cells must have at least two electrodes. The electrochemistry that occurs at a working electrode can be written in the form of a half-cell reaction. Similarly, the electrochemistry that occurs at the secondary electrode can be written with a corresponding half-cell reaction.

The electrochemistry at one electrode cannot occur without a corresponding electrochemical reaction occurring at the other electrode; in other words no half reaction can proceed by itself. There must be an electron donor (a reducing agent) and an electron acceptor (an oxidizing agent). If the potentials of half-cell reactions could be measured directly then it would be possible to determine which half cells could act as either oxidizing or reducing agents with other cells. Fortunately we can measure the potential difference between different half cells with respect to reference electrodes (Section 10.3). Half-cell potentials may be measured with respect to a standard reference half cell, the standard hydrogen electrode (Section 10.3.1), which is assigned an arbitrary potential of 0.0 V. The potential of a half cell at ***unit activity*** with respect to the standard hydrogen electrode is known as its ***standard potential***, E^0. In practice, the activity of a half cell is dependent upon both concentration and temperature, and the potential observed or recorded, E, may be related to the standard potential by the Nernst equation, Eqn (10.1):

Activity can be thought of as a thermodynamic approach for considering concentration.

For a half-cell reaction $a\text{Ox} + ne^- \rightleftharpoons b\text{Red}$:

$$E = E^0 - \{2.302RT/nF\} \log_{10} \frac{[\text{Red}]^b}{[\text{Ox}]^a} \quad (10.1)$$

where F is the Faraday constant and R is the gas constant.

EXAMPLE 10.1 A pH 3 aqueous solution contains 1×10^{-3} M MnO_4^- and 1.5×10^{-2} M Mn^{2+}. Calculate the potential of the half-cell reaction.

Method

1. Write the half-cell reaction together with E^0.
2. Calculate E from the Nernst equation.

Step 1: $MnO_4^- + 8H^+ + 5e^- \rightleftharpoons Mn^{2+} + 4H_2O \quad E^0 = 1.51$ V
Step 2: $E = E^0 - \{2.302RT/nF\}/\log [\text{Red}]/[\text{Ox}]$
Note $n = 5$ for five electrons and there are eight protons on the left-hand side of the equation.

Taking into account the stoichiometry for the reaction we can substitute into the Nernst equation:

$$\begin{aligned} E &= 1.51 - \{1.059/5\} \log [Mn^+]/[MnO_4^-][H^+]^8 \\ &= 1.51 - 0.0118 \log \{1.5 \times 10^{-2}\}/\{1 \times 10^{-3} \times (10^{-3})^8\} \text{ V} \\ &= 1.51 - 0.0118 \log \{1.5 \times 10^{-2}\}/\{1 \times 10^{-27}\} \text{ V} \\ &= 1.51 - 0.011 (\log 1.5 \times 10^{27}) \text{ V} \\ &= 1.51 - 0.011 \times 25.176 \text{ V} \\ &= 1.51 - 0.2769 \text{ V} \\ &= 1.23 \text{ V} \end{aligned}$$

The potential difference between two half cells separated spatially but electrically connected (for example, by a salt bridge as shown in Fig. 10.2) may be determined by subtracting the standard potentials of the cathodic (most negative) from the anodic (more positive) half cell. The first step is to establish which half cell will act as the anode and which will act as the cathode.

EXAMPLE 10.2 From the two half-cell equations below determine the cell potential:

$Ce^{4+} + e^- = Ce^{3+}$ $\quad E^0 = 1.61$ V

$I_3^- + 2e^- = 3I^-$ $\quad E^0 = 0.5355$ V

Method

Assign anodic and cathodic half cells and then subtract the cathodic half cell from the anodic half cell to give the overall cell potential.

Step 1: $Ce^{4+} + e^- = Ce^{3+}$ half cell more positive than:
$I_3^- + 2e^- = 3I^-$,
$Ce^{4+} + e^- = Ce^{3+}$ half cell is therefore the anodic half cell.

Step 2: Cell potential, $V = 1.61 - 0.5355$ V $= +1.0745$ V

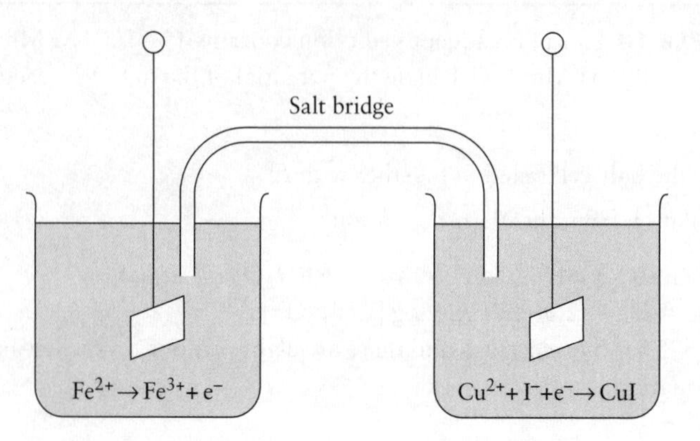

Figure 10.2 Examples of two half cells connected by a salt bridge.

10.3 Potentiometric methods and ion selective electrodes

Potentiometric methods rely on monitoring a potential to allow analytical measurements to be made.

All potentiometric methods use some form of sensing or ***working electrode*** together with a ***reference electrode***. A third ***counter electrode*** may also be employed. We shall consider the design and use of reference electrodes first.

10.3.1 Reference electrodes

Reference electrodes are extremely important in most electro-analytical determinations. A potential is always quoted with respect to another potential; potentials are therefore often referred to in terms of ***potential differences***. It is crucial to be able to both quote and identify a stable reference potential within electrochemical cells that contain many ions that are mobile and carry charge. Reference electrodes are used for this purpose. If a reference electrode defines a stable reference potential, then we can draw an analogy with describing the directions on a map to another location by quoting its position with respect to a known point. You might, for example, instruct a visitor asking for directions to travel from town A to town B, to drive 100 km in a north-easterly direction to reach town B.

Any electrode will possess a potential difference between its surface and the interfacial region of a solution in which it is immersed. In a simple model, the charge on the surface of the electrode is considered to comprise one layer of charge, while ions in solution of the opposite polarity (and close to the electrode) act as a second layer of charge; this model is known as the ***electrical double layer***. Reference electrodes are chosen because of their stable electrical double layer properties. In this way, they are able to maintain a stable potential difference that may therefore be *referred* to. Electrochemical potentials of analyte couples are normally quoted versus the reference electrode, or with respect to the reference electrode.

An ideal reference electrode should be completely ***non-polarizable***— and this means that no current should flow across the interface between the electrode and the solution in which it is immersed irrespective of the current which flows through the cell.

The two most popular reference electrodes used for electrochemical analyses are:

1. the ***silver/silver chloride*** or ***Ag/AgCl*** electrode,
2. the ***saturated calomel electrode*** or ***SCE*** electrode.

It is therefore common to find potentials quoted in terms of *mV potential differences versus Ag/AgCl* or *versus SCE*.

The first reference electrode to be reported was the ***standard hydrogen electrode*** (SHE), and it is important that we consider this reference electrode first before looking at the SCE or Ag/AgCl reference electrodes.

The standard hydrogen electrode

The standard hydrogen electrode or SHE is the reference electrode to which all other potentials are eventually related back to. The SHE is now rarely used for routine purposes since it is somewhat cumbersome and far simpler alternatives now exist. The SHE was, however, at one time extensively used for early electrochemical studies, and for this reason established itself as the defining reference electrode against which other reference electrodes (and their potentials) are still defined today. The SHE is assigned an arbitrary potential of 0.00 V at all temperatures. All other reference electrodes quote their potential with respect to this.

The silver/silver chloride and calomel electrodes are now the most widely employed reference electrodes.

Note that if a potential is quoted with respect to another reference electrode, for example, Ag/AgCl, and this potential is to be quoted with respect to the SHE, you must first take account of the potential difference between the first reference electrode and the reference electrode in question, and then correct the value quoted by this potential.

The SHE consists of a platinized platinum flag electrode immersed in an aqueous solution of H_2 and H^+ in equilibrium; Fig. 10.3. Hydrogen gas is continually bubbled through the solution at a pre-determined partial pressure. The platinized platinum surface is formed from a specially prepared platinum foil, which presents a finely divided surface and therefore

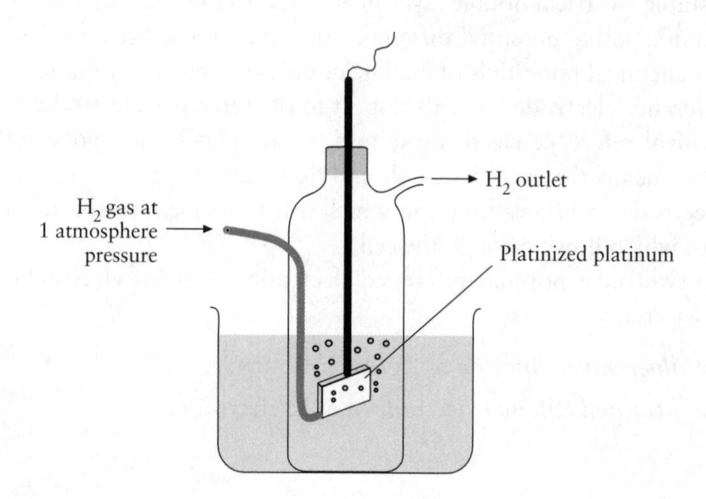

Figure 10.3 Standard hydrogen electrode.

very large surface area for the electrochemical reaction shown below, to proceed reversibly and without hindrance, Eqn (10.2):

$$2H^+_{(aq)} + 2e^- \rightleftharpoons H_{2(g)} \qquad (10.2)$$

The reference potential that the hydrogen electrode offers depends on the hydrogen *ion activity* and therefore the partial pressure at which gaseous hydrogen is bubbled through the cell. The *standard* hydrogen electrode dictates that H_2 is continually passed through the solution at exactly one atmosphere of pressure; Fig. 10.3.

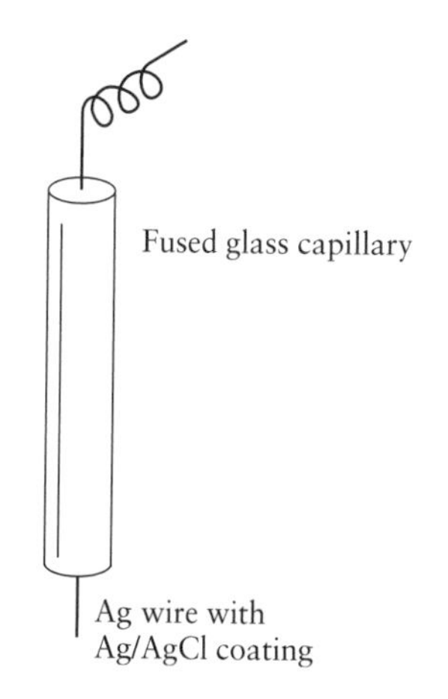

Figure 10.4 Silver/Silver chloride electrode.

The silver chloride electrode (Ag/AgCl)

The silver/silver chloride (Ag/AgCl) electrode is one of the most widely used reference electrodes within modern electro-analytical chemistry due to its simplicity and ease of use. The electrode consists of a length of silver wire, which is coated with a thin layer of silver chloride (Fig. 10.4). Ag/AgCl reference electrodes are ideal for use within almost any electrochemical cell containing chloride based supporting electrolytes.

An Ag/AgCl electrode is prepared by ***anodizing*** a silver metallic surface in a saturated solution of KCl. This is achieved by anodically polarizing the electrode within the solution of KCl, which causes the oxidation of silver to form a layer of silver chloride, Eqn (10.3):

$$Ag^+_{(aq)} + Cl^-_{(aq)} \rightarrow AgCl_{(s)} \qquad (10.3)$$

Commercial Ag/AgCl electrodes are often fabricated by embedding a length of silver wire within a glass capillary. One end of the glass capillary is fused to seal the glass–metal junction with a few centimetres of the metal left protruding from the capillary. It is not critical how much of the Ag/AgCl electrode is immersed in the solution since firstly the potential is taken from the electrical double layer of ions at the electrolyte Ag/AgCl interface and, secondly, no current flows through the reference electrode.

The Ag/AgCl electrode does have one drawback, which is that the reference potential that it provides is dependent of the chloride concentration of the analyte solution. The Ag/AgCl electrode gives a potential of +0.199 V versus the NHE in a saturated KCl solution.

The calomel electrode

The ***calomel electrode*** is another commercially available reference electrode that is widely used for electrochemical analyses. The electrode (Fig. 10.5) typically consists of a tube filled with a saturated solution of mercury(I) chloride (calomel) in the form of a paste made with mercury and mercurous chloride. The Hg_2Cl_2/Hg/KCl paste is held within a tube

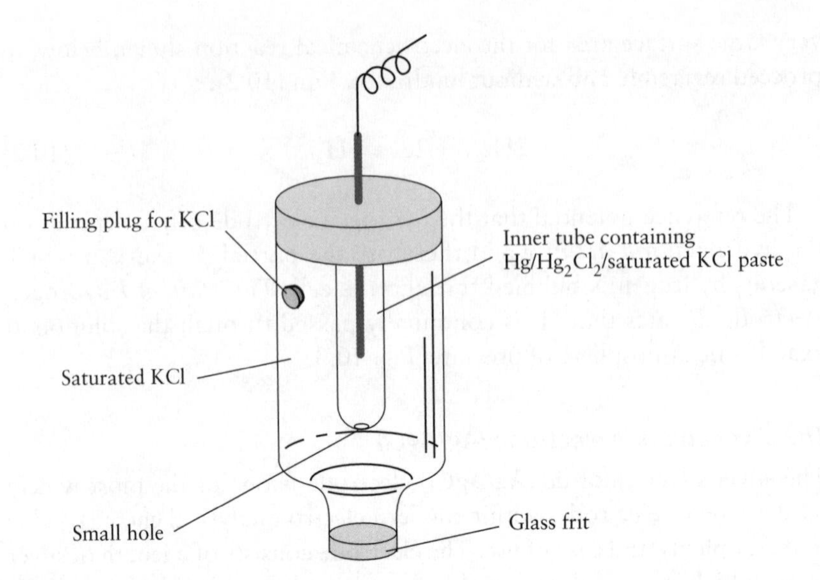

Figure 10.5 Saturated calomel electrode.

that has a small hole within its base to permit contact with a saturated KCl solution, held within an outer glass sleeve. The glass sleeve in turn has a glass frit at its base to permit electrical contact with the analyte solution.

The reference potential given by the electrode depends on the concentration of chloride within the entrapped KCl solution and the equilibrium of Eqn (10.4):

$$Hg_2Cl_{2(s)} + 2e^- \rightleftharpoons 2Hg_{(l)} + 2Cl^-_{(aq)} \qquad (10.4)$$

In order to maintain the reliability of the electrode, the KCl solution and calomel/Hg_2Cl_2/Hg/KCl paste should be periodically changed. The electrode is normally saturated with KCl to maintain a quantifiable and reliable reference potential, which at 25°C should be +0.242 V versus the SHE.

EXAMPLE 10.3 The potential of a half cell with respect to an SCE reference electrode is −0.577 V. Calculate the potential with respect to an SHE. (The cell potential using the SHE is 0.242 V less negative than SCE.)

Method

Correct for difference in reference electrode potentials:

$$\begin{aligned} E \textit{ vs.} \text{ SHE} &= E \textit{ vs.} \text{ SCE} + 0.242 \text{ V} \\ &= -0.577 + 0.242 \text{ V} \\ &= -0.335 \text{ V} \end{aligned}$$

10.4 Ion selective electrodes

There are a number of commercial potentiometric ***ion selective*** or ***ion sensitive electrodes*** (ISEs) available. The most commonly used types operate by measuring a potential difference across specially manufactured ***ion selective glasses***. The origins of all these electrodes stem from the development of the modern pH electrode, which is still the most widely used ISE today. The commercial success of the pH electrode is largely due to its remarkable reliability, robustness and selectivity towards determining the concentration of H^+ ions in almost any aqueous solution. We shall therefore consider the operation and construction of the pH electrode first and then consider a few other ISEs that have now been developed.

10.4.1 pH electrodes

The pH of a solution is defined as being equal to the negative $\log_{10}$ of the $[H^+]$. The pH electrode measures the H^+ concentration, with its output normally being displayed directly as a pH value. The pH electrode is, in fact, an H^+ ISE. pH electrodes, as we shall see, rely on the measurement of a potential across a glass membrane which varies logarithmically with the $[H^+]$ at the solution interface. This is convenient since the response of the electrode therefore follows the conventional pH scale with which we are all familiar.

A commercial pH electrode assembly typically consists of two electrodes—a reference electrode that is usually either an SCE or a silver chloride electrode and the pH sensing glass membrane electrode (Fig. 10.6). The potential is then measured between these two electrodes and correlated with a pre-determined pH–potential difference calibration profile.

The potential, E, that is measured may be predicted by the generalized and simplified form of the ***Nernst*** equation:

$$E = E^0 + \frac{RT}{nF} \ln [X] \qquad (10.5)$$

where E^0 is the potential in volts under standard conditions of temperature and concentration, R the gas constant, F the Faraday constant, T the absolute temperature, n the number of charges transferred in the reduction or oxidation process in question, and $[X]$ the concentration of the ion, which for the pH electrode is an H^+ ion. This last term should strictly be the activity of the ion, but in most cases an assumption can be taken that the concentration may simply be substituted. Now if the numerical values for RT and F are substituted at 25°C, Eqn (10.5) may be simplified to form Eqn (10.6):

$$E = E^0 + 0.059 \log_{10} [H^+] \qquad (10.6)$$

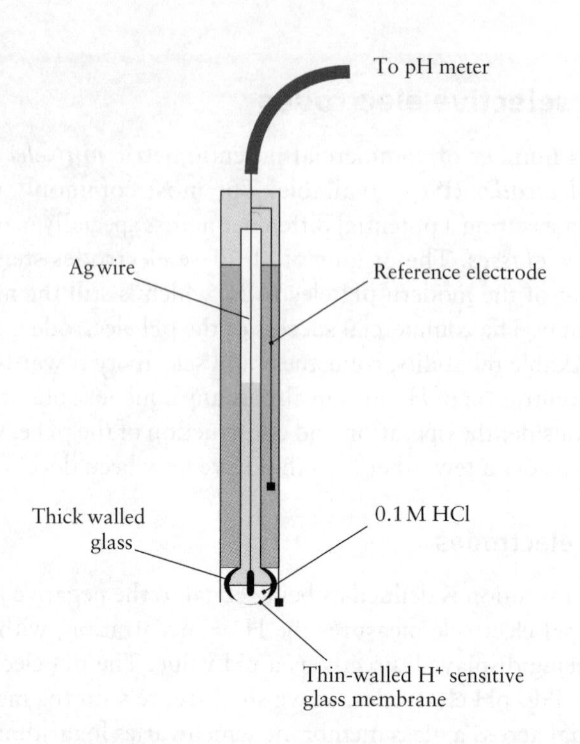

Figure 10.6 pH – Ion selective electrode.

This equation predicts the behaviour that is actually seen in practice, that is, that ***a 10-fold increase in the hydrogen ion concentration causes a change of 59 mV potential difference***. We should not forget the term T for temperature in Eqn (10.5) and that we have assumed that the electrode will be operated at 25°C. In practice, most commercial pH meters have a temperature control to correct for changes in operating temperatures.

H^+ ion selective membranes and pH electrode assemblies

The H^+ ion selective glass is specially manufactured glass that has a typical composition of approximately 63% SiO_2, 28% Li_2O, 5% BaO, 2% La_2O_3, and 2% Cs_2O. The glass is highly selective for H^+ ions as long as it remains hydrated. The electrode must therefore be kept immersed within distilled water otherwise it will lose its responsive behaviour. Electrodes that have been dried out may normally be revitalized by soaking in distilled water for a couple of days or so, prior to further use. The uptake of H^+ ions causes a potential difference across its boundaries and this is typically measured between a wire that is in electrical contact with the glass membrane (by being immersed within an HCl solution) and the reference electrode. As we have already seen, the potential difference may be interpolated as a pH reading; the greater the concentration of H^+ ions, the larger the potential difference that may be measured across the glass membrane, and this of course corresponds to a lower pH within the solution being tested.

EXAMPLE 10.4 A pH electrode is recording a pH of 4.2. Acid is added to the solution and the potential of the pH electrode increases by 118 mV. What is the pH of the new solution?

Method

An increase of 59 mV corresponds to a 10-fold increase in the $[H^+]$; from this calculate the change in pH of the solution. The pH decreases with increasing response of the pH electrode.

Step 1: Potential of pH electrode increases by 118 mV. This corresponds to 118/59 pH units = 2 pH units.

Step 2: Original pH was 4.2. Therefore new pH = 4.2 − 2.0 = 2.2.

10.4.2 Other commercially available ion selective electrodes

There are a number of other ISEs for a range of ions such as F^-, Na^+, K^+, NH_4^+, and Li^+. The pH electrode remains, however, the most reliable of all the ISEs currently available due to the small size of the H^+ ion; electrodes for larger ions invariably suffer from some interference effects from ions of similar size and charge. Many commercial pH meters are designed to allow ISEs to be substituted in the place of the pH electrode, since the instrumentation required is identical. The Nernst equation (Eqn (10.5) again relates the measured potential difference to the $\log_{10}$ of the ion concentration. An empirical calibration curve may be plotted by recording the response of the electrode (in millivolts) in certified reference solutions to ensure that the ISE is performing correctly. Ion concentrations may then be interpolated from the calibration profile.

10.5 Linear sweep and cyclic voltammetry

Linear sweep voltammetry and ***cyclic voltammetry*** are both dynamic electrochemical techniques since they involve the variation of an applied (or ***polarizing***) potential. The current is then measured with respect to either the applied potential or time. Voltammetric determinations are often used as diagnostic techniques; a very large number of electrochemical analyses (such as ***adsorptive stripping voltammetry, ASV***, and ***polarography***) are based upon linear sweep voltammetry or cyclic voltammetry.

The technique of linear sweep voltammetry is based on a potential being ramped from one potential to another. The potential sweep rate may vary from a few millivolts per second through to several hundred volts per second. The current is monitored throughout this potential sweeping—and if plotted with respect to the potential, the current–potential profile is known as a ***voltammogram***. If the potential is swept from one potential to

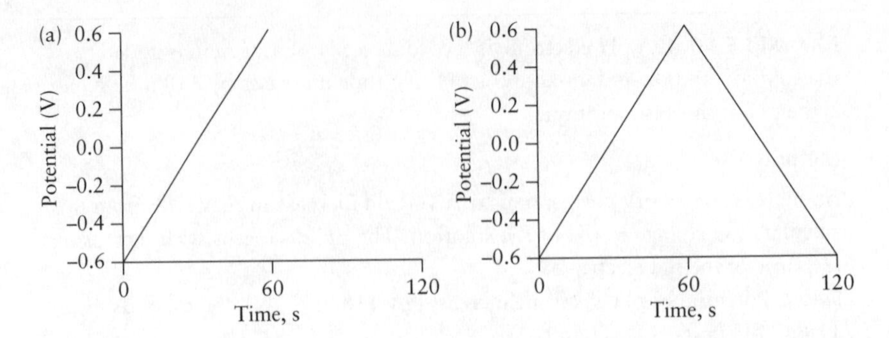

Figure 10.7 (a) Potential sweep for linear sweep voltammetry; (b) potential sweep for cyclic voltammetry.

another and stopped, the technique is known as linear sweep voltammetry; a potential profile for linear sweep voltammetry is shown in Fig. 10.7(a). If the direction of the potential ramp is then reversed at the end of the sweep until the original starting potential is reached once more, the technique is known as ***cyclic voltammetry***; Fig. 10.7(b). The potential cycle may be repeated several times if desired. The nature of the voltammogram (i.e. the shape, size, and potential of peaks) may give a great deal of information relating to the electrochemistry at the working electrode.

A three-electrode arrangement should be used for all *dynamic* electrochemical measurements. Dynamic voltammetric techniques involve varying the polarizing potential with time. The electrochemistry that occurs at a working electrode (WE) is monitored when polarized with respect to the reference electrode. A ***counter electrode*** *(*CE*)* is included to complete the circuit, as charge flows to and from the working and counter electrode. Charge is carried by the ions in the electrolyte solution and it is for this reason that the ionic concentration of the electrolyte must be monitored and preferably be controlled. For example, 0.1 M KCl is often added to the solution to act as the supporting electrolyte. The area of the counter electrode should also be at least ten times larger than the working electrode (as shown in Fig. 10.1) so that it never becomes rate limiting to the electrochemistry that occurs at the WE.

10.5.1 The shape and nature of linear and cyclic voltammograms

The electrochemical process *to be monitored* occurs at the surface of the working electrode. The analyte must therefore travel through the solution to the electrode in order for it to be either oxidized or reduced. The analyte must therefore reach the working electrode under the influence of ***electrical migration*** (charge attraction via Columb's law), ***convection*** (e.g. physical stirring of the solution) or ***diffusion***. Within an unstirred solution, ***diffusion*** is normally the principal mechanism for mass transport to and from the working electrode. If the electron transfer rate occurs at a sufficiently rapid rate between the electrode and the analyte, then diffusion may become the rate limiting factor and the electrode reaction is said to be ***under diffusional control***. If the oxidation/reduction process is reversible then it may again be shown from the ***Nernst*** equation that the forward and reverse peaks will have a $59/n$ mV separation, (where n once more denotes the number of electrons involved per molecule within the electron transfer process).

It is convenient to consider the shape of linear sweep and cyclic voltammograms together since a linear sweep voltammogram comprises the forward current/time profiles of a cyclic voltammogram. When a cyclic voltammogram is recorded, the potential is reversed at the end of the forward potential sweep, and the current is recorded during the reverse potential sweep.

If we consider the idealized voltammogram in Fig. 10.8, we see why the voltammogram takes the shape and form that it does. In our discussion we shall consider the reduction and re-oxidation of a reversible

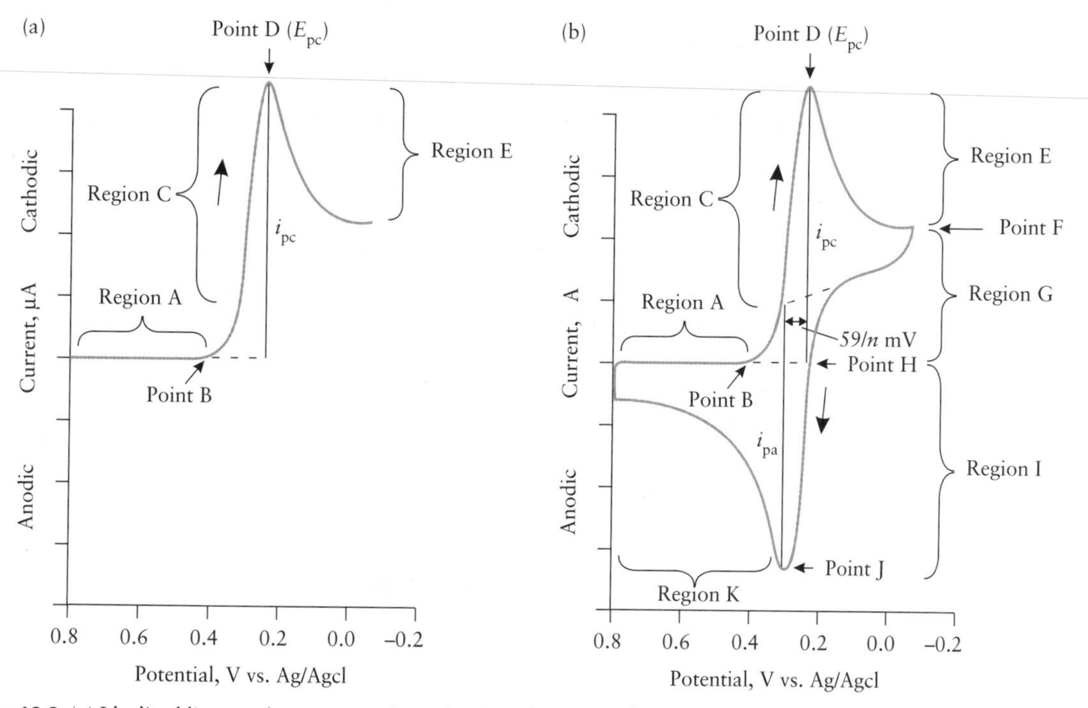

Figure 10.8 (a) Idealized linear voltammogram for reduction of $[Fe(CN)_6]^{3-}$. (b) Idealized cyclic voltammogram for a reversible redox couple such as $[Fe(CN)_6]^{3-/4-}$.

redox couple such as potassium ferricyanide which may be reduced and re-oxidized via a simple one-electron process, Eqn (10.7):

$$[Fe^{(III)}(CN)_6]^{3-} + e^- \rightleftharpoons [Fe^{(II)}(CN)_6]^{4-} \quad (10.7)$$

A linear sweep voltammogram for the reduction of $[Fe^{(III)}(CN)_6]^{3-}$ is depicted in Fig. 10.8(a). A cyclic voltammogram for the reduction of $[Fe^{(III)}(CN)_6]^{3-}$ to $[Fe^{(III)}(CN)_6]^{4-}$ and subsequent re-oxidation back to $[Fe^{(III)}(CN)_6]^{3-}$ is shown in Fig. 10.8(b). The potential is first swept from approximately +0.8 to −0.2 V versus Ag/AgCl. (*Please note in some countries it is electrochemical convention to plot more negative potentials to the right along the X axis.*) No current is observed between approximately +0.7 and +0.4 V in the absence of any electrochemical reaction occurring within this region (***region*** **A**). At around +0.4 V, a cathodic current commences due to the reduction of $[Fe(CN)_6]^{3-}$ to $[Fe(CN)_6]^{4-}$ (***point B***), Eqn (10.7). The current increases as the potential is increased due to the increased electron transfer rate (***region*** **C**). $[Fe(CN)_6]^{3-}$ is being consumed and its surface concentration decreases, and this causes a diffusion gradient to form between the surface of the electrode and the bulk solution. When the surface concentration of $[Fe(CN)_6]^{3-}$ approaches zero, the cathodic current peaks at E_{pc} (***point D***) and then falls as the diffusion gradient extends further into the solution (***region*** **E**). The rate of mass transport to the working electrode now becomes rate limiting, and the current now approaches a new equilibrium plateau until the direction of the potential sweep is reversed (***point F***).

If a linear sweep is stopped at this point a linear sweep voltammogram is recorded; Fig. 10.8(a). If the potential is reversed at this point a cyclic voltammogram is recorded.

A reduction current is observed (***region*** **G**), until the potential at which the peak cathodic current was observed is once more passed (***point H***). At this point the current momentarily passes through zero. The $[Fe(CN)_6]^{4-}$ now begins to re-oxidize (***region I***), and an oxidative (***anodic***) current is seen, which again rises until in a similar manner the surface concentration of $[Fe(CN)_6]^{4-}$ approaches zero; a peak current E_{pa} is again observed (***point J***). The current once again decreases as the original potential is once more approached (***region*** **K**).

10.5.2 Surface bound reactions

If the analyte is bound in some way to the electrode surface, then it does not have to travel to the electrode in order to undergo a charge transfer reaction. The separation between the forward and reverse voltammetric peaks due to the diffusional behaviour of the analyte will now not occur; indeed if the redox reaction is reversible, the forward and reverse peaks are expected to reside at exactly the same potential (one above the other). Voltammograms of this type are known as ***surface bound voltammograms***.

10.5.3 Instrumentation

If a current is drawn from, say, a battery, the potential across its terminals will fall as current begins to flow. It follows that if a potential source like this was used to apply the polarizing potential, the current measured would cease to be a direct function of the applied potential. An instrument called a ***potentiostat*** is used, which ***maintains the potential across the electrodes—irrespective of the current drawn by the circuit***. Note that the term *potentiostat* implies the maintenance of a potential and not a static potential *per se*.

There are many commercial potentiostats. Some electroanalytical chemists choose, however, to build their own specialized yet simple battery powered instruments from modern operational amplifiers, which may have very low signal to noise characteristics. Most potentiostats are computerized and allow for the storage of data digitally, which may then be plotted or manipulated at will, although less expensive instruments are available for recording voltammograms on *X–Y* plotters although care should be taken that the slew rate of the plotter does not distort the current or potential profiles.

10.6 Polarography and associated techniques

Polarography is a specialized form of voltammetry that employs mercury as the working electrode. Mercury has two distinct advantages over other metals for use as an electrode material: firstly, it resists the electrochemical evolution of hydrogen at potentials that could cause problems at other metallic or indeed carbon electrodes. The second advantage comes from its liquid state at room temperature since new (and therefore very clean) electrode surfaces may be readily formed for each analytical test. Solid electrodes normally require cleaning, which can be both time consuming and difficult. A specialized electrode, called the ***dropping mercury electrode*** or ***DME*** (Fig. 10.9), allows a new mercury surface to be formed every second or so *during* an analysis. Liquid mercury from a reservoir is passed through a fine capillary tube at a fixed rate. A mercury drop is formed and this continues to grow with time until it becomes too heavy and drops from the capillary tip. The next drop starts to form and the process repeats itself in a cyclical manner.

The mercury within the capillary (and therefore the drop) is progressively polarized throughout the experiment by a potential ramp with time

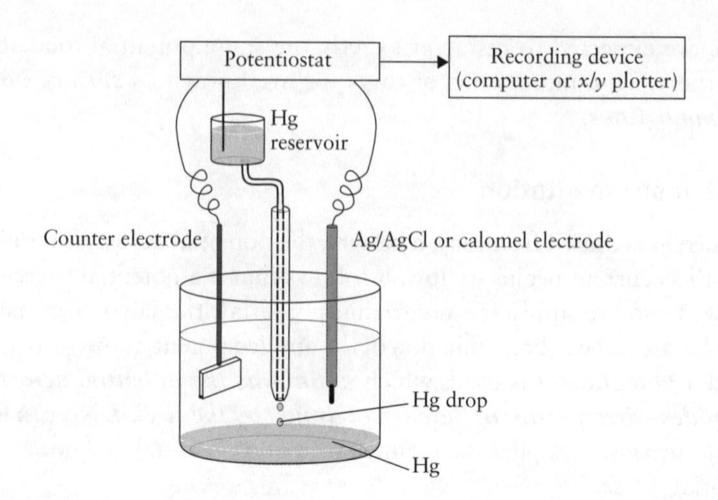

Figure 10.9 Dropping mercury electrode.

(as in Fig. 10.7(a)); the potential is again maintained by means of a potentiostat. As before, a reference (normally an SCE) and a counter electrode (usually Pt) are required to complete the circuit. The complete apparatus is known as a ***polarograph***.

The continually changing surface area of the drop causes a rather strange yet characteristic form of the voltammogram, known as a ***polarogram***. As we might expect, a potential will be reached as the potential is swept that will allow for the reduction (or indeed oxidation) of an analyte. The current is then recorded as a function of potential (and/or time); Fig. 10.10. The first feature you will notice is that the current follows a rhythmical increase and decrease with time. The current is directly proportional to the area of the working (mercury) electrode, and this is, of course, continually changing with time as the drop grows and then falls from the capillary tip. When the drop is first formed its surface area is very small; this increases over the lifetime of the drop as the drop grows in size. The current that is observed therefore also increases. When the drop becomes too large and therefore heavy it falls from the capillary tube and a new, very small, drop is formed. The process repeats itself in a similar manner. The current profile which follows this cycle therefore undulates with the frequency of the drop lifetime.

The *mean* current will rise sharply once the potential has reached a value that allows for the analyte to be measured. Polarography is used most widely for the determination of heavy metal ions that may be reduced, and, in this case, the potential ramp (or sweep) will be in a cathodic (negative) direction. In Fig. 10.10, we see that once a potential has been reached that allows for the reduction or oxidation of the analyte, the mean current increases in a staircase like manner until a new ***diffusion limited*** current is reached. Potentials in excess of this value are known as ***overpotentials***. The cathodic or anodic current therefore continues to be

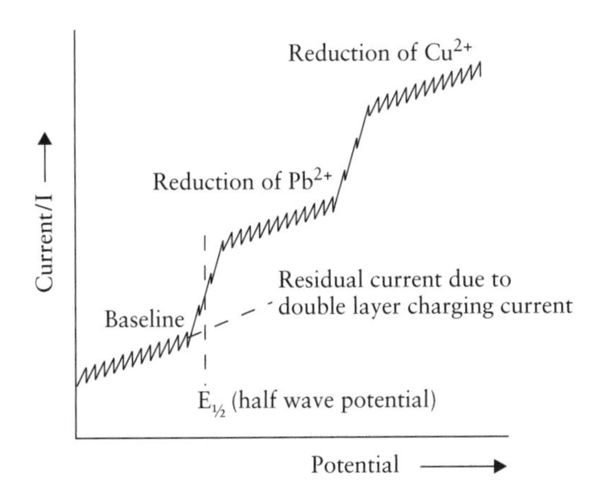

Figure 10.10 Polarogram for the reduction of Pb^{2+} and Cu^{2+}.

seen during the rest of the potential sweep. The potential that corresponds to the current half way between the baseline current and the steady state diffusion limiting current is known as the ***half-wave potential***, or $E_{1/2}$. If a potential is reached that allows a second analyte to be reduced or oxidized, then this current becomes superimposed upon the first. An example may be seen again in Fig. 10.10, which shows how Pb^{2+} and Cu^{2+} may be both reduced at potentials in excess of ~ +600 mV versus Ag/AgCl.

In each case, however, the technique may be quantitative since the current is directly proportional to the concentration of the analyte and this may be predicted by the ***Ilkovich equation***, Eqn (10.8):

$$i_d = kc \tag{10.8}$$

where i_d is the diffusion-limited current, c the concentration of the analyte, and k a constant relating conditions such as the temperature and the rate of flow of mercury.

One other physico-chemical factor embedded in the shape of the polarogram should be noticed; the mean current will slowly rise *irrespective* of whether or not a detectable analyte is present. The current due to the reduction of an analyte is therefore superimposed on top of this baseline. The drifting baseline is due to the charging of the interface between the mercury drop and solution in the manner of a capacitor; the effect is known as a ***double-layer charging current***, I_{dl}, and may limit both the sensitivity and lower limit of detection for the technique. You might think that if the drop is being continually replenished that the charging current should not increase with time. This argument is partly true, yet some mercury remains in contact with the electrolyte throughout the experiment, and this remains polarized.

The detrimental effects associated with the double layer charging current may largely be overcome using ***differential pulse polarography or DPP***, which we shall consider next.

Polarography has its disadvantages, which include the difficulty associated with the handling and toxicity of elemental mercury. The instrumentation is also somewhat cumbersome, and aqueous electrolyte solutions must be de-aerated (de-oxygenated) before analyses are attempted to remove molecular oxygen, which could be electrochemically reduced at the dropping mercury electrode (DME). Despite these drawbacks, polarography is still a widely used laboratory technique due to its sensitivity and the very low limits of detection that may be achieved. Many heavy metal ions, for example, may be determined down to ppm levels, and sensitivities of this kind may even challenge the performance of techniques such as atomic absorption spectroscopy at a small fraction of the cost—or indeed instrumental complexity.

10.6.1 Differential pulse polarography

As we have seen already, polarography employs a potential ramp to polarize a DME. Since a dc potential causes the mercury drop–solution interface to charge, a double-layer charging current is observed, which decreases the sensitivity of the analysis. It is also difficult to accurately determine (or indeed follow) the mean current through a series of rhythmical current waves.

Differential pulse polarography (DPP) largely overcomes these problems by an elegant approach that involves superimposing a series of small potential pulses on top of the linear potential sweep; Fig. 10.11.

The pulse is timed towards the end of the drop's lifetime when it is close to its largest volume before being knocked from the capillary. A fraction of a second *after* the potential pulse has ceased, the drop is mechanically knocked from the capillary tip by a small bar that physically jolts the column.

The current is measured *just before* and then *just after* the potential step. The *difference* between these two currents is then plotted as a function of potential. The potential pulse is timed towards the end of its lifetime, because this is the period in which the surface area changes at the slowest rate with time. The drop volume will increase linearly with time as the mercury flow rate remains constant, however, the rate of change of the drop's surface area decreases with time since the surface area of a sphere is given by $4\pi r^2$. The double-layer charging current is directly proportional to the surface area of the mercury drop, and it is during the later stages in the drop's lifetime that the double layer charging current increases at its slowest rate. The current due to the ***amperometric detection*** of the analyte will be dependent upon the potential that has been reached by the linear potential sweep. Since the current is measured at the beginning and the end of each pulse, and the charging current is at a very low value its contribution is minimized. Another very important effect is also

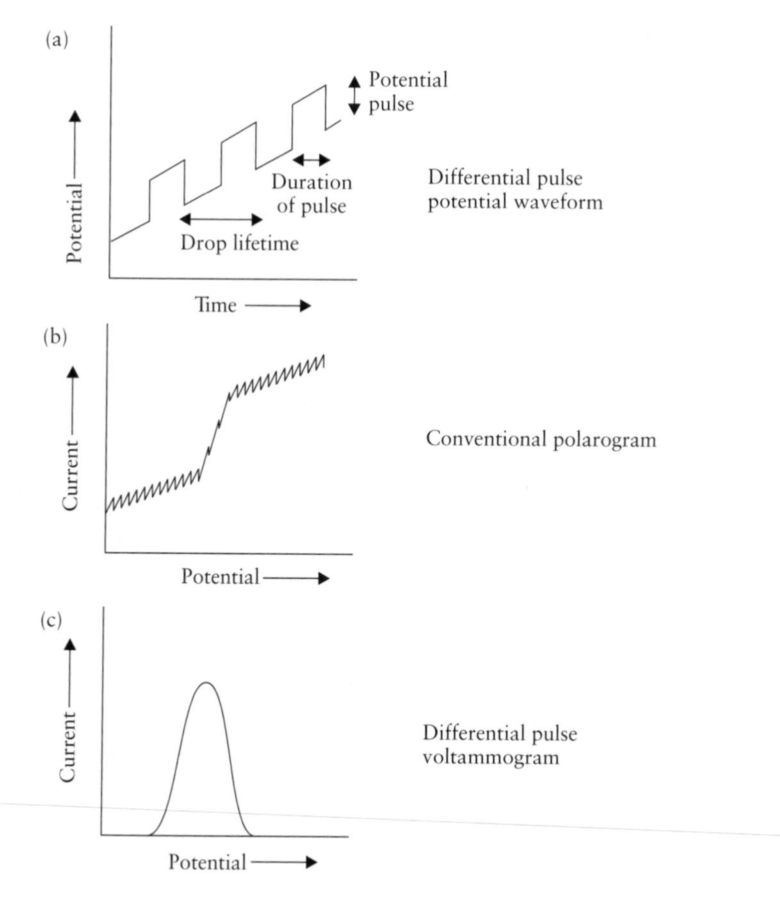

Figure 10.11 Comparison of differential pulse polarography and conventional polarography.

seen; since we now measure a *differential* current (which follows the *rate of change of current with time*), this value will increase as the rate of the redox current increases, but will then fall again as the current approaches a steady state value. The potential of the DPP current peak, corresponds to the midpoint of the *voltammetric redox wave*, and this is known as the ***half-wave potential***. The relationship between a standard polarographic redox wave and a DPP peak is shown in Fig. 10.11 for clarity. The interpretation of differential pulse voltammograms is very much easier than for standard polarograms, particularly if we wish to ascertain the accurate position of the half-wave potential. The determination of a half-wave potential often helps identify the analyte that gives rise to the polarographic wave. DPP may also help differentiate two partially overlapping polarographic waves. Two separate peaks are easier to distinguish, (even if they are shouldering each other), rather than trying to interpret one staircase-like current superimposed upon another that, in turn, has an embedded drifting baseline.

10.6.2 Adsorptive stripping voltammetry

Adsorptive stripping voltammetry, or ASV, is an extremely powerful technique that permits vastly lowered limits of detection in comparison to standard voltammetric analyses. The technique involves the *accumulation* of the analyte at the working electrode. Only after accumulation is the actual analysis attempted. Determinations may be frequently performed at parts per million or even parts per billion concentrations. In this way, extremely low analyte concentrations are analysed that would otherwise defy identification.

So how is this actually performed? Many analytes dissociate in solution; the solute ions are made to traverse (or ***migrate***) across a solution under the influence of an electric field. Anions (negative) migrate towards an anode (positive) and cations (positive) migrate towards the cathode (negative). The WE is polarized to attract and thereby accumulate the ionized analyte over some period of time. This time interval is known as the ***pre-accumulation*** or ***adsorption*** step, and may last from 1–2 s up to several minutes. The potential that is applied to the working electrode is similarly known as the ***pre-accumulation*** or ***pre-adsorption*** potential.

Once a sufficient quantity of the analyte (or indeed all the analyte in the solution), has been pre-accumulated and adsorbed at the working electrode, the potential is reversed through a linear potential ramp with time; Fig. 10.12. The current is then voltammetrically recorded as function of

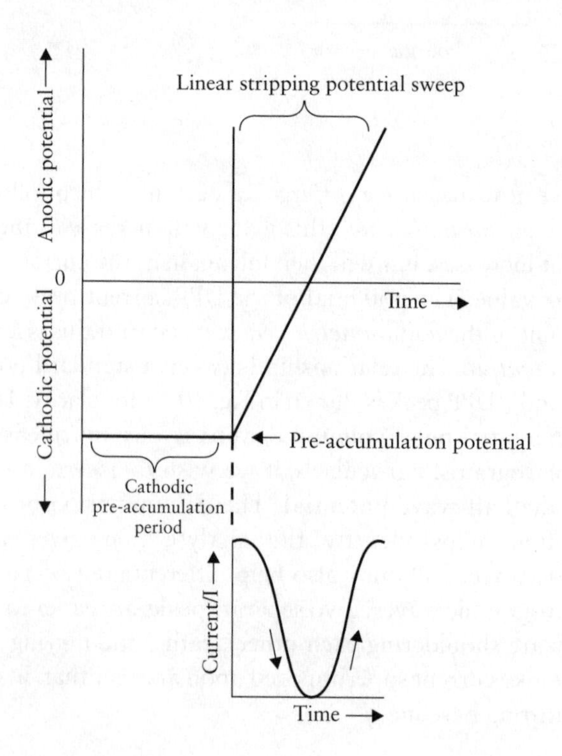

Figure 10.12 Adsorptive stripping voltammetry.

potential. The voltammogram that is recorded is known as an ***adsorptive stripping voltammogram*** since the analyte is stripped from the WE surface. Anions will be oxidized and cations reduced during the ***stripping step***. Adsorptive stripping voltammetry which involves the accumulation of cations at an anode employs the ***anodic*** stripping of the analyte, known as ***anodic stripping voltammetry***. In a similar manner, if anions are accumulated at anodes and subsequently stripped cathodically, the approach is known as ***cathodic stripping voltammetry***. These two processes are often abbreviated as **CSV** and **ASV** although the later acronym sometimes causes confusion with the terminology for adsorptive stripping voltammetry. It is therefore always wise to define precisely what is meant if we are going to use these abbreviations.

Many different electrodes are used for adsorptive stripping voltammetry, but probably the most common is the ***hanging mercury drop electrode*** (**HMDE**). The HMDE is essentially very similar to the DME, except that mercury is not continually forced through the capillary but rather a static drop is formed that hangs at the end of the capillary for the duration of the analysis. The drop is knocked off before a new analysis is performed by pushing a small amount of mercury through the capillary via a vernier screw plunger situated at the top of the column.

The HMDE again offers the advantage of providing a very clean Hg surface for each new drop that is formed. Heavy metal atoms may also dissolve in the mercury to form an amalgam thereby permitting an even greater quantity of the analyte to be accumulated prior to its analysis. Other electrodes are used for adsorptive stripping voltammetry and in particular chemically modified electrodes offer many opportunities for the enhanced analysis of a range of differing analytes. We shall briefly consider chemically modified electrodes in Section 10.7.

10.7 Chemically modified electrodes

Chemically modified electrodes or CMEs have, in recent years, attracted much research interest and it seems likely that they will continue to do so for some time. Until now, we have only considered *bare* electrodes that are metallic but might be made of other conductive materials such as, for example, carbon. In each case, the electrode is essentially untreated prior to its use. A *chemically modified electrode*, by contrast, has been treated in some way to alter or modify its surface, and therefore its analytical performance.

There are several ways that electrodes may be chemically modified, and these involve electrolytic pre-conditioning, chemically coating the electrode by covalently bonding molecules to its surface, or coating the entire electrode surface with an adsorbent polymer film. We shall only consider the latter two types since they are the most widely encountered forms of CMEs.

10.7.1 Anchorage of molecules with well-characterized surface properties

The surface chemistry of an electrode surface may sometimes be radically altered by chemically anchoring molecules that have some very well-characterized properties while still permitting the desired electrochemistry. Polyethylene glycol (PEG) is, for example, frequently used to coat the surface of biomedical devices in order to resist protein sorption. PEG may be chemically anchored to an electrode surface to improve its biocompatibility, for use within a clinical analysis. Electrode modifications of this kind are particularly useful if we wish to increase the useful lifetime of an electrode for multiple analyses in hostile environments. PEG, like many other organic molecules, may be anchored to a metallic electrode (Au is particularly good in this respect) by means of a thiol or (–SH) linkage. Redox enzymes can also be made to undergo direct electrochemistry by *electrically wiring* them to an electrode surface. A linkage is made between the enzyme's redox active centre and a metallic electrode via a highly unsaturated hydrocarbon chain which permits electron transfer along its length. Many other means of anchoring tailored molecules with specific moieties or functional groupings are presently being explored, and the reader is referred to Chapter 14 for a fuller coverage of these areas.

10.7.2 Polymer modified electrodes

By far, the most widely encountered types of CMEs are those that have been modified by the deposition of some form of polymer. Polymers are extremely versatile and may be made with hydrophobic, hydrophilic, insulating, or electrically conducting properties. Polymers may be used to entrap other molecules within their structure, such as enzymes, or in some instances to prevent chemical interferents from reaching the working electrode. The so-called *conducting* polymers are those that permit some electrical conductivity in at least one redox state. The polymer may usually be oxidised or reduced, so the conductivity of the polymer depends on its redox state. Polyaniline has three redox states, for example, which permit the polymer to exhibit either *reasonable* conductivity or insulating behaviour via its polarization. The polyaniline is easily electropolymerized from an aqueous solution of aniline (hydrochloride) monomer. Further redox molecules such as enzymes may be immobilized within the polymer acting as a matrix. In this way, a number of redox state dependent conductimetric properties may be monitored—and, in this way, a number of so-called 'chemical transistors' or 'enzyme transistors' have been developed. There are a number of other conducting polymers that may be electropolymerized at an electrode surface with the most popular including polypyrrole, polyindole, and poly-*N*-methylthiophene.

Another important use for the polymer modification of an electrode is to provide a polymeric ***permselective*** film across its surface to exclude electrochemically active interferents from reaching a working electrode surface. This approach therefore helps improve the specificity of the technique. Most permselective films rely on the charge exclusion of ionized or at least highly polar molecules via the repulsion of like charges. For example, the commercial polymer Nafion® from Du Pont, has anionic (negative) groups in a polymer that should readily permit the passage of cationic or neutral solutes across a thin film but which should greatly impede the access of anions to the electrode surface.

10.8 Micro-electrodes

Until now we have considered processes which occur at large (planar) electrodes. We shall now see how *very* small (of the order of micrometre dimensions) or ***micro-electrodes*** may be exploited by the analytical chemist. You may well ask why you should wish to use a micro-electrode since the size of any (e.g. amperometric) signal will be directly proportional to the electrode area. The answer lies in the way in which solutes travel to and from the electrode surface. Large electrodes experience linear diffusion profiles, which means that the solutes (in this case the reactants and products) may be considered to diffuse to and from the electrode in straight lines (Fig. 10.13). In fact, the solutes may diffuse in any direction, but steric hindrance means that the net effect is one in which the solutes do behave overall as if they travel in straight lines. If the rates of electron transfer at the electrode surface are fast in comparison to diffusion of the analyte, then the rate of diffusion may be the rate limiting step for the electrode reaction. The reaction is now said to be ***under diffusion control***. We have already discussed the implications of this situation when considering cyclic voltammetry (Section 10.5) and polarography (Section 10.6). This diffusion-controlled behaviour may be very troublesome for analyses since stirring the analyte solution will disrupt the diffusion gradients in the solution, and in this way, alter the electrode responses.

Micro-electrodes, by contrast, are not subject to these constraints since they experience ***hemispherical solute diffusional profiles*** as long as they are sufficiently small to behave as a single point source. In this case,

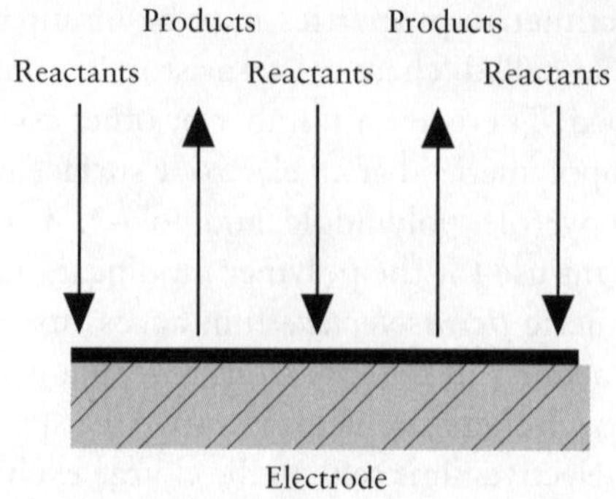

Products Products
Reactants Reactants Reactants

Electrode

Figure 10.13 Linear diffusion to and from a planar electrode.

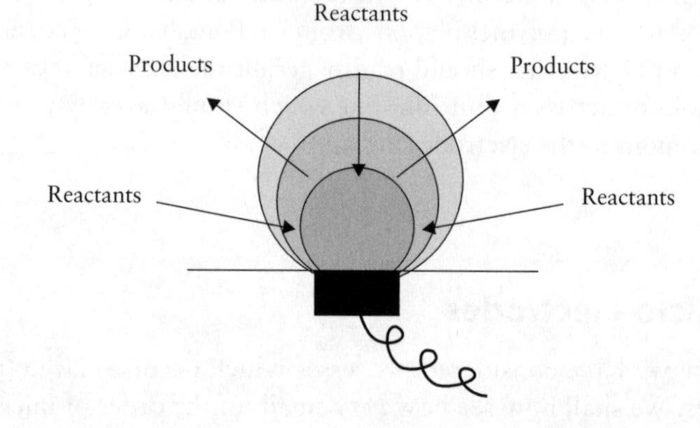

Figure 10.14 Hemispherical diffusion profiles to and from a micro-electrode.

solutes can approach or leave the electrode in a two-dimensional 180° arc, drawing a hemisphere; Fig. 10.14. The rate of solute mass transport is usually considerably greater than you could achieve at a large planar electrode, and many analyses may be taken out of diffusional control and thus rendered ***stir independent***. It is beneficial if an electrode ceases to be susceptible to stirring, or other physically induced movement within the analyte solution (i.e. convection), since it would not be necessary to control solution flow patterns, for example, in non-laboratory (remote) settings.

Micro-electrodes do suffer from some drawbacks. The most obvious of these relates to the small size of the signal (usually amperometric), which is proportional to the area and therefore very small. You may need to protect the electrode from radiated or other sources of electrical noise, so that the signal does not become buried within a baseline of noise and so rendered useless. One other approach might be to use several micro-electrodes with their signals being coupled together in order to boost the composite measured signal. Ironically, the very small size of the signal may be exploited to some advantage since it is frequently possible to perform the analysis without having to use a potentiostat. This is because: (a) the currents due to the electrochemical reaction; and (b) the ohmic drop contributions across a solution (iR) are very small, and therefore it is not necessary to use a potentiostat in order to maintain the polarizing potential, V.

The 'ohmic drop' (iR) is the drop in potential difference experienced across the solution caused by current passing through the solution and is equivalent to V = iR from Ohm's law.

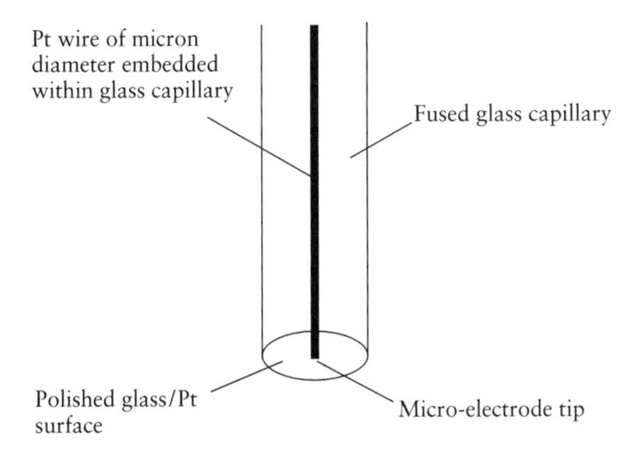

Figure 10.15 Micro-electrode.

One common method of producing micro-electrodes is to embed a wire (usually platinum) within a glass capillary which is fused at one end. The tip of the electrode is then polished with fine alumina (Al_2O_3) powder to produce a flat surface; Fig. 10.15.

Multiple micro-electrodes are often fabricated via photolithographic techniques in a similar manner to the production of printed circuit boards and silicon micro-chips. Micro-electrodes are certainly attracting ever more interest, for example, in biomedical applications (see Chapter 14), and from the wider research community. It is clear that micro-electrodes are likely to be used ever more widely for an increasing number of electrochemical analyses.

10.9 Organic phase electro-analytical chemistry and the electro-analysis of organic compounds

10.9.1 Organic solvents and the solubility of organic analytes

We have assumed until now that the analyte will be water soluble and that sufficient conductivity will be provided across the solution by the addition of a simple supporting electrolyte such as KCl. Unfortunately, many organic molecules that might otherwise lend themselves to voltammetric or polarographic analysis are either partially soluble or even totally insoluble in water.

The analyte can sometimes be solubilized by introducing a small quantity of a water miscible organic solvent such as acetic acid, dioxane, acetonitrile, or one of the alcohols. If this approach fails to solve the problem, and the analyte is only truly soluble within aprotic organic solvents, then electrochemical analysis *may* still in some instances be possible as we shall see below (Section 10.9.2).

10.9.2 Organic phase polarography

Organic phase polarography may be performed in solvents such as diethylamine or dimethyl formamide. Aprotic solvents of this type unfortunately have very high resistivities, and will totally preclude any electrochemistry without the addition of an ionisable electrolyte salt which can permit charge transfer through the solution. Lithium or tetra-alkyl salts are frequently used for this purpose since they may dissolve and ionically dissociate in completely aprotic environments.

We have already seen that polarography is most commonly employed for the determination of heavy metal ions within aqueous solutions. Polarography is, however, becoming ever more widely used for the analysis of organic compounds that may either be soluble in water or perhaps only soluble in non-polar organic solvents.

Organic phase polarography can be used for either reducing or oxidizing an analyte. The analysis of heavy metal ions involves their reduction but the determination of many organics demands their oxidation. Unfortunately, we can only use a polarizing potential of up to approximately +0.4 V versus Ag/AgCl, since above this potential mercury may itself become oxidized. Nevertheless, this problem may be overcome by using another electrode material such as carbon, platinum, or gold.

10.9.3 Electrochemical determination of organic compounds

Many organic compounds as detailed below contain functional groups that will give rise to electrochemical behaviour that may be exploited for analytical purposes.

1. *Reduction reactions*
   - organic halo groups may be reduced with the replacement of the halogen by a hydrogen;
   - alkenes and some unsaturated cyclic organics;
   - nitro, nitroso, amine oxide, and azo groups.
2. *Oxidation reactions*
   - hydroquinones and mercaptans;
   - peroxides and epoxides.

10.10 Electrochemical titrations

The most commonly used approach for determining the equivalence point of a redox titration is to use an indicator. Electrochemically based titrations are, however, becoming more popular since the equivalence point may be determined more accurately than by conventional titrimetric methods.

Electrochemical titrations are most typically followed potentiometrically, though some equivalence points may be determined amperometrically.

10.10.1 Potentiometric titrations

Potentiometric titrations are normally simple to perform. The potential, E, of an indicating electrode is measured with respect to a reference electrode, and E is plotted as a function of the volume of titrant added. The potential difference changes rapidly near the equivalence point. The equivalence point is determined with far greater accuracy than would normally be possible with a standard indicator based volumetric titration for two reasons. Firstly, we remove the human error associated with estimating a change in colour. More importantly, however, the potential difference we monitor, directly follows the change in the chemical activity of the reaction throughout the course of the titration. Since we are trying to estimate the equivalence point of the titration, the absolute potential of the indicating electrode need not be accurately known; it is the *change in potential* that is important and this potential will describe the shape and therefore the equivalence point for the titration curve. Potentiometric titrations may either be performed manually or by means of a commercial auto-titrator. Indeed, most commercial auto-titrators use some form of potentiometric indicating electrode to follow reactions.

pH Titrations

We have already seen how the pH of an acid-based titration changes dramatically when the equivalence point has been reached (Chapter 3). An easy way to follow a titration of this type is via the use of a standard potentiometric glass membrane pH electrode. By plotting the pH as a function of the titrant volume, a titration curve may be easily plotted. The equivalence point corresponds to the point which gives the steepest slope of the pH/volume titration curve; Fig 10.16. An SCE, Ag/AgCl, or other suitable reference electrode may be used. The pH and reference electrodes should ideally be kept as close to each other as possible in order to minimize the solution resistance between the two electrodes.

Ion selective electrodes for titrimetric analyses

In Section 10.3, we saw that a range of potentiometric ISEs is commercially available for a range of anions and cations. ISEs of this type may be used to follow titrations in which one of these ions is consumed within a redox reaction. For example, a Ca^{2+} ISE may be used to follow the titration of calcium and EDTA, Eqn (10.9):

$$2Ca^{2+} + EDTA^{4-} \rightarrow Ca_2(EDTA) \tag{10.9}$$

Another suitable example might be to use an Ag^{+} ISE for silver nitrate titrations.

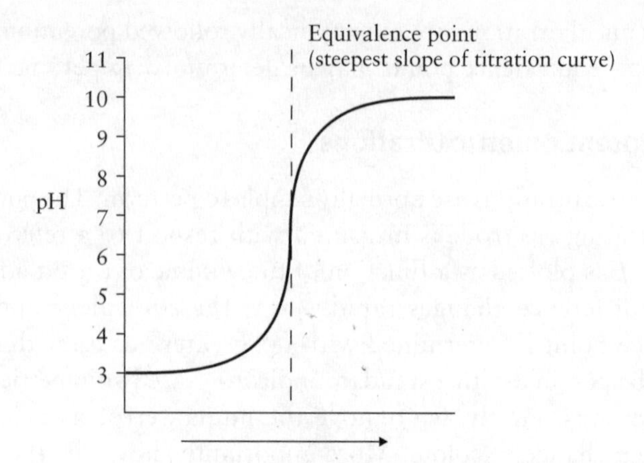

Figure 10.16 pH titration profile.

Accurate determination of potentiometric titrimetric equivalence points

The equivalence point of a potentiometric titration of the type in Fig. 10.16 corresponds to the steepest point of the curve, that is, the inflection point. Measuring this manually can easily lead to errors, although computers can be used to continually measure the slope of the line (as the first derivative of the potential with respect to the volume of titrant). The slope may then be plotted as a function of the volume of titrant used. The resultant plot follows the rate of change of potential with titrant volume added ($\Delta emf/\Delta vol$ versus volume of titrants added); Fig. 10.17. The 'S'-shaped ΔpH/volume profile now becomes a peaked curve; with the mid-peak position corresponding to the value of titrant added at the equivalence point.

Taking the first derivative of the potentiometric titration profile offers two distinct advantages. First, it is *visually* easier to follow the potentiometric end point by monitoring the rise and fall of a peak. More importantly, however, the equivalence point may be more accurately detected since the first derivative of the titration slope is continually monitored and the point of inflection does not have to be estimated by eye.

10.10.2 Amperometric titrations

Amperometric titrations monitor a current when a polarizing potential is imposed at a WE. The potential is generally fixed. This type of voltammetry is known either as ***hydrodynamic voltammetry*** or ***steady-state amperometry***. The term hydrodynamic refers to the current being a function of ***mass transport*** within the solution if the process is under ***diffusional mass transport control.***

Hydrodynamic voltammetry is used to determine the equivalence point of a titration providing that at least one of the reactants and/or products

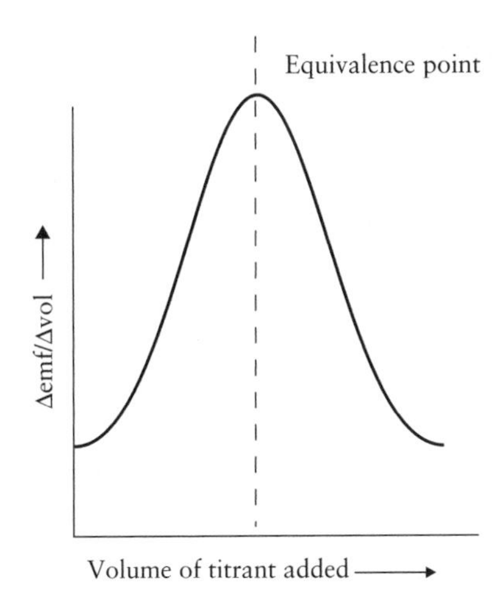

Figure 10.17 Potentiometric titration equivalence point determination via monitoring the first derivative of the titration curve.

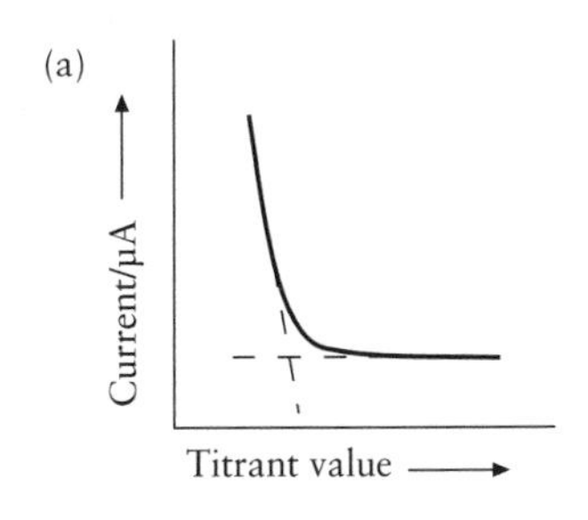

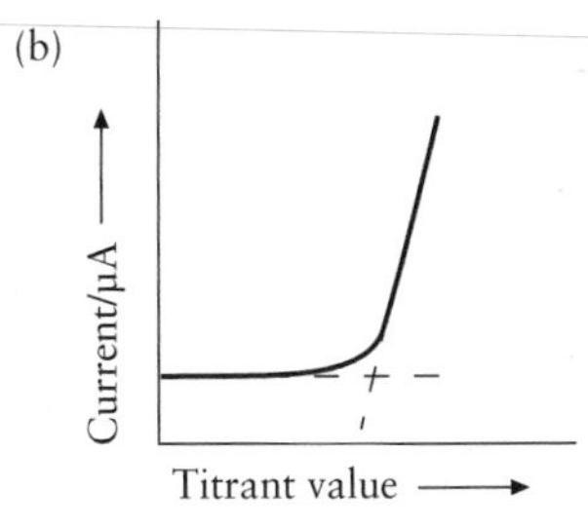

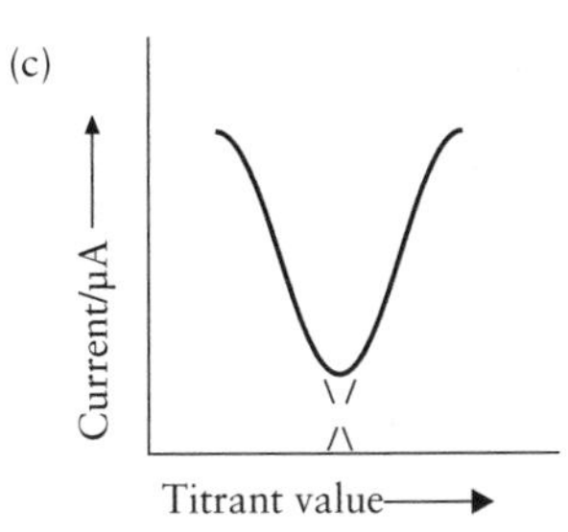

Figure 10.18 Amperometric titrations.

can be electochemically oxidized or reduced. The current is therefore measured as a function of the titrant volume. If one or more of the reactants is electrochemically active then the current due to the oxidation or reduction of this species will fall as it is consumed, Fig. 10.18(a).

Mass transport is a term used to describe the movement of a reactant or analyte (or any product formed during an electrochemical reaction) to and from an electrode surface.

Conversely, if the product (or one of the products) is electrochemically active, then the current due to the electrochemical oxidation or reduction of this species will rise, Fig. 10.18(b). If a reactant and a product are *both* electrochemically active then the current will fall in a linear fashion towards the equivalence point and then linearly rise again as more of the titrant is added, see Fig. 10.18(c). In this instance it should be remembered that the current due to either the reactants or the products is directly proportional to their respective concentrations. In each case the equivalence point can be estimated by extrapolating the two straight line sections of the current/volume plots.

Hydrodynamic voltammetry always involves some consumption, however small, of one or more of the reactants and/or products. Microelectrodes are therefore often used in order to minimize the extent of the consumption.

Amperometric titrations are often employed where titration reactions involve the formation of either a stable complex or precipitate. EDTA titrations, for example, form stable complexes with many heavy metal ions such as Fe^{2+}, Cu^{2+}, and Pb^{2+}. Other precipitating agents include silver nitrate for the precipitation of halide ions and lead nitrate is used for the precipitative titration of sulphate ions.

10.11 Oxygen electrodes

The oxygen electrode was first described by Leyland Clark in the 1960s and is now widely used to monitor the oxygen concentration of aqueous solutions. The electrode relies on the amperometric reduction of oxygen at a cathodic WE polarized at approximately −600 mV versus Ag/AgCl. The WE is normally Pt over which is stretched a thin oxygen permeable membrane Teflon; Fig. 10.19. The Teflon film is designed to readily permit oxygen diffusion to the underlying working electrode whilst preventing the passage of any ionised solutes. A counter electrode (frequently in the form of a concentric ring around the WE) completes the circuit. A small wick soaked in electrolyte may sometimes be placed in between the working and counter electrodes to ensure electrical conductivity is maintained between them. In some designs, the counter and reference electrodes are combined together to form one electrode.

Oxygen is reduced according to Eqn (10.10):

$$O_2 + 4H^+ + 4e^- \rightarrow 2H_2O \qquad (10.10)$$

The current is proportional to the partial pressure of oxygen within the sample. The electrode must be pre-calibrated with both de-aerated and oxygen saturated samples. O_2 saturated water contains approximately

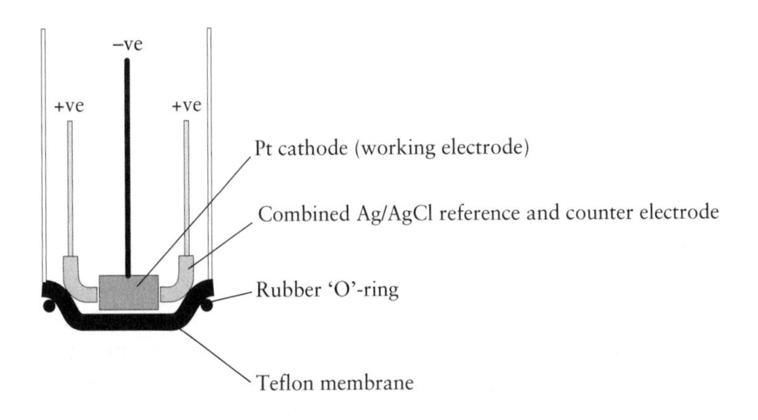

Figure 10.19 Oxygen electrode.

28 μg cm^{-3} at normal atmospheric pressure at 298 K. The currents for the de-aerated (baseline) and oxygen saturated samples have both to be empirically determined. A two-point linear calibration profile may then be determined, from which O_2 concentrations of unknown samples may be estimated.

Some halogen gases such as chlorine (Cl_2) may dissolve in water ($Cl_{2(g)} + H_2O_{(l)} \rightleftharpoons HCl_{(aq)} + HOCL_{(aq)}$) and can cause interference to the O_2 electrode since they may both traverse the Teflon permselective membrane.

The oxygen electrode is now routinely incorporated into clinical multi-gas analysers for the determination of oxygen within blood and serum samples (see Section 14.3) and is also sometimes used as a component within an amperometric biosensor in which an oxygen consuming enzyme is employed.

10.12 The scope of electrochemically based sensors

There is an ever-burgeoning research effort currently being devoted to the development of many different sensors with applications spanning medical, environmental, and industrial process control sectors.

Electrochemically based sensors represent one of the most important contemporary classes of sensor, and we shall consider the scope of their application in this section. Further discussions of biomedical sensors and biosensors are given in Chapter 14.

There are several obvious examples. A diabetic patient may simply test his or her own blood for its glucose concentration using a hand-held biosensor. Automated portable hand-held pH electrodes represent another excellent example of a sensor that has now reached the stage where it can be used by almost anyone without any specialist training. Although the pH electrode has been in existence for some years, it still represents an almost idealized sensor due to its robustness, selectivity and simplicity, which the majority of contemporary sensors still struggle to match.

Despite the vast research effort that is presently being focused on developing new sensors, comparatively few devices have yet realized commercial success. While this situation is undoubtedly set to change, many sensors fail to satisfy a number of key performance criteria, particularly in terms of the specificity, sensitivity, longevity, robustness, or indeed financial viability. Technological inertia within the scientific community should also not be overlooked. Sensors primarily offer a technology that seeks to provide alternative approaches towards a number of well-established techniques. In order for an analyst to switch techniques they must be satisfied that: (a) the new approach is at least as reliable as the approach already being used; (b) the new sensor matches and probably exceeds all of the main performance criteria of the technique already being used; and possibly (c) some cost savings may be made. If the new sensor meets all the requirements, the possible customer might still have to justify an initial capital re-investment for a new technology when they already possesses the analytical capacity they require.

10.12.1 Modes of operation

We shall first consider the differing modes of operation for electrochemical sensors and then consider some of the range of applications they may be used for.

Electrochemical sensors may operate in: (a) a potentiometric; (b) an amperometric; or (c) a conductimetric mode of operation.

Potentiometric sensors

Potentiometric sensors have one principal advantage over many techniques in that their measurement does not involve the destruction or consumption of the analyte. A potential is measured in response to the analyte, but since no current flows, the analyte is neither oxidized nor reduced and therefore is not consumed. As the analyte is not consumed at the WE surface, diffusional gradients are not set up between the bulk analyte solution and the sensor surface. Potentiometric sensors are therefore inherently stir independent in their behaviour, which certainly facilitates their ease of use. This contrasts with amperometric sensors whose response behaviour is normally strongly dependent on the stir rate.

ISEs are by far the most widely used form of potentiometric sensor which are currently commercially available. ***Chemically selective field effect transistors*** or *CHEMFETs* (Fig. 10.20) represent another development

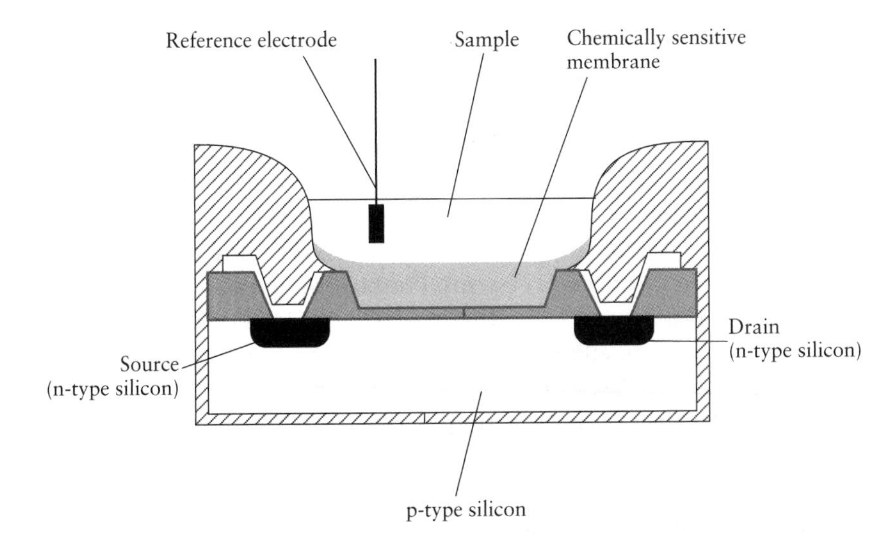

Figure 10.20 A CHEMFET.

and are based on a modified form of the ***field-effect transistor***. The regulation of current in a field-effect transistor is related to the potential at which the gate region is held. A CHEMFET uses an ion-selective membrane, which is held in contact with the gate region. In this way the potential at the ion selective layer modulates the current between the source and drain. For some time it has been predicted that CHEMFETs would quickly dominate the sensor market; to date this certainly has not happened. The principal problem CHEMFETs encounter relates to the ease with which the transistor's gate region is poisoned. While the production of CHEMFETs may not be excessively expensive, disposable CHEMFETs would not prove commercially viable for widespread use.

Amperometric sensors

Amperometric sensors enjoy widespread applications and, at the time of writing, account for the largest share of the chemical and biosensor markets. Amperometric techniques, in many instances, represent the most robust of the electrochemical techniques, although this approach does inherently consume some of the analyte. Though the consumption of a very small quantity of the analyte may not significantly affect the bulk analyte concentration, solute diffusion gradients are created, which imparts an unwanted stir-dependent response. Amperometric sensors are therefore often unsuitable for on-line monitoring of industrial processes and require specialist care in use. Micro-electrodes often overcome many of these problems, yet bring with them their own drawbacks associated with the small currents that have to be monitored. Micro-electrodes are also frequently far costlier to fabricate than larger planar electrodes.

Conductimetric sensors

Ions transport the charge between the electrodes of an electrochemical cell. Normally, we wish the ion conductivity of an electrolyte to be sufficiently high that the electrochemistry of the sensing electrode is not impeded in any way. Sometimes, however, the ionic conductivity of the solution may actually be the parameter that we wish to measure.

Conductivity sensors are widely used for monitoring the quality of laboratory water purification systems. Freshly distilled water should have an $[H^+]$ and $[OH^-]$ each equating to 10^{-7} M. Carbon dioxide from the air will slowly dissolve within the water causing the conductivity of the solution to increase with the dissolved concentration of hydrogen carbonate $[HCO_3^-]$. Ionizable impurities dramatically increase the conductivity of the water.

Conductivity meters may also be used to follow the titration of a weak base and a strong acid (or indeed strong base and weak acid). The differences in the extents of dissociation of a weak acid or base and that of a strong acid or base gives rise to a very large change in conductivity of the solution once the equivalence point is reached.

10.12.2 Conducting polymer based sensors

A number of bio- and chemical sensors monitor the conductivity of a ***conducting polymer*** that bridges two closely spaced electrodes. A so-called conducting polymer can be defined as a polymer that permits the passage of charge in at least one of its redox states. Most conducting polymers are electrically insulating in another redox state, and the change of conductivity between the redox states may be used to follow the redox state of the polymer. If the conductivity of the polymer can, in turn, be modulated by the redox activity of a chemical reaction, then we may use this as the basis of a ***conductimetric*** sensor. A recent approach has been to incorporate a redox enzyme (e.g. glucose oxidase) within a conducting polymer such as polyaniline or polypyrrole. The conductivity between two closely spaced electrodes bridged by a conducting polymer that contains an enzyme, may be directly modulated by the redox activity of the enzyme and in turn by the concentration of the enzyme substrate. Sensors of this type are discussed more fully in Chapter 14.

10.12.3 Applications

Electrochemical sensors are used for a variety of applications, which might include:

- sensors for monitoring industrial processes
- sensors for health care

- sensors for monitoring environmental pollution
- sensors for use as research tools.

It should be remembered (but is, however, frequently overlooked at the design stage) that individual sensor requirements vary considerably according to the final sensor applications. A clinical glucose sensor intended for use in a hospital pathology laboratory, for example, might be expected to perform repeated tests for a number of days without requiring maintenance; by contrast, a glucose sensor for use by the diabetic at home will probably, for the sake of simplicity, use disposable electrode strips, which only perform one test. Industrial glucose sensors for the on-line monitoring of a food production line may, however, require the greatest longevity, since maintenance of such sensors would frequently result in disrupting the manufacturing process, with the inevitable loss of productivity. Broadly speaking, sensors may be classified into four operating modes:

- *disposable*—for single analyses.
- *serial analysis mode*—intended to operate for a finite length.
- *on-line monitoring mode*—which provides real-time information. In this arrangement, it is common for a small quantity of analyte to be passed via the sensor and then fed back to the bulk reaction/analyte mixture.
- *in-line arrangement*—where the sensor is placed *in situ* for providing real-time monitoring.

The scope for sensor-based analyses is increasing continually. In most Western countries, legislation is becoming ever more stringent for the determination of atmospheric and water-borne pollutants, so sensors may help to meet these analytical and regulatory challenges. In a similar manner, we all seek assurance that residual pesticides are monitored within our food and that the air we breathe is of an acceptable quality.

Sensors will never wholly replace existing analytical techniques yet if they can help decrease the number of more complicated and costly analyses they will certainly perform a useful role. One way in which this may be achieved is for sensors to act as a front-line screening tests. If a sensor can identify a possible problem such as confirming the presence of an industrial pollutant in, say, a river, then further more elaborate testing can be performed. Screening of this type may dramatically decease the number of critical tests which need to be performed, which can simplify the testing operation, reduce costs and indeed may help focus the analytical effort towards tests that truly require closer scrutiny.

Exercises and problems

10.1. Why must all electroanalytical cells contain a reference electrode?

10.2. Why is the standard hydrogen electrode now very rarely used?

10.3. Describe three forms of mass transport to electrode surfaces.

10.4. From the two half-cell equations below determine the standard cell potential:

$Fe^{3+} + e^- = Fe^{2+}$ $\quad E^0 = 0.771$ V
$I_3^- + 2e^- = 3I^-$ $\quad E^0 = 0.5355$ V

10.5. From the two half-cell equations below determine the standard cell potential:

$Fe^{3+} + e^- = Fe^{2+}$ $\quad E^0 = 0.771$ V
$Zn^{2+} + 2e^- = Zn$ $\quad E^0 = -0.763$ V

10.6. From the two half-cell equations below determine the standard cell potential:

$I_{2(aq)} + 2e^- = 2I^-$ $\quad E^0 = 0.6197$ V
$Sn^{4+} + 2e^- = Sn^{2+}$ $\quad E^0 = 0.154$ V

10.7. A pH 3 aqueous solution contains 1×10^{-3} M CrO_7^{2-} and 1.5×10^{-2} M Cr^{3+}. Calculate the potential of the half-cell reaction.

($E^0 = 1.33$ V:
$Cr_2O_7^{2-} + 14H^+ + 6e^- \rightleftharpoons 2Cr^{3+} + 7H_2O$)

10.8. The potential of a cell with respect to an SCE reference electrode is −0.845 V. Calculate the potential with respect to a SHE. (The cell potential using the SHE is 0.242 V less negative than SCE.)

10.9. The potential of a half cell with respect to an SCE reference electrode is −0.793 V. Calculate the potential with respect to an Ag/AgCl electrode. (The cell potential using the Ag/AgCl is 0.014 V less negative than SCE.)

10.10. A pH electrode is recording a pH of 6.1. Acid is added to the solution and the potential of the pH electrode increases by 177 mV. What is the pH of the new solution?

10.11. Explain why ion selective electrodes exhibit logarithmic responses with respect to analyte concentration.

10.12. Why are electrolytes often added to analyte samples prior to electroanalysis?

10.13. Explain what is meant by a chemically modified electrode.

10.14. What advantages can be gained by using micro-electrodes? What possible drawbacks can be encountered using micro-electrodes?

10.15. Sketch a cyclic voltammogram for a diffusion-controlled reversible one-electron process; how will the shape of this voltammogram differ for a two-electron process?

10.16. What are the advantages and disadvantages associated with using a dropping mercury electrode in comparison to solid electrodes?

Summary

1. Electrochemical cells must contain at least two electrodes to complete a circuit. In many cases, three electrodes are used and these are known as the working, counter (or secondary or auxiliary) and reference electrodes.

2. The counter/auxiliary/secondary electrode should always have a surface area of at least 10 times that of the working electrode to ensure that the electrochemistry at the working electrode is never rate limiting.

3. Double-layer charging is a phenomenon that occurs when two layers of charge on either side of the electrode/solution interface accumulate when the electrode is polarized.

4. The two most widely used reference electrodes are the silver/silver chloride (Ag/AgCl) and calomel electrodes (SCE); all reference electrode potentials are quoted with respect to the standard hydrogen electrode.

5. pH electrodes operate via a potentiostatic approach and their responses increase in proportion to the $\log_{10}$ of the H^+ ion concentration as dictated by the Nernst equation:

$$E = E^0 \pm \ln \frac{RT}{nF}[X]$$

where E^0 is the potential in volts under standard conditions of temperature and concentration, R the gas constant, T the absolute temperature, F is the Faraday constant, n the number of charges transferred in the reduction or oxidation process in question, and $[X]$ the concentration of the ion, for which the pH electrode is a H^+ ion.

6. There are a number of other ion selective electrodes; for example, F^+, Na^+, K^+, Li^+, and NH_4^+.

7. Linear sweep voltammetry involves imposing a potential upon a working electrode that is ramped linearly with time and recording the current; cyclic voltammetry is similar to linear sweep voltammetry but involves reversing the potential at the end of a potential sweep.

8. Polarography is a technique that is based upon the polarization of a dropping mercury electrode and may be used for the analysis of, for example, metal ions in solution.

9. The potential is typically ramped linearly with time. The current that is recorded is known as a polarogram and possesses a saw-tooth profile as each mercury drop grows and falls from the capillary tip.

10. Differential pulse polarogrphy involves superimposing a series of small potential pulses on top of the linear potential sweep to overcome limitations associated with double layer charging effects; this approach also often allows a lowering of the minimum detection limit.

11. Adsorptive stripping voltammetry is an analytical approach involving adsorbing the analyte onto an electrode surface during a pre-accumulation step. Accumulation occurs by polarizing the electrode so that the analyte moves towards the electrode by electrical migration. The analyte adsorbs at the electrode surface during the pre-accumulation. The analyte is quantified via a stripping process that involves reversing the potential at the working electrode (normally in the form of a potential ramp). The current that flows is recorded in the form of an adsorptive stripping voltammogram.

12. Chemically modified electrodes possess surfaces modified in some manner, for example, by coating with a polymer to impart some property to the electrode.

13. Micro-electrodes are electrodes with dimensions of the order of micrometres and offer advantages such as lowering of the minimum detection limits as well as imparting stir-independent response characteristics. These benefits occur as a consequence of the hemispherical diffusional mass-transport profiles micro-electrodes experience in comparison to linear diffusion experienced at planar surfaces.

14. Electrochemical techniques are routinely used for the monitoring of titrations via both potentiometric and amperometric means.

15. Oxygen electrodes may be used for monitoring aqueous oxygen concentrations and find application within both clinical and biological analyses.

16. Electrochemical biosensors are used in a number of clinical, biological, and environmental applications. The most widely used biosensors are those for the determination of blood glucose for the treatment of diabetes.

17. Other types of electrochemical sensors include, for example, devices for chemical determinations based on field effect transistors and are known as 'CHEMFETs'.

Further reading

Bard, A. J. and Faulkner, L. R. 2nd edition (2003). *Electrochemical methods: fundamentals and applications*. Wiley.

Fischer, A. C. (1996). *Electrode dynamics*. Oxford Chemistry Primers, Oxford University Press.

Monk, P. M. S. (2001). *Fundamentals of electroanalytical chemistry*. Wiley.

Riley, T. and Watson, A. (1987). *Polarography and other voltammetric methods*. Wiley.

Wang, J. (2000). *Analytical electrochemistry*. Wiley.

Nuclear magnetic resonance spectroscopy

11

Skills and concepts

This chapter will help you to understand:

- What is meant by the magnetogyric ratio for a magnet precessing in a magnetic field and the Lamor frequency for a nucleus.
- What is meant by high-resolution NMR spectrometry and how this differs from wide-line NMR spectroscopy.
- What is meant by the chemical shift and why tetramethysilane (TMS) is widely used as a reference compound in NMR spectroscopy.
- The meaning and origins of spin–spin couplings of neighbouring chemical moieties and how the splitting patterns of peaks can be used for structural elucidation of unknown compounds.
- How integration peaks can be used to determine the number of nuclei within a functional group.
- Why ^{1}H NMR spectroscopy is most widely used as the nucleus for NMR determinations.
- How the frequency of magnetic fields in an NMR instrument may be modulated and controlled.
- The nuclear Overhauser effect and how this may be used for resolution enhancement.
- Some applications for NMR spectroscopic approaches using nuclei other than the proton.
- How NMR spectroscopy may be used for a number of practical analytical applications.

11.1 Introduction to nuclear magnetic resonance phenomena

Nuclear magnetic resonance spectroscopy (or NMR*)* is an instrumental technique based on monitoring how spinning nuclei with magnetic dipoles interact with applied magnetic field(s) and absorb radiation.

The theoretical basis underpinning NMR effects dates back to the work of Pauli, who in 1924 suggested that certain atomic nuclei should possess spin and magnetic moment properties. Pauli postulated that this should allow splitting of their energy levels on exposure to appropriate magnetic fields. It was not until 1946, however, that Purcell and Block (at the Universities of Harvard and Stanford), independently demonstrated that certain nuclei could indeed absorb electromagnetic radiation in this manner. Research during the next decade was directed towards understanding how neighbouring nuclei affect the chemical environment in which individual nuclei reside. In turn, this led to rules that are still used for structural identification of molecular species.

It was in 1953 that the first commercial NMR instrument was marketed. This machine was specifically designed for structural determination and analytical applications. Since then, the use of NMR spectroscopy has burgeoned in an unprecedented way. It is not an understatement to say that NMR has had truly profound influences on many areas of organic, inorganic, and bio-chemistry.

11.1.1 The absorption of electromagnetic radiation via NMR

Nuclei can only absorb energy from the electromagnetic field when they possess a magnetic dipole moment. Magnetic dipoles occur when the spin quantum number, I, is not an integer. Isotopes such as ^{12}C, ^{16}O, or ^{32}S cannot for this reason absorb radiation from the magnetic field and it follows that these nuclei cannot be observed or studied via NMR. By contrast, maximal absorption peaks are observed for isotopes that possess a spin quantum number of $I = \frac{1}{2}$, for example, ^{1}H, ^{19}F, ^{31}P, or ^{29}Si.

See Section 6.2.1 for a discussion of spin quantum numbers.

NMR instruments designed to measure the absorption of electromagnetic radiation have one electromagnet, and are known as single-coil instruments. Electromagnetic absorption in instruments of this type is typically monitored over a frequency range of 4–900 MHz.

The magnetic dipoles of the nuclei interact with a static magnetic field, H, by precessing in a manner analogous to a gyroscope spinning in a gravitational field. In this instance, however, each nucleus behaves like a tiny magnet spinning within the magnetic field.

To understand the nature of NMR effects we shall first consider a model of how a magnet can spin in an applied magnetic field. Imagine a simple bar magnet (such as a small compass needle) at rest and aligned with a magnetic field. If it is displaced perpendicularly to the field then the ends of the magnet will swing first one way and then the other about the axis of the field until some external force (e.g. friction) brings the magnet to rest.

If, by contrast, the magnet spins rapidly around its north–south axis, then a displacement of the magnet from the axis of the magnetic field will cause the nucleus to move or ***precess*** in a circular direction around the

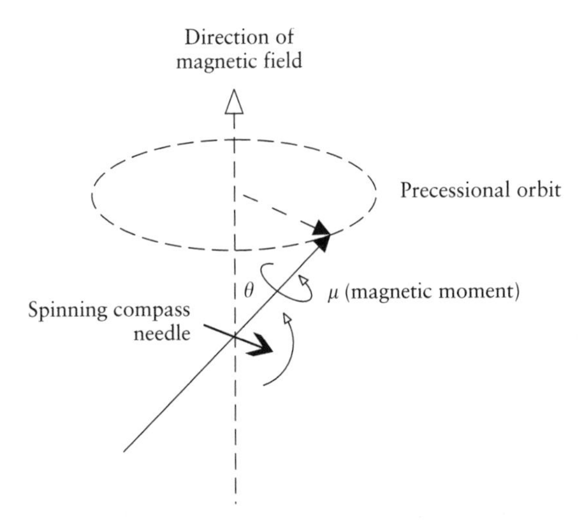

Figure 11.1 Precession of a magnet in an applied magnetic field.

magnetic field as depicted in Fig. 11.1. This motion is analogous to the motion of a gyroscope when displaced by some external force from the vertical whilst spinning under the influence of gravity. The angular frequency, ω_0, of this motion depends on both the strength of the magnetic field, B_0, and the ***magnetogyric ratio***, γ, for a particular magnet, and can be easily predicted according to Eqn (11.1):

$$\omega_0 = \gamma B_0 \tag{11.1}$$

We can now extend this model to the spinning of nuclei within magnetic fields. It follows that each orbit of spin of a nucleus has a particular energy and, in order to change this to another orbit, either absorption or emission of a certain quantity of energy is required.

Quantum mechanics dictates that nuclei spinning in a magnetic field will have $2I + 1$ possible orientations (and hence energy levels). A hydrogen nucleus (or the proton) has a value of $I = \frac{1}{2}$, so the nucleus of H has two possible energy levels. The energy difference between these energy levels may be predicted from Eqn (11.2):

$$\Delta E = \frac{\mu H}{I} \tag{11.2}$$

where μ is the magnetic moment of the spinning nucleus.

The energy difference between these differing energy levels is quantized since each separate energy level is fixed. The energy absorbed corresponds to the energy required to excite the nucleus from one energy level to another, and this may be calculated according to Eqn (11.3):

$$\Delta E = h\nu \tag{11.3}$$

where h is the Planck constant.

It follows that we can predict (by re-arranging Eqn (11.2)) which frequencies of the electromagnetic radiation will be absorbed by a given nucleus, Eqn (11.4):

$$\nu = \frac{\Delta E}{h} \tag{11.4}$$

The characteristic frequency at which nuclei absorb radiation is known as the ***Larmor*** frequency. It should be noted that this frequency depends on the strength of the magnetic field and that the Larmor frequency will be higher with stronger magnetic fields.

The ratio of the angular frequency of a nucleus, ω, to the strength of the static magnetic field, H, is known as the *gyromagnetic* or *magnetogyric* ratio, γ, Eqn (11.5):

$$\gamma = \frac{\omega}{H} \tag{11.5}$$

In practice, nuclei are irradiated with an electromagnetic field from an oscillator coil oriented at 90° to the fixed magnetic field. Irradiation causes circularly polarized radiation, which, in turn, causes the magnetic moment of the nucleus to flip. This process gives rise to an absorption. The frequency of the alternating magnetic field may typically range from a few MHz to 900 MHz, or more.

The energy difference between two different energy levels, as predicted by Eqn (11.2), is typically very small, meaning that the radiofrequency source need not be large, although the detector must be sensitive.

11.1.2 NMR emission spectroscopy

All nuclei capable of nuclear resonance will act as nuclear oscillators and, consequently, will also radiate energy. At the Larmor frequency, all the nuclei spin in phase and so collectively act as a coherent radiant source. This emitted radiation may be detected by placing a second coil perpendicular to the oscillator coil and the fixed magnetic field. Instruments of this type are designed to monitor resonant emitted energy and are known as two-coil instruments.

It is worth noting that it is customary to plot *both* absorbance and resonant emission NMR spectra as peaks above the baseline. A NMR spectrum itself does not directly inform the observer whether absorption or emission is being measured, but this should be obvious from the type of instrument being used.

11.2 Wide-line or low-resolution NMR spectroscopy

Wide-line (or low-resolution) NMR spectroscopy measures either the absorption or resonant emission of any isotope possessing spin quantum numbers of $I > 0$.

Figure 11.2 shows a simplified schematic of a wide-line NMR spectrometer. Samples are prepared in glass tubes and placed in the fields of the static and alternating magnetic fields. The absorption or emission of electromagnetic radiation is then monitored with an appropriate detector.

Spectra are obtained by sweeping the intensity of the alternating magnetic field from a strength of around 0–1 T (0–10^4 Gauss).

The spectrum produced in this way will possess a peak for each isotope for which $I > 0$; a typical spectrum of water containing trace quantities of copper and silicon is shown in Fig. 11.3. The relative abundance of the differing isotopes is reflected in the area under each peak. It, therefore, follows that integrating each area allows for quantitative determinations provided calibrations are performed with respect to standardized (e.g. certified) reference materials (see Chapter 2).

Wide-line (low-resolution) NMR has not been widely exploited, however, for quantitative determinations, which is partly due to the extremely high running costs and initial outlay associated with NMR instrumentation.

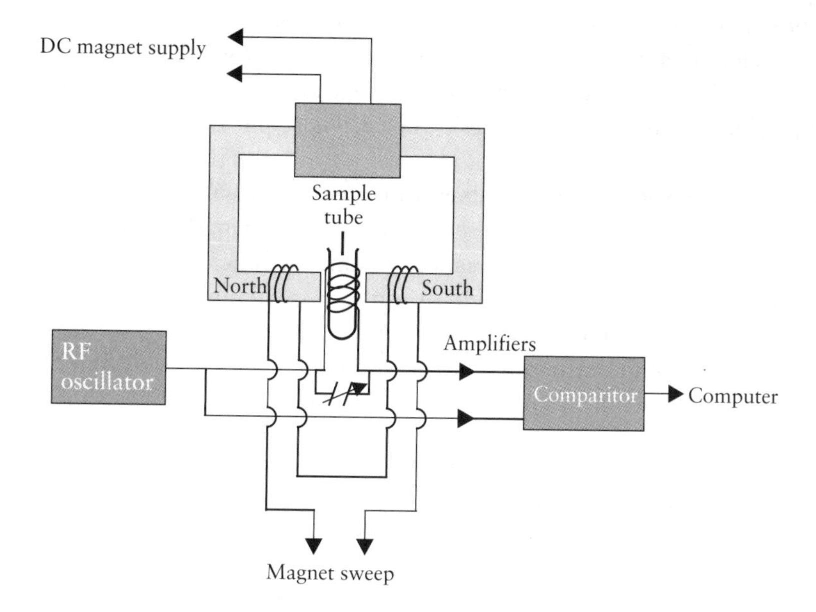

Figure 11.2 Schematic of a wide-line (low-resolution) single-coil NMR spectrometer.

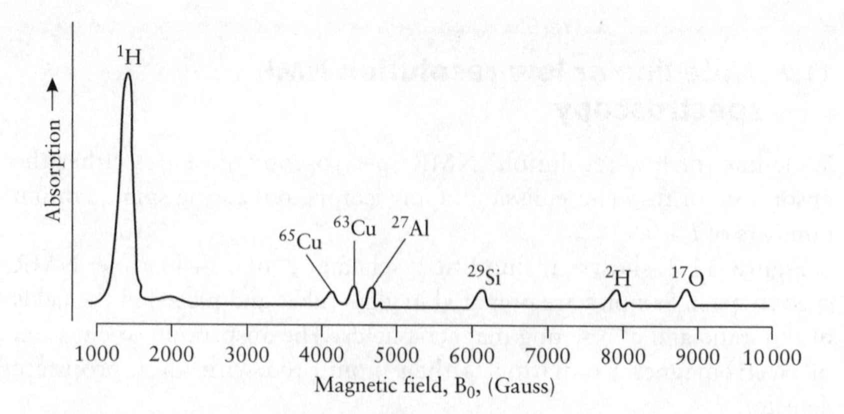

Figure 11.3 NMR spectrum of water containing trace quantities of copper, silicon, and aluminium.

11.3 An introduction to high-resolution NMR spectroscopy

NMR spectrometers are designed to monitor the nuclear magnetic resonance of a single nuclear species, since in practice this yields maximum chemical information. Any real sample will almost always contain many different atomic components. By far the most widely used target nucleus within NMR spectroscopy is the proton and we often refer to this approach as *proton NMR spectroscopy*. Proton NMR is particularly useful for determining the structure of organic compounds that ubiquitously contain hydrogen. NMR spectroscopy not only allows the quantification of a particular isotope, but also provides information relating to the chemical environment in which the nucleus resides. The information in the NMR spectrum that relates to the chemical environment of the nucleus resides is known as the ***fine structure*** of the NMR spectrum.

Nuclei are shielded to differing extents by electrons that either reside within orbitals associated directly with a particular atom or within bonding orbitals that are partially de-localized. These shielding effects give rise to the fine structure of an NMR spectrum. It follows that if similar nuclei reside within slightly differing chemical environments they are shielded by varying amounts and so different magnetic field strengths will be needed to bring these nuclei into resonance. Resonance is achieved by sweeping the intensity of the magnetic field over a very small range to bring each of the nuclei of a given isotopic configuration into resonance.

Two types of fine structure may be observed in a high-resolution NMR spectrum: the chemical shift and spin–spin coupling effects. We shall now consider each of these in turn.

11.3.1 The chemical shift

We have just seen that different nuclei experience differing shielding effects due to the cloud of electrons that surrounds them. The ***chemical***

shift, δ, describes the change in frequency of the magnetic field that must be applied in order to bring specified nuclei into resonance with the applied field. The chemical shift represents the *ratio* of the change in magnetic field with respect to the magnetic field required to bring nuclei with no shielding effects into resonance; as such the chemical field is a dimensionless parameter and is normally expressed in terms of parts per million (ppm) for the required change in the applied field.

It is impossible to obtain free protons that experience no shielding from electrons; it is, therefore, necessary to assign a standard environment for an electronic (chemical) environment for protons against which the chemical shift may be measured. The protons in ***tetramethylsilane (TMS)*** are normally used for this purpose since all 12 protons reside in a chemically equivalent environment. A small quantity of TMS is, therefore, often added to samples prior to their analysis. A peak at $\delta = 0$ in these cases corresponds to the protons of TMS. Different nuclei reside in differing chemical environments and experience differing degrees of electronic shielding and, therefore, possess individual chemical shifts in an NMR spectrum.

The chemical shift of a spectrum is expressed along the x-axis of an NMR spectrum. We recall that different chemical shifts correspond to nuclei residing in different chemical environments in a particular molecule. Examples here could include protons in CH_2 or CH_3 groups in a molecule of ethyl iodide.

Before we can interpret how individual peaks are represented in an NMR spectrum, we need to consider the effects of electronic spin–spin coupling on the fine structure of NMR spectra.

11.3.2 Spin–spin coupling

Spin–spin coupling is a process in which the spinning nuclei of two or more atoms (in different chemical environments) interact to generate a further feature of the fine structure of the NMR spectrum. The effects of spin–spin coupling are seen by the splitting of a peak at a given chemical shift to give a multiplet peak.

The protons in one functional group undergo mutual spin–spin coupling with the protons of an adjacent functional group. There is a simple rule for predicting how spin–spin coupling between adjacent functional groupings will split peaks into multiplets: *If a functional group, A, with N protons undergoes spin–spin coupling with another (or multiple) proton containing group(s), then the NMR peak due to the protons on neighbouring groups will be split into a multiplet containing* N + 1 *peaks*. This rule holds for all functional groups in a molecule, so the effects of all spin–spin interactions by adjacent groups should be considered if we are either going to predict the presence of peaks that should be seen within a

NMR spectrum for a particular molecule, or to interpret spectra using the number of peaks within a multiplet.

If two nuclei reside in equivalent chemical environments, then the peak due to these nuclei will not be split and it follows that spin–spin coupling effects are only seen in NMR spectra if protons residing in differing chemical environments are present.

The $N+1$ spin–spin coupling rule is a consequence of the number or orientations in which protons of a particular field can align themselves relative to the magnetic field. These effects are, however, evidenced by the splitting of NMR peaks that correspond to the protons of an adjacent proton-containing group. The detailed reasoning of why this occurs is embedded in quantum theory and is beyond the scope of this book.[1] We shall consider how we arrive at $N+1$ orientations for the alignment of protons with the magnetic field.

Let us now consider a full high-resolution NMR spectrum of a simple representative molecule. We shall choose ethyl iodide in the first instance, CH_3CH_2I (a ^1H NMR spectrum of which is shown in Fig. 11.4) since it represents one of the simplest molecules to posses protons in differing chemical environments that are capable of undergoing spin–spin coupling with each other. This molecule possesses five protons in total within its methyl (three proton) and methylene (two proton) groups. The ^1H NMR spectrum consists of two groups of peaks (the peak at 0 ppm is due to the tetramethylsilane—TMS standard). In this example, the protons of the CH_3 group couple with the protons of the CH_2 group, and so the CH_3 peak is split into a triplet since $N+1 = 2+1 = 3$. The CH_2 peak is split into a quartet (i.e. $N+1 = 3+1 = 4$).

Integration curves are also shown in Fig. 11.4. The heights of each describe the relative areas under the peaks. We can thus see that the total

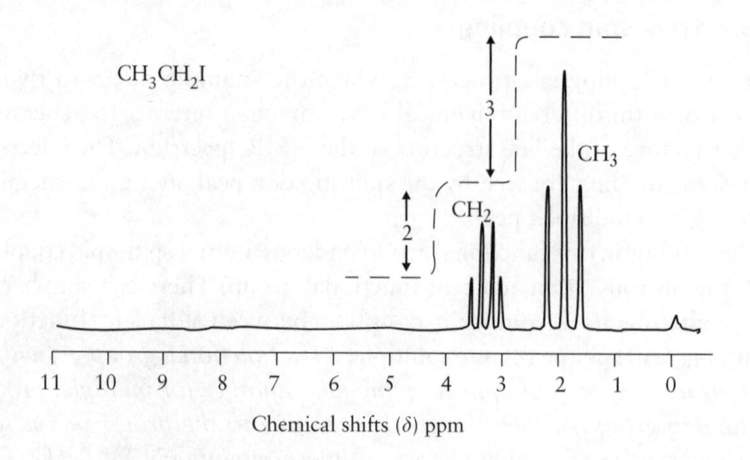

Figure 11.4 NMR spectrum for ethyl iodide (CH_3CH_2I).

[1] Refer to Atkins, P.W. and Friedman, R.S. (1997). *Molecular Quantum Mechanics* (3rd edn). Oxford: Oxford University Press.

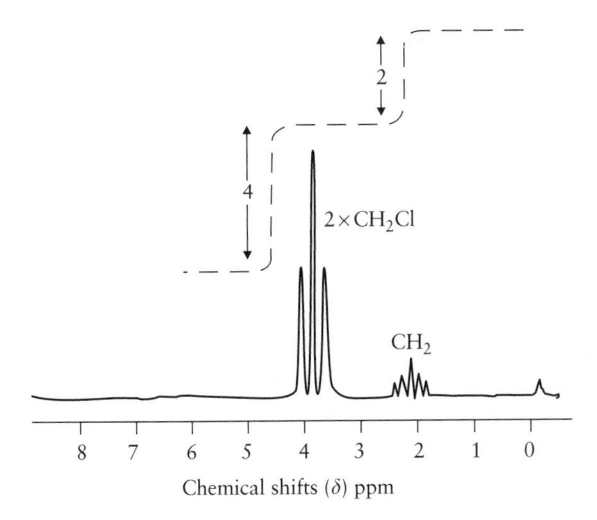

Figure 11.5 NMR spectrum of 1,3-dichloropropane.

area for the CH_3 protons is 1.5 times that for the CH_2 protons. The areas of these peaks are, therefore, in the ratio 3 : 2, reflecting the number of protons in each functional moiety.

It follows that if we can predict the structure for an NMR spectrum, then we ought to be able to interpret NMR spectra with regard to both the assigning of protons in functional groups as well as the relative positioning of these groups in a molecule.

Another compound we shall consider is 1,3-dichloropropane, $ClCH_2CH_2CH_2Cl$, the spectrum of which is shown in Fig. 11.5. We can see the protons form two groups, the central methylene group and the two chloromethyl groups that are identical. The spectrum shows two groups of peaks in the expected (4 : 2) or (2 : 1) ratio. The group centred at 3.9 ppm is due to the CH_2Cl groups and is split by the two protons of the methylene group into a triplet. The methylene (CH_2) group centred at 2.1 ppm is split by four equivalent protons into a quintet.

11.4 NMR instrumentation

11.4.1 Frequency and magnetic field control devices

Since an exact field-to-frequency ratio is required to bring a particular nucleus to resonance, any small variations in either the strength of the magnetic field or the frequency of oscillation must be corrected for.

The magnetic field into which the sample is placed must be extremely uniform and normally maintained to within 1 part in a million per MHz frequency; it follows that a 100 MHz machine should be able to produce a homogeneous magnetic field to within 1 part in 100 million. This is achieved in practice by using *shimming coils*, which take the form of auxiliary windings around the main electromagnet. A DC current is passed through the shimming coils to counteract any inhomogeniety in the magnetic field. Samples are also often spun in the magnetic field to overcome additional problems associated with the fluctuations in the magnetic field caused by the sample holder or the sample itself.

The alternating excitation in the magnetic field is generated using a radio-frequency crystal oscillator. The resolution obtainable increases with the frequency of oscillation; however, higher frequencies necessitate the use of stronger magnetic fields, which, in turn, increases the cost of the instrumentation. The simplest high-resolution NMR machines operate at 60 MHz whereas 900 MHz or more may be obtainable in state-of-the-art instrumentation.

An automatic frequency controlling mechanism known as a *frequency lock* is normally included in the form of specialized circuitry and/or software. The frequency lock may either take reference to the same nucleus under investigation (known as a *homonuclear locking approach*) or, alternatively, may make reference to a different nucleus, in which case the approach is said to be using a *heteronuclear lock*. The frequency lock must take a reference from nuclei from some molecular species within the sample (known as an *internal lock approach*) or, alternatively, may be taken from a separate reference sample (which is known as using an *external lock* approach).

11.4.2 Double resonance spin decoupling of complicated spectra

Spin–spin coupling techniques as described in Section 11.3.2 greatly help in the identification of resonances. This approach can, however, lead to spectra that are so complicated that they become unintelligible, especially in the case of larger molecules. In cases such as these, a spin decoupler is often used. This device consists of an auxiliary coil through which an alternating current is passed to generate a radio-magnetic field that is superimposed on the stimulating magnetic field. The second signal is tuned to the resonant frequency of one set of the coupled protons and causes the undesired interfering protons to rapidly equilibrate between their two energy levels ensuring that the protons being observed cannot differentiate between the separate states. In this way, the secondary magnetic field causes the multiplet in the spectrum (due to the other set of coupled protons) to collapse—thereby leaving a sharp identifiable peak of the same area as the original (multiplet) peak.

11.4.3 NMR signal and resolution enhancement via the nuclear Overhauser effect

The ***nuclear Overhauser effect*** describes a phenomenon that occurs as a result of another form of double resonance and can be induced by means of an auxiliary coil. Excited nuclei naturally relax with time. As they do so, the NMR signal they generate diminishes; in practice, the nuclei are continually exited and relax and in this way a steady-state signal can be recorded. If one of a pair of closely spaced and spin–spin coupled nuclei is saturated, then the rate of relaxation of the other diminishes. In this way, the magnitude of the signal may be enhanced. This increase in the magnitude of specific peaks may clearly help to increase the signal-to-noise ratio of specific parts of the spectrum, thereby facilitating its identification. The nuclear Overhauser effect diminishes with an inverse sixth-power relationship between the distance between two spin–spin coupled nuclei. The effect can, therefore, also be used to help identify peaks due to nuclei in close proximity to each other and this, in turn, may help elucidate the structure of the most complicated of molecular structures.

11.5 Fourier transform NMR

Fourier transformation NMR spectroscopy involves irradiating a sample with a wide range of frequencies (the so-called white radiation) to excite all possible resonances simultaneously. A Fourier transform mathematical operation is then used to transform the signal from a time-dependent to a frequency-dependent domain. In practice, samples are exposed to a short burst (or pulse) of radio-magnetic energy, which is equivalent to a band of radiation centred around a particular frequency—the bandwidth depends on the duration of the pulse with; for example, a 10 μs pulse covering a band width of about 10^5 Hz. Exited nuclei stimulated in this way emit radiation at their characteristic Larmor frequencies as they relax via a process known as ***Free Induction Decay*** (FID). A probe coil monitors emissions between each successive pulse. In practice, multiple signals are collected for each sample for signal averaging prior to Fourier transform signal processing.

11.6 NMR shift reagents

NMR shift reagents are used to help identify the location of specific protons that belong to specific moieties in a molecule. Their use may be extremely useful for the structural elucidation of more complicated molecules.

One common approach involves using either Eu or Pr containing organic reagents to form a Lewis acid type complex with functional moieties that contain lone pairs of electrons. The effect of reagents of this type is to shift downfield the resonance of protons thereby providing structural information relating to the molecule as well as helping to resolve overlapping peaks where these occur.

11.7 NMR of nuclei other than the hydrogen (proton) nucleus

The principles of NMR spectroscopy when applied to elements other than hydrogen are exactly the same as those already encountered. Apart from hydrogen, 18 other elements have naturally occurring nuclei with $I = \frac{1}{2}$, which could, in theory, lend themselves to NMR analysis. In practice, however, only ^{19}F, ^{31}P, and ^{13}C are commonly utilized due to the lower sensitivities associated with many other isotopes as well as their relative abundance in nature.

In general, the chemical shifts increase with atomic number, which allows spectra to be obtained with instruments of lower resolution than would be required for proton NMR spectroscopy. (This does depend on the relative abundance of a particular element in a sample.)

^{19}F NMR spectroscopy may be performed in a near identical manner to that of ^{1}H NMR spectroscopy—albeit with a lower sensitivity. ^{19}F NMR techniques allow the analysis of: (i) fluorine-containing organic compounds within complex mixtures; or (ii) the identification of 'labelled' compounds that have been deliberately reacted with fluorine-containing reagents; this again may be especially useful for the analysis of organic compounds in complex mixtures.

^{31}P NMR spectroscopy offers lower sensitivity than either ^{1}H or ^{19}F NMR approaches, although in all other ways the technique is similar. ^{31}P NMR spectroscopy is especially useful for the analysis of phosphorous-containing compounds such as phosphates, thiophosphates, and phosphines.

^{13}C NMR spectroscopy, unfortunately, offers extremely low sensitivity due to the low natural abundancy of ^{13}C within the environment, although modern instrumentation is now allowing ^{13}C NMR to be used for quantitative determinations in some instances.

^{2}D, ^{15}Na, and solid-state NMR may also all be used for analytical purposes but are beyond the scope of this book.

11.8 Analytical applications for NMR

It should be stated at the outset of this section that the primary applications for NMR are normally for structural identification and confirmation, and not analytical chemistry *per se*, although NMR can be used as a useful tool for both qualitative and quantitative determinations.

11.8.1 Qualitative determinations—structural determination

NMR spectroscopy is typically used for qualitative analyses in one of two ways. First, for applications in which the presence of a particular compound needs to be identified in a mixture. Second for structural identification of specific functional groups or elucidation of the entire structure of a compound.

11.8.2 Quantitative determinations—methodology and practical applications

NMR uniquely provides signal peaks that are directly proportional to the number of nuclei from which they originate. For this reason, quantitative determinations do not require pure samples. Having said this, the more complicated a sample mixture becomes, the greater will be the probability for overlapping of some or all of the peaks and this can ultimately render a spectrum so complicated as to be unintelligible.

The signal area per proton can be easily identified in many cases, however, via the use of an internal standard such as (tetrachloromethane or cyclohexane) being added to the sample at a known concentration. The signal due to the standard should obviously not overlap with that of the analyte, and for this reason, silicon derivatives are often chosen for use as internal standards due to the high upfield location of their proton peaks.

If a suitable internal standard is chosen, then the areas of the peaks for such a compound may be used to determine the concentration of the analyte directly.

One of the greatest problems associated with analytical applications for NMR spectroscopy relates, unfortunately, to the running costs associated with instruments of this type. Economic considerations become particularly relevant if less-expensive approaches can suffice (see Chapter 16—relating to critical choice of technique, where factors such as these are discussed further).

Quantitative analysis of multi-component mixtures

NMR may allow for quantitative determinations of a number of different components in relatively simple mixtures. NMR spectra of pure samples of the suspected components are normally first run to individually identify the peaks due to each of the proton environments in the sample, prior to recording the composite NMR spectrum of the analyte mixture. The key to such determinations is that at least one uniquely identifiable peak may be established for each compound in order to permit qualitative determinations.

Quantitative elemental analyses in mixtures

NMR spectroscopy can be used to quantify the total concentration of a particular nucleus in a sample. ^{1}H NMR, for example, allows the analyst to determine the total hydrogen content in an organic compound or sample of unknown origin or composition. ^{19}F NMR techniques may, in a similar way, be used to determine the fluorine content in organic fluorocompounds, which is especially useful since quantitative fluorine compositional analyses of this type invariably prove to be particularly difficult and troublesome via almost any other approach.

Quantitative analysis of functional groups in organic samples

NMR can be used to quantify the total concentration of functional moieties such as hydroxyl or carboxyl groups in a sample. Analyses of this type are always easier if the sample only contains one molecular species from the particular class of compound in question (e.g. an alcohol or a carboxylic acid), since the protons from the appropriate functional group will all reside in similar chemical environments. Quantitative determinations of this type may also be possible with samples containing a mixture of different molecular species from a particular class of compound—provided the identity of each of the components can be established in order to permit known standard sample concentrations to be quantitatively compared with unknown samples.

Exercises and problems

11.1. Two bottles, both labelled trichloroethane, contain liquids that have different boiling points. When their NMR is checked, each bottle gives a different spectrum.

Bottle A has two groups of peaks, a doublet at 4.0 ppm (relative intensity 2) and a triplet at 5.9 ppm (relative intensity 1).

Bottle B gives a much simpler spectrum, with only a singlet at 2.9 ppm being visible.

Identify the formula for two compounds.

11.2. In NMR spectrometry, explain what are the advantages of using a magnet with as large a field strength as possible.

11.3. Sketch the expected form of a high-resolution NMR spectrum for: (i) acetic acid; (ii) acetone; (iii) cyclohexane.

11.4. Explain what is meant by a frequency lock system for an NMR spectrometer.

11.5. What are shims in NMR spectrometers and what are they used for?

11.6 How could NMR be used to distinguish the three isomers of pentane (C_5H_{12})?

11.7 A compound has been isolated from a reaction and shown to have the empirical formula C_2H_4O. The NMR spectrum shows peaks at 1.3 ppm (triplet), 2.0 ppm (singlet) and 4.1 ppm (quartet) in the ratio 3:3:2. Give possible structures for the compound.

Summary

1. Nuclear magnetic resonance (NMR) is an instrumental technique based on monitoring how spinning nuclei with magnetic dipoles interact with applied magnetic fields and absorb radiation.

2. Nuclei can only absorb energy from an electromagnetic field when they possess a magnetic dipole moment. Magnetic dipoles occur when the spin quantum number, I, is not an integer.

3. Magnetic dipoles of nuclei interact with a static magnetic field, H, by precessing in a manner analogous to a gyroscope spinning in a gravitational field.

4. The angular frequency, ω_0, of the motion depends on both the strength of the magnetic field, B_0, and the magnetogyric ratio, γ, for a particular magnet as described by: $\omega_0 = \gamma B_0$

5. Frequencies of electromagnetic radiation, ν, that will be absorbed may be predicted by: $\nu = \Delta E/h$, and are known as Larmor frequencies.

6. Frequencies of alternating magnetic fields range from a few MHz to 900 MHz or more.

7. Wide-line or low-resolution NMR spectroscopy measures the absorption or emission of isotopes possessing spin-quantum numbers of >0. This is acheived by sweeping the intensity of an alternating magnetic field from a strength of 0–1 T or more.

8. High-resolution NMR is more versatile than low-resolution NMR spectroscopy and finds applications for structural identification of unknown compounds.

9. The resonance of a number of differing nuclei may be exploited, although proton (^1H) NMR spectroscopy is the most widely used.

10. A feature of high-resolution NMR spectroscopy is the splitting of peaks due to interactions with nuclei on neighbouring functional groups; this is known as the fine structure of the spectrum.

11. The chemical shift, δ, describes the change in frequency of the magnetic field that may be applied in order to bring specific nuclei into resonance with an applied field.

12. Chemical shifts are often measured with respect to standards. The four protons of tetramethylsilane are normally used for this purpose. The peak due to TMS is normally designated a $\delta = 0$.

13. Spin–spin coupling describes the interaction of protons in two or more chemical environments (normally on neighbouring functional groups in a molecule) to give rise to the fine structure of the NMR spectrum.

14. If a functional group, R, with protons undergoes spin–spin coupling with another (or multiple) proton possessing N protons, then the NMR peak due to the protons on the R group will be split into a multiplet with $N + 1$ peaks.

15. Integration peaks describe the relative areas under NMR peaks and this may be related to the number of protons in a particular chemical environment.

16. Frequency and magnetic field control devices such as shimming coils are often used in NMR instrumentation.

17. The nuclear Overhauser effect describes a phenomenon that occurs as a result of another form of double resonance induced via the use of an auxiliary coil.

18. NMR spectroscopy can be performed using nuclei such as $^{2}H(D)$, ^{13}C, ^{15}Na, ^{19}F, and ^{31}P.

19. Analytical applications of NMR spectroscopy include: (i) quantitative analyses of multi-component mixtures; (ii) quantitative elemental analyses in mixtures; and (iii) quantitative analyses of functional groups in organic compound containing samples.

Further reading

Abraham, R. J. and Fischer, J. (1988). *Introduction to nuclear magnetic resonance spectroscopy*. Wiley.

Callaghan, P. T. (1993). *Principles of nuclear magnetic resonance microscopy*. Clarendon Press.

Hore, P. J. (1995). *Nuclear magnetic resonance*. Oxford Chemistry Primers, Oxford University Press.

Jackson, L. M. (1969). *Applications of nuclear magnetic resonance spectroscopy in organic chemistry*. International Series of Monographs in Organic Chemistry, Pergamon Press.

Infrared techniques 12

Skills and concepts

This chapter will help you to understand:

- The origins of molecular oscillations and how these can give rise to absorptions within the infrared (IR) region of the electromagnetic spectrum.
- How to calculate the number of degrees of freedom for given molecular oscillations.
- The convention of using wavenumbers instead of wavelengths.
- How the presence of some functional moieties may be inferred by absorption at characteristic wavenumber regions.
- The use of differing IR sources with IR spectroscopy.
- The advantages and disadvantages associated with the use of a number of IR detectors including thermopiles, bolometers, as well as pyroelectric and photo-conducting detectors.
- The operation of conventional dispersive grating-based and Fourier transform multiplex IR instrumentation.
- How gas sampling approaches are used for the IR analysis of gaseous samples.
- The use of IR emission approaches for analytical applications.

12.1 Introduction

IR radiation forms part of the electromagnetic spectrum and covers a range of wavelengths from approximately 800 to 1 000 000 nm (0.8–1000 μm), although analytical IR techniques in practice normally only exploit radiation in the wavelength range of 2500–16 000 nm (2.5–16 μm).

IR analytical spectroscopy utilizes the molecular absorption of radiation and in this way shows similarities to UV–visible spectroscopy although this is where the comparison ends.

UV–visible spectroscopy can be described using a photon capture model (Chapter 4) in which photons of exactly the correct energy are adsorbed by molecules to excite electrons from one quantized energy level to another. The model we shall use to describe the absorption of IR radiation by molecules is based, however, not on the particulate nature of electromagnetic radiation, but rather on its more wave-like properties.

IR wave-like radiation can be thought of as passing through a molecule in which the individual atoms or functional groups are free (to a greater or lesser extent) to move, flex, or vibrate around their molecular bonds. In this way, atomic motion in molecules occurs in a manner that can be thought of as being analogous to the motion of a spring or guitar string that will vibrate at a specific and predictable frequency. Molecules oscillate in a predictable manner around molecular bonds due to electromagnetic dipoles within the molecule; these dipoles are formed as a result of atoms of differing mass being bonded together via valence electrons residing within bonding orbitals.

If the wavelength of the incident IR radiation coincides with the frequency of oscillation of a molecular bond, then resonance between the bond and the incoming radiation absorption can occur. When this occurs, energy can be imparted from the radiation to the molecule and in this way absorption occurs. The frequency of the molecular vibration is not altered as energy is absorbed, although the amplitude of the molecular vibration *does* change. It is in this context that the absorption of IR radiation may be thought of as being analogous to a guitar string absorbing energy as it is plucked.

Molecular oscillation can occur via the simple ***stretching and contraction*** of molecular bonds as well as ***rocking, wagging, scissoring, or twisting*** motions of functional groups (see Section 12.3).

Since each motion involves oscillation at a specific frequency, absorption can help characterize the type of molecular oscillation occurring and so often the functional groups present in a molecule. IR spectroscopy can, therefore, be used to help with structural identification by organic chemists, and this probably represents it largest application within chemistry (see Section 12.4).

The absorption of IR radiation follows the Beer–Lambert law (Chapter 4) and, for this reason, the absorption is directly related to the concentration of specific bonds (or functional groups) within a sample and, hence, to the concentration of specific analytes. Unfortunately, the molar absorptivities for the absorption of IR radiation via molecules are typically very much smaller than those encountered for the absorption of UV or visible light, and this normally prevents IR techniques being routinely used for the determination of trace compounds.

12.2 Presentation of IR spectra

By historical convention IR spectra are normally displayed in a somewhat different manner to UV–visible spectra. An IR spectrum of octane is shown in Fig. 12.1, which we shall use to illustrate our discussion. The y-scale (ordinate) of an IR spectrum is normally plotted in terms of percentage transmittance instead of absorbance. This means, in practice, that the baseline of spectra lie towards the top of the spectrum (maximal transmittance), and IR absorption peaks are directed down the page.

You will see also that the x-scaling is not displayed in terms of either wavelength or frequency but rather 'wavenumbers'. Wavenumbers represent the reciprocal of the wavelength ($1/\lambda$) and have units of cm^{-1}. The use of the wavenumber rather than the wavelength owes its origin to historical convention, but does allow for the use of rather more manageable numbers. It should be noted that increasing wavenumbers correspond to increasing frequency and, therefore, to progressively more energetic radiation.

12.3 Molecular oscillations and vibrational coupling of differing oscillations

We have already briefly described in Section 12.1 how individual atoms within molecules are not rigidly fixed relative to their neighbours but, in fact, oscillate and move via flexibility in their molecular bonds. The frequency at which a molecular oscillation takes place will be largely governed by: (i) the nature of individual bonds within a molecule together; and (ii) the types of moiety that are directly associated with the oscillation.

As we have already seen, molecular oscillations or vibrations can be classified as typically involving stretching, bending, and a number of other motions. Stretching motions involve continuous change of the interatomic

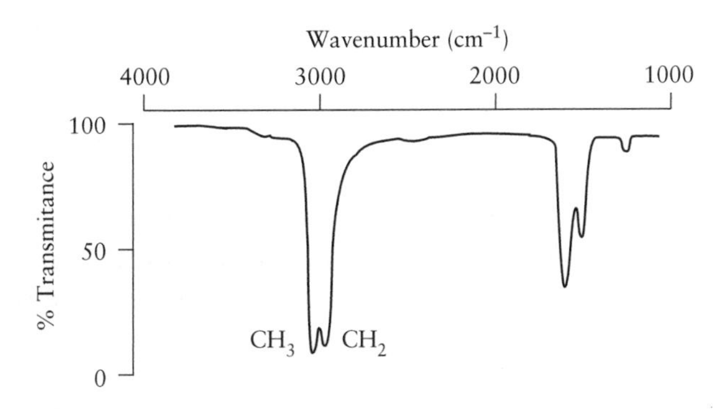

Figure 12.1 IR spectrum for octane.

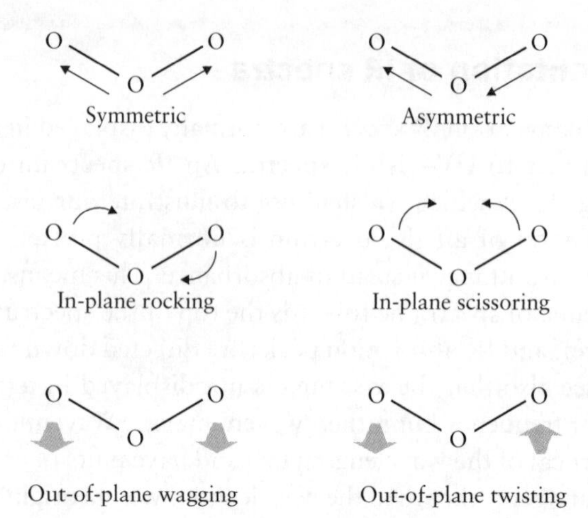

Figure 12.2 A number of differing bending molecular oscillations.

distance along the axis of the bond, between two atoms or groups within a molecule. By contrast, bending vibrations involve changes in the angle between bonds; the main types of bending oscillations are known as ***scissoring, rocking, wagging***, and ***twisting*** vibrations as depicted in Fig. 12.2.

Since molecular oscillations occur in three dimensions, different oscillations can couple to produce further oscillatory motions that themselves can give rise to absorption of IR radiation at characteristic wavelengths (wavenumbers). Some vibrations can also occur at overtone frequencies (as simple multiples of the fundamental vibrations), while the overtone frequencies can, of course, themselves couple to give further resonant frequencies.

An overtone frequency is a multiple of the fundamental frequency, that is, $\times$ 2, 3, 4, . . ., the lowest frequency of vibration.

The number of modes of vibration of a *non*-linear molecule of *n* atoms can be predicted to be equal to $3n - 6$ and for a linear molecule $3n - 5$ modes of vibration.

These values are calculated according to the following argument: three co-ordinates are required to locate the position of each atom within a polyatomic molecule—and since each co-ordinate defines one degree of freedom we now have theoretical $3n$ directions of possible movement. Some degree of freedom must be subtracted from the total since they will be accounted for by other modes of movement. Three degrees of freedom are taken up by movement of the centre of gravity of the whole molecule in space (translational motion), and this now gives us a revised total of $3n - 3$ modes of vibration. A further three degrees of movement must also be subtracted to account for the rotation of the entire molecule around its centre of gravity: we now have a further and final total of ***$3n - 6$ modes of freedom for a non-linear molecule***. The linear molecule represents a special case; one mode of vibration must be omitted since rotation around the central bond(s) axis is not possible and this now gives us a new ***total of $3n - 5$ for a linear molecule***.

In practice, there are often fewer peaks observed than would be predicted that can be accounted for by four main factors as described below:

1. Fewer peaks are observed when the symmetry of the molecule is such that no change in a dipole occurs via one or more possible molecular oscillations.
2. If a molecule can exhibit two or more differing vibrations that are nearly identical to each other in energy, then IR absorption peaks for these processes can sometimes merge and appear to be one.
3. If a molecular vibration gives rise to an extremely small absorption, then in some instances it may be too small to be detected.
4. The absorption of a peak may not be monitored if it is either outside the wavenumber (wavelength) range being monitored or beyond the range of the instrument.

The frequency of vibration, and hence the wavenumber corresponding to the absorption, can be influenced by other molecular vibrations within the same molecule that interfere with or 'couple' with neighbouring vibrational oscillations. The extent of the coupling of two molecular vibrations is influenced by a number of factors, and these include:

1. Coupling of two vibrational modes, although this requires a common bond to exist between the vibrating groups if this is to be significant.
2. Little or no coupling occurring between groups due to separation by two or more bond lengths.
3. Vibrational coupling—and this tends to be strongest when there is an atom common to the two vibrational moieties.
4. Coupling occurs between a bending vibration and a stretching vibration if the stretching bond forms one side of the angle that alters during the bending motion.
5. An increase in coupling when moieties have individual energies that are almost equal in energy to each other.
6. For coupling to occur, vibrations must be of the same class of symmetry, as defined by group theory.[2]

12.3.1 Examples of IR absorption by simple molecules

In order to examine how absorption processes such as coupling effects are expressed within IR molecular absorption spectra, we shall now consider two simple examples.

[2] For a treatment of group theory, see Atkins, P.W. and Friedman, R.S. (1997). *Molecular Quantum Mechanics* (3rd edn). Oxford: Oxford University Press.

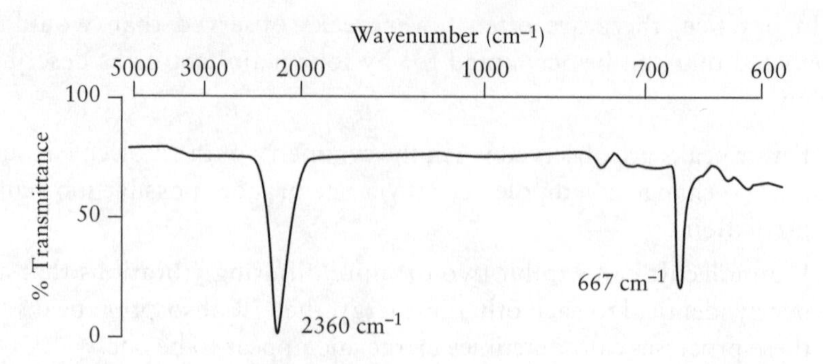

Figure 12.3 IR spectrum for CO_2.

Example of absorption of a linear tri-atomic molecule—CO_2

We shall consider the carbon dioxide molecule, CO_2, as an example of a molecule that exhibits IR absorption coupling effects. CO_2 (O=C=O) is a linear molecule and so in the absence of coupling effects the molecule would be expected to exhibit four normal modes of vibration ($3n - 5$) (see Section 12.2)—and, therefore, four absorption peaks. Experimentally, CO_2 only exhibits two absorption peaks at 667 and 2360 cm^{-1}. The reason for this discrepancy is as follows: There are two stretching vibrations that can occur, as depicted in Fig. 12.3. One of these is symmetric whereas the other is asymmetric. It is only the asymmetric stretching, however, that changes the dipole of the molecule and, therefore, is capable of giving rise to an absorption peak (at 2360 cm^{-1}).

Two scissoring vibrational modes are also possible, however, these are energetically equivalent to each other and said to be ***degenerate***.

Example of absorption of a non-linear tri-atomic molecule—H_2O

Water, H_2O, is a non-linear molecule since the two hydrogen atoms are linked to the oxygen via only one molecular bond each. It follows that the two hydrogens are not in line with each other, which gives rise to a permanent dipole within the molecule. Theory predicts that the molecule will possess three normal modes of vibration ($3 \times 3 - 6$) (see Section 12.2). In this example, three absorption peaks are indeed observed, corresponding to one symmetric stretching, one asymmetric stretching, and one scissoring vibrational mode of vibration. The scissoring vibrational mode gives rise to an absorption peak at 1595 cm^{-1}, the symmetric stretching of the molecule gives rise to an absorption peak at 3650 cm^{-1}, and the asymmetric stretching vibrational mode gives an absorption peak at 3760 cm^{-1}, Fig. 12.4.

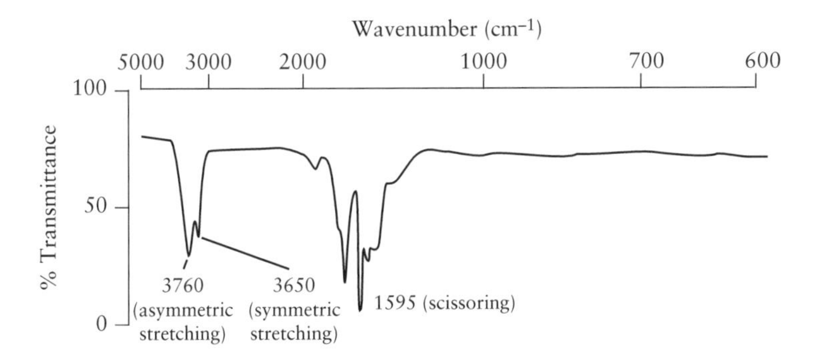

Figure 12.4 IR spectrum for H_2O.

12.4 Characteristic absorption spectra for commonly encountered functional groups and normal vibrational modes

We have already seen that the stretching, scissoring, or bending of molecular bonds in many cases gives rise to IR absorptions at characteristic frequencies. Most of these absorptions are found within a wavenumber range of 1250–3600 cm^{-1}, which, for this reason, is known as the ***group frequency region***. The exact energy of the vibration is often modulated via coupling effects, and this gives a *range* of wavenumbers over which absorptions can occur and are known as the ***group frequency regions*** of the spectra. The identification of absorption peaks can often be used to help identify a class of molecule; for example, alcohol, aldehyde, etc. The exact wavenumber at which an absorption occurs is, therefore, very often highly specific to an individual molecule and this may help in the identification of unknown compounds.

Characteristic absorption regions for molecular vibration are tabulated in ***correlation charts*** to aid structural identification of spectra relating to unknown chemical species. A number of commonly encountered molecular vibrations, together with the relevant molecular groupings and the group frequency regions for absorptions, are shown in the correlation chart of Table 12.1.

Absorption peaks within the group frequency region often overlap or shoulder each other, and this can unfortunately complicate structural elucidation and identification. Further variations in the spectrum can moreover also occur depending on: (i) how a sample is prepared (e.g. within a mull, pellet, powder, etc.); and (ii) whether the sample is in the gaseous, liquid, or solid form. It is rarely possible to identify a compound unambiguously from the identification of peaks from correlation coefficients alone—although confirmation may often be achieved via the fingerprint region of the spectrum (see Section 12.5).

Table 12.1 Table of group frequencies for some molecular vibrations associated with commonly encountered organic bonds and groups

Bond	Molecular bond/ grouping	Group frequency absorptions (wavenumbers, cm^{-1})
C–H	Alkanes	2850–2970 and 1340–1470
C–H	Alkenes	3010–3095 and 675–995
C–H	Alkynes	3300
C–H	Aromatic rings	3010–3100 and 690–900
O–H	Monomeric alcohols and phenols	3590–3650
O–H	Monomeric carboxylic acids	3500–3650
O–H	Hydrogen bonded carboxylic acids	2500–2700
N–H	Amines and amides	3300–3500
C=C	Alkenes	1610–1680
–C≡C–	Alkynes	2100–2260
–C–N	Amines and amides	1180–1360
–C≡N	Nitriles	2210–2280
–C–O (with two further single bonds on C)	Alcohols, carboxylic acids, ethers, and esters	1050–1300
–C=O (with one further single bond on C)	Aldehydes, ketones, carboxylic acids, and esters	1690–1760
$–NO_2$	Nitro compounds	1500–1570 and 1300–1370

EXAMPLE 12.1

A chemist attempted to synthesize benzoic acid by the oxidation of benzyl alcohol. However, a mixture of products was obtained and when chromatography was used to separate them, three organic compounds were obtained. The compounds gave the spectra below. Attempt to identify the compounds.

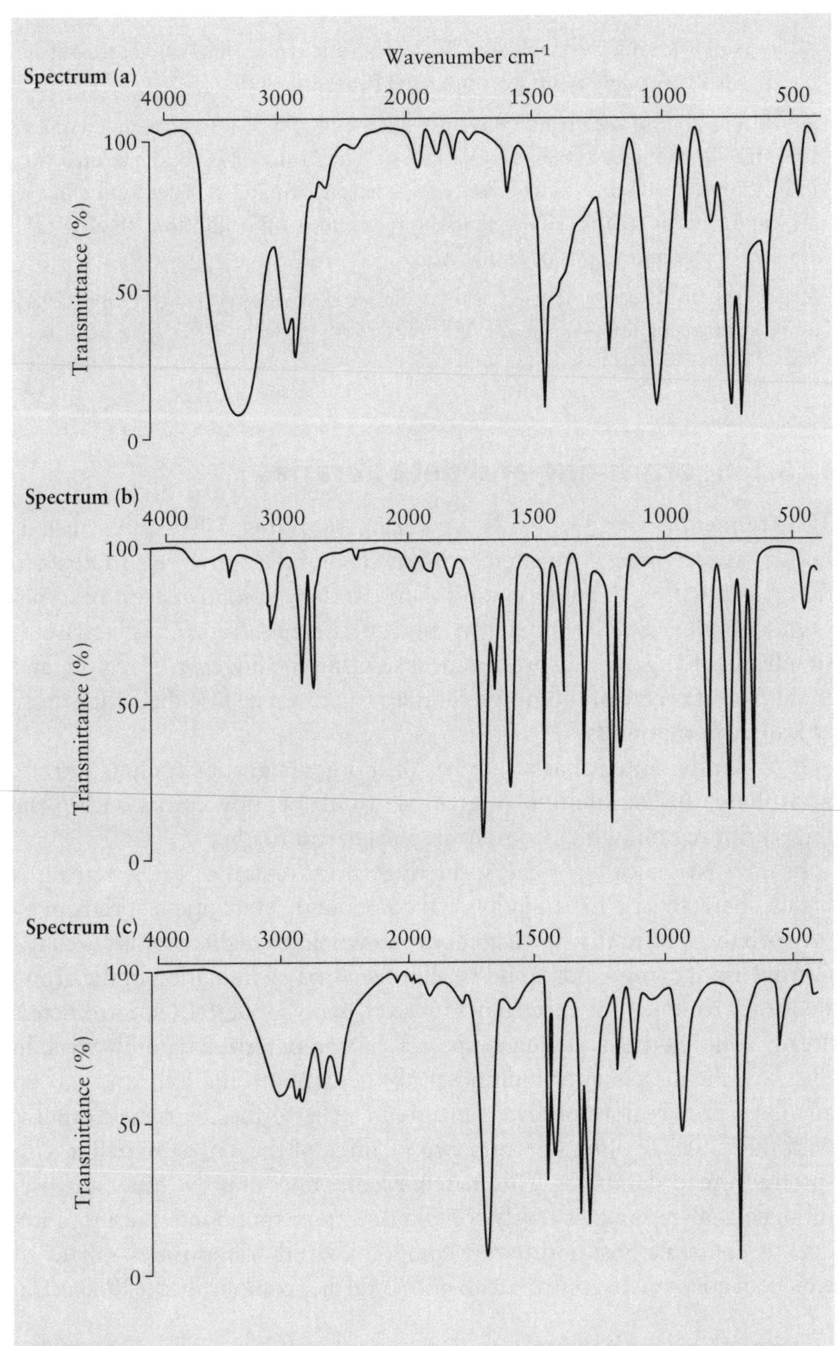

Method

Step 1: The first spectrum, spectrum (a), shows a strong broad peak in the region above 3000 cm^{-1}, indicative of an $-OH$ stretch, so it could be a compound such as an alcohol or a carboxylic acid. There are also strong peaks in the spectrum at 1600–1500 and 800–600 cm^{-1}, which indicates the presence

Confirmation of these materials could be obtained by NMR or mass spectroscopy. Also, as we now have an idea of which material is which, simple physical techniques such as mixed melting points/boiling points could be used to confirm the structures.

of an aromatic ring. No carbonyl peak is visible, ruling out an acid, so it is likely that the product is simply unreacted benzyl alcohol.

Step 2: The second spectrum, spectrum (b), does not appear to have much of an OH stretch, ruling out an alcohol or acid and has a strong carbonyl stretch, so indicating the presence of an aldehyde or ketone. Again the spectrum shows strong aromatic peaks. Given that the reaction is an oxidation, the logical choice for this material is benzaldehyde.

Step 3: The third spectrum, spectrum (c), has strong carbonyl and OH vibrations along with strong aromatic peaks. This is probably the product benzoic acid.

12.5 Fingerprinting and data libraries

The frequency range of an IR spectrum spanning 700–1200 cm^{-1} is known as the ***fingerprint region***. Most single bonds give rise to absorption peaks within this region, and their overlap, in many cases, becomes quite complex. Small changes in molecular structure often give rise to significant changes in absorption peaks within the fingerprint region, and in this way, spectra of unknown samples can often be matched with those of known compounds.

It should be noted that a number of inorganic groups such as nitrate, phosphate, and sulphate also give rise to absorption peaks within the fingerprint region, which complicate the spectra further.

In the vast majority of cases, the fingerprint region of the spectrum is highly characteristic for an individual compound. Matching a spectrum by eye with the spectra of known compounds would be highly tedious even if a narrow list of compounds could be shortlisted via examination of the group frequency region of the spectrum. However, many modern IR spectrometers greatly simplify the matching of spectra via computerized data libraries. In this way, the matching of multiple peaks across both the frequency group and fingerprint regions of the spectrum can be performed extremely rapidly, with the software often selecting two or three of the closest matches with spectra from its databases. This matching software can prove especially useful since analyte samples rarely contain one pure compound, but more frequently contain a host of differing compounds and/or impurities—many of which, of course, can contribute to and so further complicate the IR spectra.

12.6 Sample handling for infrared techniques

Samples for IR analysis may be solids, liquids, solutions (with dissolved solutes for analysis), or gases. We shall consider how infrared analyses may be performed on each.

12.6.1 Solid and liquid samples analysed by ATR approaches

The infrared spectrum of solid samples may be obtained by either (i) dissolving the solid in a suitable solvent, (ii) preparing a slurry or 'mull' of the sample within an inert liquid, or (iii) using an attenuated internal reflectance technique as described in Section 12.10. The approach that is chosen depends on the nature of the sample and the equipment available for the analysis.

Attenuated total internal reflection (ATR) techniques have transformed the way in which sample handling is routinely performed. Solid samples of materials such as leather or paper may be analysed directly by ATR techniques by placing the sample in direct contact with an ATR crystal. Many liquids and solutions ranging from environmental water through to biological samples (e.g. blood) may also be analysed by ATR approaches.

12.6.2 Preparation of 'mulls' for dispersive IR analyses

Prior to the advent of Fourier transform and ATR instrumentation, many solids required a 'mull' to be prepared. These approaches may be still used with older dispersive-type instruments (Section 12.9.1). The technique requires mixing a ground sample of the solid with an inert liquid to form a slurry or 'mull'. The most popular liquid used for preparing mulls is 'Nujol', a heavy hydrocarbon oil. Another liquid, 'Fluoroprobe', a halogenated polymer, is sometimes used as an alternative. Water is not suitable for the preparation of mulls because (i) it displays strong IR absorption, (ii) it is insufficiently viscous to hold the salt plates together by surface tension, and (iii) it would dissolve the NaCl discs.

Having prepared the mull, a small sample is smeared on one of a pair of NaCl discs (again chosen to avoid IR absorption). The second disc is gently placed on the first and these are subsequently held together by surface tension/capillary action. The two salt discs holding the sample mull are then placed within the path of the IR beam to allow absorption measurements to be performed directly. This procedure is depicted in Fig. 12.5.

The preparation of mulls and the use of salt discs do not allow reproducible absorbances to be determined, since the 'mull' thin film thickness is clearly extremely variable.

12.6.3 The use of compressed KBr discs

An alternative method to using a mull is to disperse the solid within a KBr disc. A small amount of a solid sample is ground together with KBr to form an intimate disc. This mixture is then put into a press and compressed to form a disc that can be placed directly into the IR beam. This has the advantage that there is no problem with adsorption peaks due to the mulling liquid and also it provides for a much better quantitative technique since it is possible to weigh the amounts of sample and KBr present in the disc.

12.6.4 Solutions for IR spectroscopy

Solutions of samples may be analysed directly in dedicated cells to allow direct quantification of IR absorption. The single most important factor is to ensure that the chosen solvent does not absorb in the IR region of interest. A number of organic solvents may be used within IR spectroscopy, although the usefulness of this approach has been limited following restrictions in recent years in many countries involving the routine use of solvents such as benzene, carbon tetrachloride, and chloroform.

Another simple approach for obtaining non-quantitative IR spectra of liquid samples involves placing a drop of the liquid or solution (providing this is not water containing) between two NaCl plates. The same approach may be used, however, for water-containing samples with the use of water-insoluble CaF_2 discs.

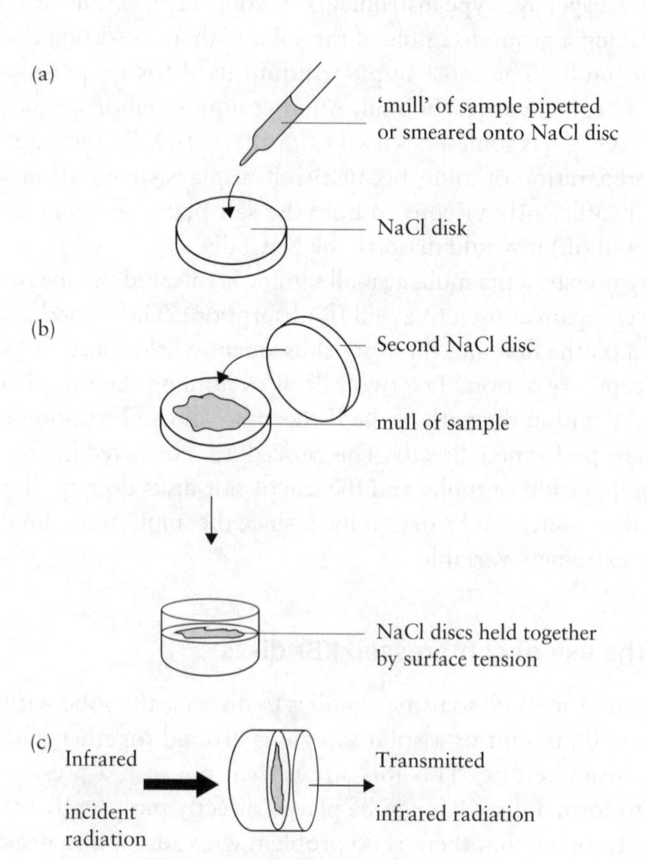

Figure 12.5 Preparation of 'mulls' for dispersive IR analysis. (a) Preparation of mull and application to one NaCl disc. (b) Two NaCl discs brought together. (c) Placement of NaCl discs in IR spectrometer.

12.6.5 Solids

Thin films of solids of some polymeric samples (e.g. polystyrene, polyethylene) may be placed in the path of an IR beam directly for IR transmission analyses. The most widely used approach for the determination of IR spectra is now, however, via the use of ATR approaches, Section 12.10.

12.6.6 Gases

Dedicated gas handling cells may be used for obtaining infrared spectra of gaseous samples and these are described in Section 12.11.

12.7 IR sources

IR sources comprise an inert solid that is electrically heated to a temperature in excess of 1500°C. The radiation emitted at these temperatures approximates to that from a black body, although the intensity of the radiation is not uniform across all frequencies of the IR range. A maximum intensity of the emitted radiation occurs at frequencies equivalent to approximately 5000–6000 cm^{-1}, with the intensity of emitted radiation falling away either side of these limits—although this is most pronounced at shorter rather than longer wavelengths.

12.7.1 The Nernst glower IR source

The ***Nernst glower*** source comprises a thin cylinder fabricated from rare earth oxides of approximately 20 mm length and 1–2 mm diameter. Current is passed through the cylinder for heating via electrical connections at either end. At operating temperatures, the cylinder glows dull red.

12.7.2 The globar source

The ***globar*** is an electrically heated silicon carbide rod of around 5 cm in length and 4 mm in diameter. The globar source typically provides a more intense source of IR radiation than the Nernst glower below wavenumbers of 2000 cm^{-1}.

12.7.3 The incandescent wire IR source

The ***incandescent wire*** is an IR source formed from a tightly wound and electrically heated spiral wire. It offers a greater longevity than many other sources although it cannot provide the same intensity as the Nernst or globar alternatives.

12.7.4 The mercury arc IR source

Far-IR radiation is radiation of the electromagnetic radiation with wavelengths of approximately $>50\ \mu m$ (or wavenumbers $< 200\ cm^{-1}$)

High-pressure ***mercury arc*** sources are used for far-IR applications where the intensity of the radiation provided via alternatives such as the Nernst, globar and incadescent sources diminishes to unacceptably low levels. The mercury arc source consists of a quartz-jacketed tube containing mercury at a pressure greater than 1 atmosphere, through which current is passed to form an internal plasma that emits IR radiation in the far IR region.

12.7.5 The tungsten filament source

Conventional tungsten filament lamps (Chapter 4) can be used as IR sources for the near-IR region throughout a range of approximately 4000–12 500 cm^{-1}.

12.7.6 The carbon dioxide laser source

Tunable carbon dioxide lasers can be used as special very high intensity IR sources for the determination of analytes in aqueous solution and atmospheric airborne pollutants, and can facilitate the quantitative determination of, for example, ammonia, benzene, ethanol, and nitrogen dioxide. The radiation provided by this source only ranges from 900 to 1100 cm^{-1} wavenumbers, and rather than being continuous, is actually composed of approximately 100 closely spaced discrete lines that can be selected by tuning the laser.

12.8 IR detectors

There are three main classes of IR detectors, and these are generally classified as operating via thermal, pyroelectric, or photoconducting principles. Photometers and dispersive spectrometers typically employ either thermal and pyroelectric detectors whereas Fourier transform instruments typically employ either pyroelectric or photoconducting detectors.

12.8.1 Thermal detectors

A black body in this context is one that absorbs all incident radiation.

Thermal detectors operate via monitoring the heating effects of IR radiation upon a component that is designed to act as a black body. Materials are chosen with very low heat capacities in order to maximize temperature changes upon irradiation with IR radiation. Under the most favourable of conditions, thermal detectors can respond to changes of a few thousandths of a kelvin. Thermal detectors are normally shielded from ambient thermal radiation to help minimize signal drift and noise effects. Instruments employing thermal detectors almost always employ IR beam choppers so that the signal may be differentiated from electrical or other sources of noise.

Thermocouples are the most popular types of thermal detector. The simplest thermocouples comprise a length of a metal (such as antimony) to which two pieces of dissimilar metals such as bismuth are attached. A potential difference develops between the two junctions upon heating and this can be related to changes in temperature between the two metal/metal junctions upon heating of *one* of the junctions. The junction that is heated via IR radiation is known as the ***active junction*** while the other is known as the ***reference junction***. The later is shielded from the incident IR radiation. In order to improve the performance of the detector, the thermocouple is normally enclosed within an evacuated chamber with a window that is transparent to IR radiation, to minimize the effects of ambient thermal fluctuations.

The sensitivity of such devices can be enhanced by connecting and using together several thermocouples to form a ***thermopile***, which, in conjunction with suitable operational amplifier circuitry, can allow detectors to respond to temperature variations of as little as 10^{-6} K.

An alternative approach involves the use of ***bolometers*** that are based either on simple ***thermistors*** (e.g. made out of germanium) or specially fabricated platinum or nickel thin strips—all of which are designed to exhibit resistivity changes in response to changes in temperature.

12.8.2 Pyroelectric detectors

Pyroelectric materials such as deuterated triglycine sulphate $(ND_2CD_2COOD)_3$ are insulators that are polarizable upon applying an electric field across opposing faces—with the degree of polarization being dependent on the dielectric constant of the material, which is itself a function of temperature. Pyroelectric detectors typically comprise a layer of pyroelectric material sandwiched between two electrodes (one of which is made of an IR transparent material) to form a capacitor. IR radiation passing through the transparent window causes a heating of the pyrolectric material and this gives rise to a change in the polarization of the material and so capacitance of the detector. The capacitance of the detector can be monitored electronically and so related to the intensity of the IR radiation.

12.8.3 Photoconducting detectors

An IR mercury cadmium detector consists of a thin semiconductor film of a material such as cadmium telluride that coats a non-conducting glass surface housed within an evacuated glass envelope. Exposure of the semiconductor to IR radiation promotes valence electrons from non-conducting states to conducting states thereby increasing the conductivity of the device and this again can be monitored electronically and related back to the intensity of the incident IR radiation.

12.9 General purpose IR spectrometers and photometers

There are four main types of IR spectrometers routinely used for analytical applications. The first of these, the ***Dispersive Grating Spectrometer*** has acted as the main analytical workhorse for IR applications within many laboratories although instruments of this type are progressively being replaced by ***Fourier Transform Multiplet based instrumentation.*** The third class of instruments are the ***Non-dispersive photometers*** developed for the quantitative determination of organic compounds within the atmosphere, while the fourth and last class of instrument are the ***Reflectance photometers.*** We shall within limit our discussion to dispersive grating and Fourier transform instruments since these are the most widely used for analytical applications.

12.9.1 Dispersive grating IR spectrometers

Dispersive grating spectrometers are instruments in which the spectrum is *sequentially* analysed following the *dispersion* of multi-wavelength radiation by means of either a monochromator or diffraction grating. Dispersive IR spectrometers have for several decades formed the mainstay of IR instrumentation for analytical applications although they are progressively being replaced by Fourier transform instrumentation—as we shall discuss in the next section. Dispersive spectrometers are normally double-beam instruments that use diffraction gratings for the dispersion and frequency selection of IR radiation from a 'white' source.

A double-beam arrangement is normally used in conjunction with a beam chopper to compensate for IR absorption within the gas or a solvent in which the analyte is dissolved. A typical schematic arrangement for an IR dispersive spectrometer is shown in Fig. 12.6. In many ways a dispersive IR spectrometer can be thought of as being similar to a UV–visible instrument in the way it operates. In contrast to UV/visible instrumentation, however,

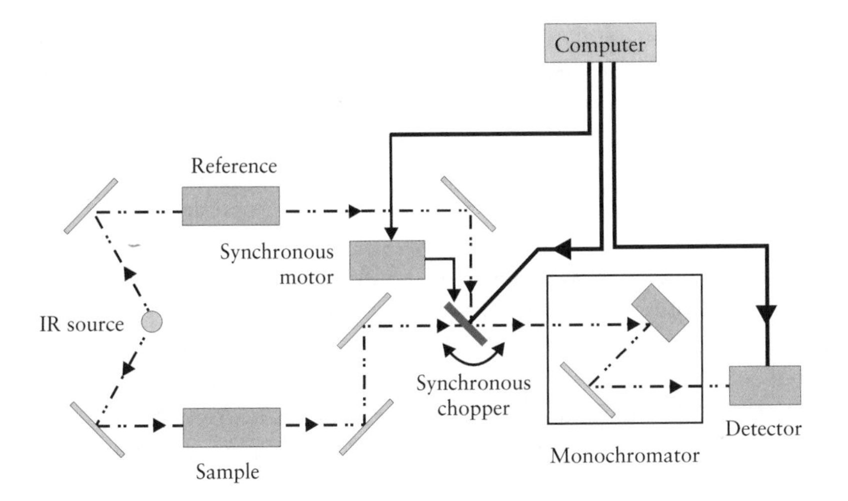

Figure 12.6 Schematic for a dispersive IR spectrometer.

it is normal in IR spectrometry to place the grating and monochromator *after* the IR beam has passed through the sample. In this way, any radiation scattered by the sample is removed by the monochromator and, therefore, does not reach the detector. This approach is possible with IR techniques since IR radiation is insufficiently energetic to damage most samples.

Instruments that employ attenuators to regulate (reduce) the power of the reference transmitted beam in relation to the intensity of the beam that emerges from the sample are known as ***null-type instruments***. The absorption of the sample is monitored in null-type instruments in the following way. The reference beam having passed through the attenuator next passes through a chopper. The chopper alternately allows the reference beam and the beam transmitted by the sample to pass to the grating and so on to the detector. If the two beams are of identical power, then the output from the detector will be constant. If the intensity of the transmitted beam is stronger or weaker than the reference beam, then the output of the detector will alternate between high and low values. In this way, the composite alternating signal can be correlated with the intensity of the transmitted beam and the absorption of the sample can be determined.

12.9.2 Fourier transform multiplet spectrometers

Fourier transform multiplex instruments operate in an entirely different approach to their dispersive counterparts since all wavelengths are detected and measured simultaneously, which explains why they are known as ***multiplex instruments***.

Fourier transform IR (FTIR) spectrometers allow unrivalled speed, resolution, sensitivity, and wavelength precision that cannot be matched by dispersive instrumentation. Early Fourier transform instruments were expensive, cumbersome, and bulky as well as requiring more frequent servicing than their dispersive counterparts. However, modern FTIR instruments, by contrast, are typically smaller than dispersive spectrometers, and also require less servicing than dispersive IR spectrometers.

Although a number of multiplex instrumental approaches other than those based on Fourier transform techniques are possible, Fourier transform spectrometers now predominate.

Since we detect all wavelengths simultaneously, we must have some means to resolve a spectrum in terms of radiant power—as a function of wavelength (or wavenumbers). This is achieved using an interferometer together with Fourier transformation of the transmitted IR radiation. The advent of modern computational powers represents one of the major factors that has helped FTIR techniques to supersede dispersive approaches since this allows the Fourier transformation to be performed rapidly.

Interferometers

All FTIR instruments employ interferometers. There are many different ways in which interferometers operate, although we shall in this book only describe one in detail: the ***Michelson interferometer***. This interferometer is the most widely used design even though the principle underpinning its operation was first described towards the end of the nineteenth century.

Normally, the sample is first irradiated with polychromatic (white) IR radiation by means of two beams that are *nearly equal in intensity* to each other. The difference in phase of the two beams reaching the detector may then be used to determine the intensity of transmitted radiation with respect to time—and as we shall see shortly frequency (or wavelength–wavenumber) by means of a mathematical Fourier transform of the data. The radiation emitted from the source radiation is first collimated and directed towards a *beam splitting mirror* so that half of the radiation is transmitted through the mirror while the other half is reflected, Fig. 12.7. The two beams of radiation are then reflected back again towards the beam-splitter by a pair of mirrors, one of which is fixed while the other can be moved. The beam splitter now directs half of each of the beams towards the sample and detector while the other half is directed back towards the source. If the two beams passing through the sample are to remain exactly in phase with each other as they reach the detector, the pathlength through which both the beams travel, must be exactly the same as each other. Since one of the mirrors may be moved, the pathlength of the two beams may be varied slightly. If the pathlength through which the two beams travel differs, then the two beams will come into and out of phase with time.

A collimator is a device for producing a parallel beam of radiation.

In such situations, constructive or destructive interference can occur depending on the phase angle of the two beams as they couple. In most spectrometers, the mirror is moved at a linear and extremely carefully

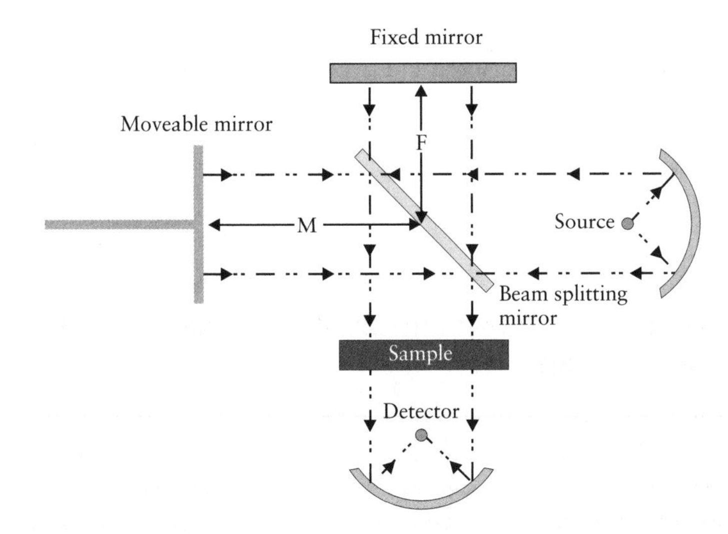

Figure 12.7 Schematic of a multiplex (Fourier transform) IR spectrometer.

controlled rate. The intensity of the coupled beam will both be a function of time and absorption by the sample. In this way, the two beams that travel through the sample are then coupled together to produce an interferogram in which the intensity of the coupled beam can be plotted as a function of time, for example, Fig 12.8(a), which, in turn, may be de-convoluted to form a conventional absorption peak, Fig. 12.8(b).

The difference between the pathlengths of the two beams, as caused by movement of one of the mirrors within the interferometer, is known as the ***retardation***, or δ. The power of the interferogram signal is denoted $P(\delta)$. The first stage in IR Fourier transform spectroscopy involves recording $P(\delta)$ as a function of δ. Interferograms of this type, therefore, contain

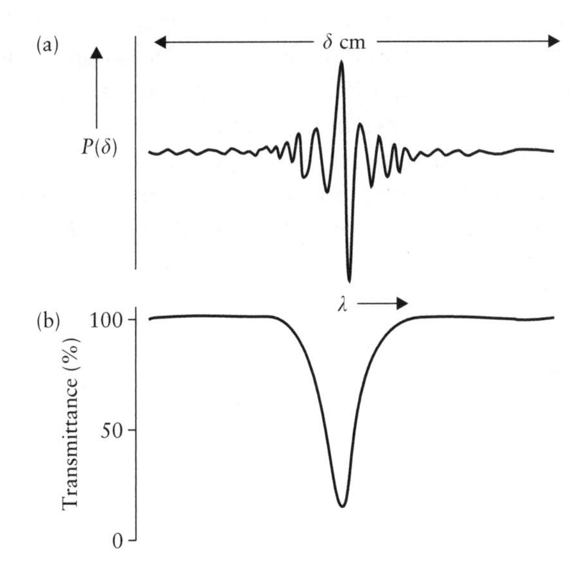

12.8 (a) IR interferogram. (b) De-convoluted IR peak.

information relating to the absorption of the sample at all wavelengths (or wavenumbers). The interferogram may be de-convoluted via a Fourier transformation using computer algorithms allowing the intensity of IR transmission to be related to wavelength (or wavenumbers), and in this way an IR spectrum may be plotted.

12.10 Attenuated total internal reflection—FTIR instrumentation

Attenuated total internal reflection (ATR) techniques allow IR spectra to be obtained for solids, pastes, and/or powders, by exploiting the reflection of IR radiation when IR radiation passes from one medium to another that possess differing refractive indices. One of the media represents the sample while the other is a crystal that forms part of the spectrometer instrumentation. The fraction of the radiation that is reflected is determined by a number of factors including the angle of incidence. As the angle of incidence increases, so does the percentage of reflected radiation until a certain angle known as the ***critical angle is reached.*** At angles greater than the critical angle total reflection occurs.

In practice, the beam of radiation penetrates the surface of the less dense medium before reflection occurs. The radiation that penetrates the less dense layer is known as the ***evanescent wave***. The depth of penetration of the evanescent wave is determined by: (i) the wavelength of the radiation; (ii) the refractive indices of the two media; and (iii) the angle of incidence, θ. Evanescent waves typically penetrate the sample by a few micrometres. If the less dense layer (e.g. the sample) absorbs IR radiation, then attenuation of the beam is seen at wavelengths (wavenumbers) corresponding to IR absorption bands of the sample. This approach is known as attenuated internal reflection IR spectroscopy. A schematic of a typical ATR cell arrangement is shown in Fig. 12.9.

In practice, all contemporary ATR spectrometers are based upon Fourier transform instrumentation and so the technique is often known as ATR-FTIR spectroscopy. Commercially available ATR accessories or

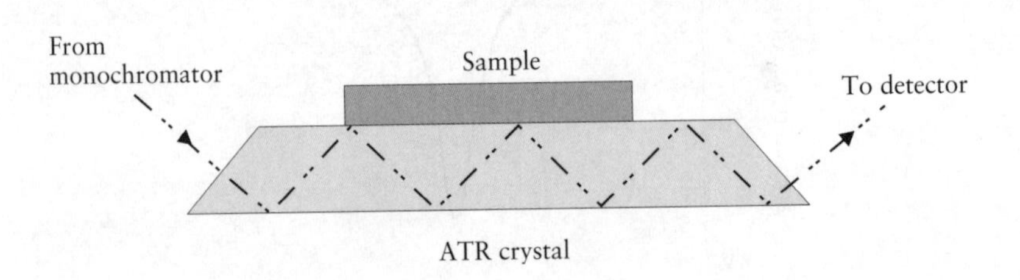

Figure 12.9 ATR cell for IR spectroscopy.

adapters can be used with many FTIR instruments and are designed so that the incident IR beam is directed through the crystal and then onto the detector. The ATR adapter is normally designed so that the crystal can be easily changed with the most commonly chosen crystals being germanium, zinc selenide, or a mixed crystal of thallium bromide/iodide. Since differing crystals are associated with differing and characteristic critical angles, the choice of crystal allows some variation of the depth of penetration of the evanescent wave into the sample. The angle of incidence of the IR beam can also normally be varied to allow further control of the depth of penetration of the evanescent wave.

ATR-FTIR techniques offer a number of advantages. Amongst the most important are ease of sample preparation and versatility towards samples that would otherwise be difficult to obtain. Powders, solids, and liquids may all be easily studied via ATR techniques provided that a good contact can be made between the sample and the crystal surface. A small lid can sometimes be used to gently compress the sample against the face of the crystal, although great care must be taken to not break the crystal by over-tightening the clamps of the lid, since they are typically both quite fragile and expensive!

This approach allows many samples such as fabrics, leather, and fibres to be analysed using IR spectroscopy with relative ease.

ATR-IR approaches yield spectra that are essentially similar to those obtained via normal transmission approaches, although the relative intensities of some peaks may differ. The absorbancies observed within reflectance spectra are normally independent of sample thickness, since the evanescent wave penetrates the sample by only a few micrometres.

12.11 Gas-sampling photometers for IR spectroscopy

A number of portable IR gas analysers for the routine quantitative determination of organic pollutants in the atmosphere are based on a filter photometer approach. A schematic diagram of a typical instrument is shown in Fig. 12.10. Instruments of this type typically contain IR sources

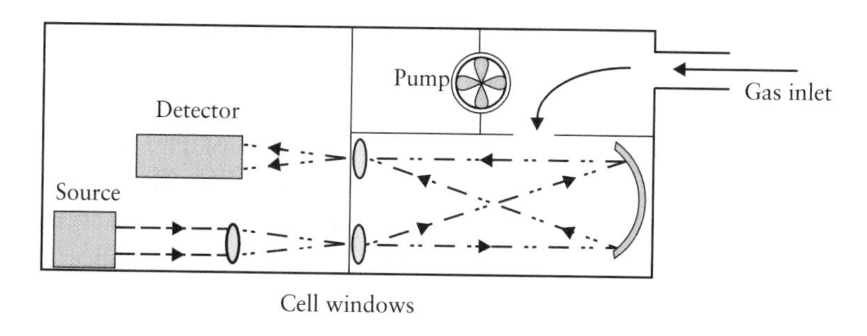

Figure 12.10 Schematic of a gas-sampling IR spectrometer.

comprising a ceramic rod around which is wound a nichrome wire and these are normally used in conjunction with a pyroelectric detector.

A number of different filters are produced—each for determining a specific atmospheric pollutant. Gaseous samples are normally collected using a small battery-powered pump. The gaseous chamber is typically a few centimetres in length, although a series of reflecting mirrors (not shown for clarity) are sometimes included at either end of this chamber to increase the effective absorbing pathlength of the instrument; in this way the sensitivity of the instrument can often be greatly enhanced.

12.12 Near-IR spectroscopy

The near-IR region ranges from approximately 700 nm (13 000 cm^{-1}) at the fringe of the upper edge of the visible region of the spectrum to wavelengths of around 3000 nm (3300 cm^{-1}). The most frequent molecular absorptions in this region are normally overtones of C–H, N–H, or O–H stretching and/or vibrational bands—or combinations of these. Unfortunately, absorbencies are characteristically smaller than those seen within the main IR region. Near-IR instrumentation typically uses tungsten lamps in conjunction with lead sulphide photoconductor-based detectors.

Optical cells of either quartz or fused silica, together with solvents such as carbon tetrachloride, must be used for the transmission of radiation within this wavelength range. The instrumentation used is, in principle, very similar to that of many UV–visible spectrometers and since the near-IR region borders the visible spectrum, some commercial UV–visible spectrometers have been built so as to allow near-IR capability.

12.13 Far-IR spectroscopy

As with most other IR techniques, the majority of IR spectrometers with far-IR capability are now based on interferometers and Fourier transform signal processing. The far-IR region of the spectrum normally extends from wavelengths of approximately 10^6 nm (200 cm^{-1}) down to 5×10^4 nm (10 cm^{-1}). Far-IR spectroscopy is most widely used for inorganic analyses, since absorption within this region largely corresponds to bending or vibrational motions between metal ions and ligands. The specificity of analyses such as these is often extremely favourable in comparison to many other analytical approaches, since the wavelength at which absorption occurs is normally dependent upon both the metal atom and the ligand in question.

Some compounds containing only low molecular weight atoms can absorb in the far-IR region due to skeletal bending motions, provided that the molecule possesses at least two or more atoms other than hydrogen. These types of absorption can be particularly useful for determining a number of substituted aromatic organics. The spectra of such compounds are often quite complicated, although this can facilitate finger-print identification for specific compounds.

The rotational movements of gases possessing permanent dipoles can also cause absorptions in the far-IR region, which permit determination of, for example, gaseous H_2O, O_3, or HCl.

12.14 **IR emission spectroscopy**

Molecules capable of absorbing radiation may, in theory at least, also emit IR energy upon heating. There are, however, a number of problems associated with the determination of compounds via IR emissions especially relating to issues such as signal-to-noise ratios and the analysis of thermally labile analytes that can decompose on heating. Samples capable of IR emission analysis can be heated electrically and analysed within a gaseous cell as long as the sample is not vaporized. The widespread use of interferometers and Fourier transform techniques has now greatly helped the capture and signal processing of very weak signals.

Exercises and problems

12.1. Why are wavenumber notations used instead of wavelengths within IR spectroscopy?

12.2. IR spectrometers typically operate over a wavelength range of 3–15 μm; express this range in terms of frequency and wavenumbers.

12.3. How many degrees of freedom does the gas NO_2 possess?

12.4. Why are mercury arc IR sources sometimes used in preference to alternatives such as Globar sources?

12.5. Discuss the relative advantages of using thermopiles, bolometers, pyroelectric, and photo-conducting detectors in IR instrumentation.

12.6. Why are Fourier-transform-based IR spectrometers superseding dispersive-based instruments?

12.7. Why may attenuated total internal reflection IR approaches allow spectra to be recorded for samples that would otherwise be difficult to study via IR spectroscopy?

12.8. Solvent bottles a–f have all lost their labels. The bottles are known to contain acetone, acetonitrile, chloroform, ethanol, hexane, and toluene. From the IR spectra, assign the solvents to their bottles (Fig. Q8).

12.9. Describe the molecular vibrations associated with infrared absorptions.

12.10. The IR spectrum of gaseous HCl shows a series of peaks centred at about 2900 cm^{-1} corresponding to

Fig. Q8 IR spectra for six bottles containing solvents to be identified.

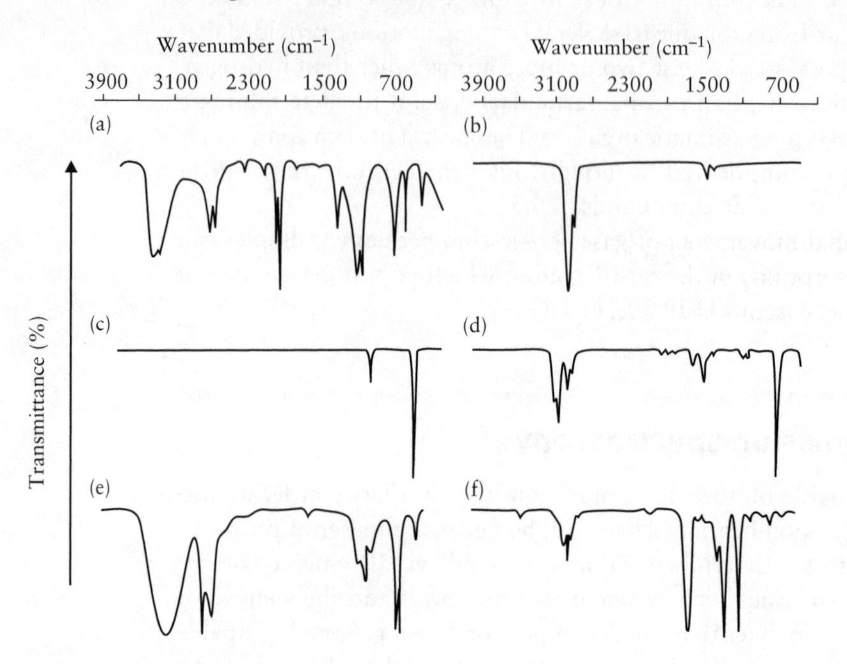

rotational transitions of the HCl molecules. Closer examination of each adsorption peaks shows it to consist of two peaks with a size ratio of approximately 3 : 1. Explain this behaviour.

12.11. Discuss how the IR spectra of a carboxylic acid in liquid form may well differ from that in dilute solution in chloroform.

Summary

1. IR radiation forms part of the electromagnetic spectrum and covers a range of wavelengths from approximately 0.8 to 1000 μm.

2. IR absorption occurs via the stimulation of a number of molecular vibrations including rocking, wagging, scissoring, and twisting motions.

3. IR absorption follows the Beer–Lambert law.

4. IR spectra are normally presented with the *Y*-scale representing transmitance ($1/A$) (or sometimes percentage transmission) and the *X*-scale in terms of wavenumbers ($1/\lambda$) cm^{-1}.

5. Specific functional groups such as C–H, C=C, C–N, NO_2 and OH groups give rise to absorptions at characteristic wavenumbers and these allow compounds to be matched by 'fingerprinting' approaches with reference to data libraries.

6. A number of IR sources are used with IR instruments and include Nernst glower, globar, mercury arc, tungsten filament, carbon dioxide, and laser sources.

7. Detectors used in IR spectrometers include thermal, pyrolectric, and photoconducting based devices.

8. IR spectrometers operate either via dispersive grating or multiplex principles.

9. Multiplex instruments employ an interferometer; the Michelson-type interferometer is the most commonly encountered design. Multiplex instruments rely on de-convoluting the signal by a mathematical treatment of the data such as a Fourier transform. Many IR spectrometers are simply known as Fourier transform IR (or FTIR) spectrometers.

10. Attenuated total internal reflection (ATR) techniques allow IR spectra to be obtained for many solids, pastes, and powders that would otherwise be impossible to study using dispersive instrumentation.

11. ATR techniques exploit the reflection of IR radiation that occurs when IR radiation passes from one medium to another. The radiation that penetrates the less dense layer is known as the evanescent wave.

12. Gas sampling devices coupled to portable IR spectrometers allow sampling of air for environmental applications.

13. Near-IR spectroscopy ranges extend from approximately 770 to 3330 cm^{-1} and may be used for a number of specialist applications.

14. Far-IR regions of the electromagnetic spectrum extend from 200 to 10 cm^{-1} and may be used for a number of inorganic analyses such as the determination of H_2O and HCl.

15. IR emission spectroscopy is performed by heating a sample until it emits IR radiation. This can only be achieved, however, if (a) the sample is thermally stable and (b) vaporization of the sample will not occur at the temperatures required.

Further reading

Colthup, N. B. (1989). *Introduction to infrared and Raman spectroscopy*. Academic Press.

Gunzler, H. and Heise, M. H. (2002). *IR spectroscopy: an introduction*. Wiley-VCH.

Stuart, B. (2004). *Infrared spectroscopy – experimentation and applications*. Wiley.

Further reading

Part IV

Analytical chemistry in practice: contemporary analytical science

Radiochemical analytical methods

13

Skills and concepts

This chapter will help you to understand:

- The nature of α, β, and γ emissions from radioactive decay processes.
- What is meant by a half-life for a nuclide and how this concept can be used for radioactive counting and quantitative measurements for radioactive labelling analyses.
- How to perform isotope dilution analyses.
- The basis of neutron activation analyses and how nuclear reactors, accelerators, or radioactive nuclide sources can be used to achieve neutron activation.
- The basis of carbon dating techniques and how these may be used.
- The scope of radio-isotope based techniques in bio-analytical and clinical applications.

13.1 Introduction

Radiochemical analyses exploit either radioactive isotopes or radiation derived from them.

There are three main classes of radiochemical analysis classified according to the nature of the radioactivity, namely (i) analyses based upon the determination of natural radioisotopes; (ii) activation analyses; and (iii) tracer based techniques.

13.2 Radioactive isotopes and the basis of radiochemical analyses

The nuclei of all atoms (with the exception of ^{1}H) contain a number of neutrons and protons. The number of protons, Z, is also known as the

atomic number and defines the chemical properties of the atom as well as specifying a particular element. Atoms of the same element may contain differing numbers of neutrons and are known as isotopes of an element. Some nuclei decay via the emission of sub-atomic particles such as α or β particles or alternatively via decompositions associated with the release of X- and/or γ-rays. Nuclei that spontaneously decompose in this way are known as ***radioactive nuclides***, the isotopes of which are known as radioactive isotopes of an element. Radioactive nuclides are sometimes more simply just termed ***radionuclides***.

Nuclei that do not spontaneously decay via radioactive emission are said to be stable isotopes. Radionuclides continue to decompose until stable nuclides (isotopes) are formed.

α and β particles along with X- and γ-rays are all ***ionizing*** and this allows them to be readily detected and indeed quantified via photographic film or detectors such as Geiger–Müller tubes. This ability for detection forms the basis of the majority of radiochemical analytical methods.

13.3 Radioactive decay products

13.3.1 Alpha decay

The α particle ${}^{4}_{2}He$ (a helium nucleus) contains two protons and two neutrons. α particles are emitted by radioactive isotopes (typically of mass numbers greater than approximately 150) as they decompose to give daughter elements of lower atomic masses. The decay of ${}^{238}_{92}U$ to give ${}^{234}_{90}Th$ as shown in Eqn (13.1) is an example of such as process:

$$ {}^{238}_{92}U \rightarrow {}^{234}_{90}Th + {}^{4}_{2}He \qquad (13.1)$$

Processes that give rise to—and accompany—α emissions proceed via clearly defined mechanisms or pathways, which means that the energy re-distributions that accompany these processes are quantized. Energy releases via the emission of either X- or γ-rays frequently accompany α particle decay processes.

α particles are ejected from the decaying parent nucleus with considerable kinetic energy (again this is both quantifiable and predictable). This kinetic energy is capable of causing considerable ionization upon collision with the atoms or molecules of many materials. α particles progressively lose energy upon each collision; however, because of their relatively high mass, they have a low capability for penetration and may be typically stopped by a thick piece of card. An α particle emitter can often be identified by determining the path length over which the particles travel (typically a few centimetres through air).

13.3.2 Beta decay

There are three nuclear reactions that can be classified as β decay processes and we shall consider each of these in turn. β particles are either electrons or positrons.

The first of the β decay processes we shall consider involves the capture of an electron by the nucleus, followed by the release of an X-ray photon; this process is unfortunately of no analytical significance.

Both of the other two processes involve the ejection of an electron or a positron from the nucleus (along with the formation of a neutrino, ν). In one of these processes, a neutron is converted into a proton and an electron, which is subsequently expelled; an example of this type of reaction is shown in Eqn (13.2):

$$^{14}_{6}C \rightarrow ^{14}_{7}N + \beta^{-} + \nu \tag{13.2}$$

In the other process involving the formation of a positron, β^{+}, the total number of protons within the nucleus decreases by one, an example of which is shown in Eqn (13.3):

$$^{65}_{30}Zn \rightarrow ^{65}_{29}Cu + \beta^{+} + \nu \tag{13.3}$$

Positrons are ultimately annihilated via reaction with an electron and this is accompanied by the generation and emission of two γ photons.

Unlike α particles, β particles are generated with a continuous range of kinetic energies. The penetrating power of β particles far exceeds that of α particles due to their vastly smaller mass. In practice, β particles can typically penetrate up to several tens of centimetres of air, with a minimum of a piece of aluminium foil normally being required to stop radiation.

13.3.3 Gamma-ray emission

It should be realized that there is no difference between X-rays and γ (gamma)-rays except that X-rays are formed via electronic transitions, whereas γ-rays originate from nuclear processes.

Many α or β decompositions are accompanied by the release of γ-rays as excited nuclei relax via one or more quantized steps. In this way, the γ-rays themselves have specific energies (frequencies) and so can often be used in a fingerprinting manner for identifying specific radioactive decompositions.

γ-rays are highly penetrating with several centimetres of lead typically being required as a shield. γ-rays lose energy and are stopped in this way via three main mechanisms as they pass through matter. The mechanism that predominates largely depends upon the frequency (and thus the energy) of the γ-rays.

For low-energy γ-rays, energy is largely lost via the ***photoelectric effect*** in which the energy of a γ photon causes the excitation and displacement of an electron from an atomic orbital within the material through which the radiation passes. In most instances, the photon is entirely consumed so stopping the radiation.

For γ-rays of intermediate energy levels, the ***Compton effect*** may be observed as radiation passes through matter. Electrons are again displaced from atomic orbitals although not all of the energy of the photon is consumed, with the remainder of the energy continuing through the material as a γ-ray of lower energy. In some instances this can give rise to further photoelectric or Compton effects.

Extremely highly energetic γ photons (>1.02 MeV) can sometimes be completely absorbed in the vicinity of a nucleus via the formation of a positron and an electron in a process known as ***pair production***.

13.3.4 X-ray emission

As we already have seen, X-rays are generated via electronic transitions and there are two nuclear radioactive decay processes that give rise to the loss of inner shell electrons and X-rays.

The first of these processes involves the capture of an electron by the nucleus as discussed earlier, while the other occurs via an ***Auger*** electronic emission. Auger electronic emissions occur as a result of an interaction between an excited nucleus and an orbital electron, so as to cause the excitation and loss of the electron with a kinetic energy equal to the difference between the energy of the nuclear transition and the binding energy of the electron. During this process, excess energy is lost via the emission of an X-ray photon.

13.3.5 Radioactive decay rates

Radioactive decays are completely random processes in their timing and so no prediction can be made as to when a particular atom will undergo decay. The study of the statistics describing the radioactive decay of large populations of a particular nuclide *can*, however, allow us to predict with a high degree of certainty *how many* radioactive events will be expected to be observed over a given period of time. The way in which these statistics are described is in terms of ***half-lives***, $t_{1/2}$, which are highly characteristic for specific nuclides.

Radioactive decay processes follow first order kinetics with a decay constant (normally denoted λ) characteristic of individual decay processes. We can predict that during the time duration of the half life, $t_{1/2}$, half of the radionuclides will have decayed as described by Eqn (13.4):

$$t_{1/2} = \frac{\ln 2}{\lambda} \cong \frac{0.693}{\lambda} \qquad (13.4)$$

Half-lives of radionuclides range from millions of years down to fractions of a second. The activity, A, of a radionuclide has the units of s^{-1} and describes the rate at which a radionuclide decays, Eqn (13.5), so that:

$$A = -\frac{dN}{dt} = \lambda N \qquad (13.5)$$

where N is the number of atoms undergoing radioactive decay.

Confusion can sometime arise since activities are normally expressed in terms of either becquerels (Bq), where 1 Bq corresponds to 1 decay per second, or the older units of curies (Ci), where 1 Ci $\equiv 3.7 \times 10^{10}$ Bq and corresponds to the activity of 1 g radium-226.

Absolute activities are hard to determine since detectors are never 100% efficient, and for this reason, measured radioactivity is normally expressed in terms of count rates, R, that can be related to the activity by the introduction of a detector coefficient, c, Eqn (13.6), so that:

$$R = cA = c\lambda N \qquad (13.6)$$

13.4 Background corrections

Counts recorded during a radiochemical analysis will always include contributions from background sources such as cosmic radiation and/or radon gas within the atmosphere. It is therefore normally necessary to correct for these other sources of radiation by first determining the true background count so that it may be subtracted from all subsequent analytical counts. In this way we can express a new corrected count rate, R_c, that can be related to the measured count rate for the sample, R_{ms}, and the background count according to Eqn (13.7):

$$R_c = R_{ms} - R_b \qquad (13.7)$$

13.5 Instrumentation

Scintillation counters, gas-filled detectors, and semiconductor-based detectors can all be used to measure radioactive events in a very similar manner to the detection of X-rays since α and β particles, along with γ photons, are all ionizing forms of radiation. Each of these three types of detector rely on the production of photoelectrons upon absorption of radiation and this can give rise to many ion pairs in the form of a cascade, which generates a measurable electrical pulse.

13.5.1 Measurement of α particles

Analyte samples that are known (or suspected) to be α emitters are normally prepared as thin film coatings upon a host underlying material to minimize self-absorption by the material, since as we have seen, the penetrating power of α particles is very small. α spectra consist of discrete peaks that may be used for identification by using pulse height analysers.

13.5.2 Measurement of β particles

Liquid scintillation counters, Fig. 13.1, are typically used for counting low-energy β emitters such as ^{35}S, tritium or ^{14}C with the sample typically being dissolved in a solution of the scintillating compound that is placed in a vial situated equidistant between two photomultiplyer tubes in a light proof housing. The output of the two counters is fed to a coincidence counter that only registers a pulse when both detectors record a signal simultaneously. In this way, background noise associated with the detectors and amplifiers can be significantly decreased, since it is unlikely that such effects would influence both detectors simultaneously. Liquid scintillation

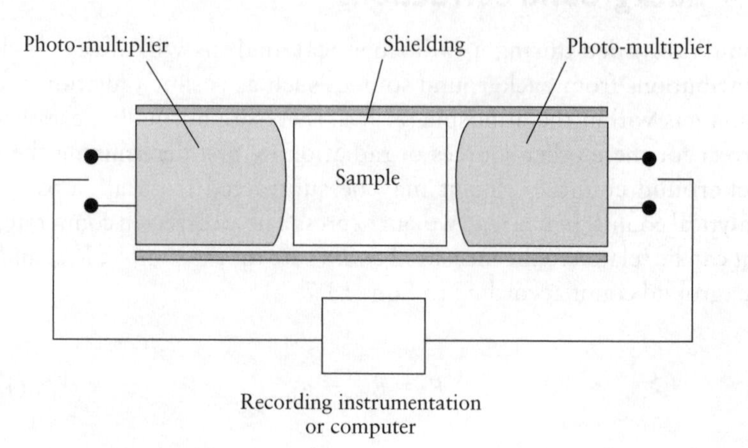

Figure 13.1 Schematic of a liquid scintillation counter.

counting is one of the most widely used radiochemical analytical approaches especially with clinical applications.

Higher-energy β emissions can be counted using a ***well scintillation*** (Fig. 13.2), ***Geiger–Müller tube*** or proportional tube counter placed at a suitable distance close to the source, which should normally be prepared so as to have a flat planar surface.

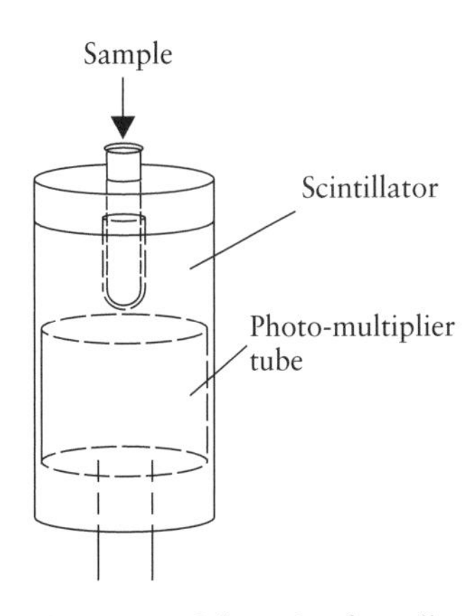

Figure 13.2 Schematic of a well scintillation counter.

13.5.3 Measurement of γ photons

γ-radiation is normally measured using Geiger–Müller tubes Fig. 13.3, the ends of which are normally protected by thin aluminium or Mylar windows so as to filter out α or β background interferences. γ-ray spectrophotometers are also often used for identification of the emitting radioisotope(s) according to the frequency of the radiation as well as allowing for quantification of the number of radioactive emissions in a given time according to the number of photons received.

13.6 Isotope dilution based analyses

Isotope dilution methods often offer high degrees of selectivity and can be used in conjunction with both stable and radioactive isotopes. Since radioactive counting simplifies quantifying specific isotopes, most isotope dilution based techniques are based on this approach and we will for this reason limit our discussions to radioactive dilution based counting approaches.

A radioactively labelled isotope form of the analyte must first be prepared and isolated to the highest level of purity possible. The count rate for this radiolabelled analyte must then be determined. A sample of this preparation is then mixed thoroughly with a weighed quantity of the analyte sample to be analysed. The analyte is then again isolated and

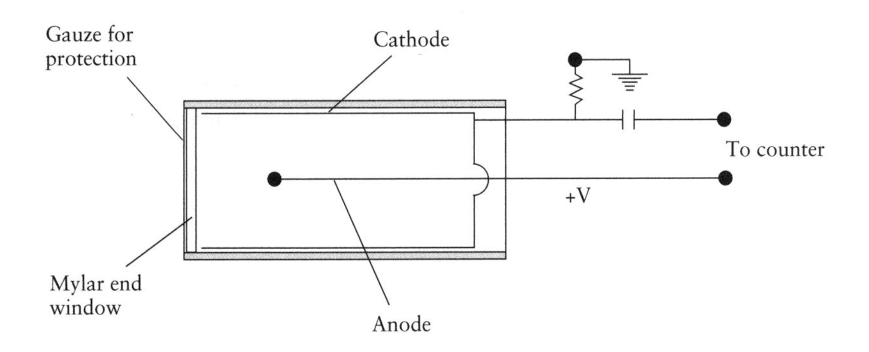

Figure 13.3 Schematic of a Geiger–Müller tube.

purified from the raw sample and a new count rate determined. In this way the level of dilution of the sample can be calculated within a weighed sample and thus related to the quantity of the non-radioactive form of the analyte, according to Eqn (13.8):

$$M_I = \frac{R_T}{R_M}(M_M - M_T) \tag{13.8}$$

where M_I, M_M, and M_T are the measured masses in grams of the isolated sample, the purified mixture, and the tracer, respectively, with R_T and R_M similarly representing the background corrected count rates for the tracer and mixture, respectively.

This procedure does not rely on quantitative recovery of the sample, but rather on the purification of samples and carefully determined weight measurements to allow accurate determinations of dilution effects; and so quantification of the analyte.

Isotope dilution methods have been used for the determination of many elements (normally in organic molecules), for a range of differing analyses. The greatest uses for these techniques are, however, within clinical or biochemical settings, for the analysis of, for example, vitamin D, insulin, a number of amino acids, thyroxine, and penicillin (as well as a number of other antibiotics).

13.7 Neutron activation methods

Neutron activation methods are based upon inducing radioactivity within a sample by irradiating it with neutrons in a reactor, an accelerator, or a radioactive nuclide source. Neutron bombardment of some stable nuclides can give rise to the release of radiation. Neutrons formed within all three of these types of source are highly energetic and sometimes require slowing down to energies of ~0.04 MeV. This can be achieved by passing them through a moderator containing a large number of protons or deuterium atoms such as water, paraffin, or deuterium oxide. Neutrons lose energy via collisions with nuclei in the moderator via a process known as ***elastic scattering*** until the neutrons reach thermal equilibrium with their surroundings. These neutrons are known as ***thermal neutrons*** and are widely used to activate samples for analytical purposes, although for a limited number of applications ***fast neutrons*** of energies of up to 10 MeV must be used.

Free neutrons are not stable and decay to give protons and electrons with a half-life of approximately 12.5 min, although in practice they are highly reactive with the nuclei of many atoms and do not normally exist in the free state for long periods of time. The reactivity of free neutrons is largely due to their ability to approach nuclei in the absence of electrostatic repulsions since neutrons are neutral.

One of the most widely exploited reactions of neutrons involves their capture by analyte molecules to produce nuclides whose atomic number stays the same but increase their atomic mass by one. The nucleus will now be in an excited state; the ***binding energy*** associated with the capture of the neutron is typically released via a radioactive decay via the emission of a γ-ray and/or nuclear particles (e.g. α particles or protons). It is the emission of radiation during the relaxation of the newly formed nuclide, that offers a route for quantitative analyses. The rate of formation of the radionuclide depends on the flux rate of neutrons as well as the irradiation time, although, the rate of decay of the newly formed radionuclide will be dictated by a characteristic half-life.

The measured radioactivity, A, of the sample will therefore increase with irradiation time until a point is reached at which the rate of decay will equal its rate of formation, as shown in Fig. 13.4. It can also be seen within this figure that higher neutron flux rates give rises to higher levels of activity within the sample as expected, although plateaus of activity are encountered with longer irradiation times.

The mass of the analyte can be determined according to Eqn (13.9), where M_A and M_S represent the mass of the analyte and the sample, respectively and R_A and R_S the corresponding decay rates.

$$M_A = \frac{R_A}{R_S} M_S \tag{13.9}$$

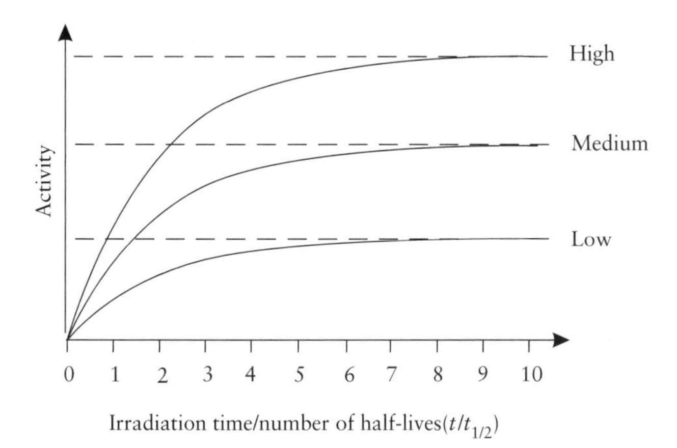

Figure 13.4 Measured radioactivity with respect to irradiation time.

13.7.1 Sources of neutrons for activation-based analyses

Nuclear reactors

Nuclear reactors produce very large numbers of neutrons (with neutron fluxes often exceeding 10^{11} neutrons cm^{-2} s^{-1}), suitable for activation-based analyses, although determinations of this type can only be performed in the immediate vicinity of a suitable reactor and this requires large-scale resources at centralized laboratories.

Accelerators

Accelerators for generating of neutrons for activation-based analyses are commercially available, and typically comprise a deuterium ion source, electrode assemblies for the acceleration of the ion through 150 kV or more and a target of tritium absorbed onto titanium or zirconium. Deuterium ions are focused (towards the target) to produce neutrons, Eqn (13.10), upon impact.

$$^{2}_{1}H + ^{3}_{1}H \rightarrow ^{4}_{2}He + ^{1}_{0}n \qquad (13.10)$$

Radioactive nuclide

Radioactive nuclides are the simplest sources of neutrons for analytical purposes, although neutron flux rates obtainable cannot match those from nuclear reactors and this will be reflected in the lower limits of detection achievable.

A number of radioisotopes may be used as neutron sources. One of the most widely used comprises a mixture of an α emitting radioisotope of, for example, americium or plutonium with a stable isotope of a lighter element such as beryllium. Neutrons in such samples can be emitted as a consequence of collisions of α particles with the atoms of the lighter element as, for example, shown in Eqn (13.11).

$$^{9}_{4}Be + ^{4}_{2}He \rightarrow ^{12}_{6}C + ^{1}_{0}n + \gamma\, 5.7\, MeV \qquad (13.11)$$

Another approach involves simply using radioisotopes, such as californium-254, that undergo spontaneous fission processes involving the emission of neutrons.

13.8 ^{14}Carbon dating

^{14}C dating is a very widely used technique for the dating of objects containing organic matter derived from plant tissue. This approach allows dating of objects over time spans of many thousands of years with applications ranging in palaeontology (the study of fossils) or the dating of historical objects such as fabrics derived from plant material such as cotton or wool (e.g. used for the much publicized dating of the Turin shroud).

^{14}C is formed in the atmosphere via the action of cosmic rays on CO_2 within the upper limits of the atmosphere. However, this isotope is unstable, possessing a half-life of 5568 years and decays via a β emission according to Eqn (13.12):

$$^{14}_{6}C \rightarrow ^{14}_{7}N + e^{-}(\beta) + \text{neutrino}\ (\nu) \qquad (13.12)$$

Radioactive ^{14}C in CO_2 becomes fixed within carbohydrates via photosynthesis. ^{14}C fixation stops (and therefore the accumulation of radioactive $^{14}CO_2$) when the plant naturally dies or when it is eaten by an animal; in this way, the age of materials incorporating plant derived material can be determined by radioactive counting in order to quantify the percentage of radioactive $^{14}C : ^{12}C$ within a sample.

Carbon dating based techniques and the accuracy of dating objects via this approach are not without controversy, however, since a number of assumptions must be made, including that the proportion of $^{14}CO_2$ within the atmosphere remains constant. In practice, fluctuations in the proportion of radioactive ^{14}C in the atmosphere do exist and are encountered, although in most cases these will largely nullify each other out over the lifetime of a plant.

13.9 The use of radioactive isotopes in medicine

Radiochemical techniques are also used within medicine for analytical, diagnostic, and research purposes and these are discussed in Chapter 14.

Exercises and problems

13.1. What detectors are suitable for monitoring: (i) α; (ii) β; and (iii) γ emissions?

13.2. What are the relative advantages and disadvantages of using neutron activation analyses for analytical purposes?

13.3. A neutron activation approach is used to determine the concentration of a pesticide within a soil sample. If the count rate for a sample of 1.5 g following activation is 750 counts s^{-1} and the analyte gives 234 counts s^{-1}, what is the mass of the analyte?

13.4. A mixture is to be analysed for the antibiotic oxytetracycline by a ^{14}C isotope dilution method. A 10 mg sample of pure oxytetracycline has an activity of 5000 counts min^{-1} is added to the sample. A total of 0.5 mg of the antibiotic is isolated from the sample. What is the content, in grams, of the antibiotic in the original sample?

Summary

1. Some nuclei decay via the emission of sub-atomic particles such as α or β particles; some decompositions are accompanied by the release of γ-rays.

2. Radioactive nuclides are sometimes simply termed radionuclides.

3. α and β particles as well as X- and γ-rays are all ionizing and this allows their detection by techniques such as Geiger–Müller tubes or photographic films.

4. α particles may be thought of as helium nuclei (${}^{4}_{2}H$) and because of their relatively high mass and slow velocity, may be stopped by a thick piece of card.

5. β particles are electrons or positrons and may penetrate a few tens of centimetres in air; typically they can be stopped by a thin piece of aluminium foil.

6. X- and γ-rays are similar, except that X-rays originate from electronic transitions, whereas γ-rays originate from nuclear processes.

7. γ-rays require several centimetres of lead to stop them.

8. For low-energy X-rays, energy is lost as they travel through matter via the photo-electric effect; the photons cause excitation and displacement of electrons from atomic orbitals.

9. Intermediate energy X-rays lose energy as they travel through matter by both the Compton and photo-electric effects. In the Compton effect, some but not all of the energy of a photon is lost as electrons are displaced from atomic orbitals. The photon may then continue to lose energy by further Compton or photo-electric interactions.

10. X-rays are generated via electronic transitions related to radioactive decay processes. The first of these involves capture of electrons by the nucleus; the second is via an Auger electronic transition following an interaction between an excited nucleus and an orbital electron giving rise to the loss of the electron and the emission of an X-ray photon.

11. Radioactive decay processes follow first order kinetics with well-characterized half-lives, $t_{1/2}$, in which:

$$t_{1/2} = \frac{\ln 2}{\lambda}$$

where λ is a decay constant specific for a particular radioactive decay process.

12. Half-lives range from fractions of a second to millions of years.

13. Radioactive nuclides may be utilized in isotope dilution based analyses. First, a radioactively labelled isotope form of the analyte must be prepared. The count rate for this radio-labelled analyte is next determined. A sample of this preparation is then mixed with a weighed quantity of the analyte. The analyte is again isolated and purified from the raw sample and a new count rate determined.

14. The level of dilution of the sample can be calculated in a weighed sample and thus related to the quantity of the non-radioactive form of the analyte, that is:

$$M_I = \frac{R_T}{R_M}(M_M - M_T)$$

where M_I, M_M, and M_T are the measured masses of the isolated sample, the purified mixture, and the tracer, respectively, with R_T and R_M representing the background corrected count rates for the tracer and mixture, respectively.

15. Neutron activation analyses are based on inducing radioactivity within a sample via its irradiation with neutrons from a reactor, accelerator, or radioactive nuclide source.

16. Carbon dating relies on monitoring the decay of ^{14}C with the emission of a β particle and a neutrino, and may be used for the dating of objects such as fossils containing organic matter derived from plants.

17. Radioactive nuclides are also used in medicine as well as for research purposes.

Further reading

Alfassi, Z. B. (1994). *Chemical analysis by nuclear methods*. Wiley.

Geary, W. and James, A. (1986). *Radiochemical methods*. Wiley.

Newton, G. W. A. (ed.) (1999). *Environmental radiochemical analysis*. Special Publications Series, Royal Society of Chemistry.

Bio-analytical methods 14

Skills and concepts

This chapter will help you to understand:

- How whole blood samples are collected and how blood serum and blood plasma are prepared for clinical analyses.
- The principles of blood gas analysers and their clinical significance.
- The clinical significance of blood electrolyte levels and how these may be determined.
- The scope of immunological analytical approaches.
- The principles of radio-immunological, fluorescent immunoassay, and ELISA tests.
- What is meant by first- and second-generation biosensors.
- The operation of a membrane-based first-generation biosensor.
- The operation of an H_2O_2-based glucose oxidase biosensor for glucose.
- What is meant by a mediator and how these may be used in second-generation biosensors.
- The advantages and disadvantages associated with first- and second-generation biosensors.
- The operation of simple thermometric, optical, and mass-based biosensors.
- How carbohydrate determinations may be performed based on HPLC, gas chromatography, and a number of chemical approaches including reduction-based analyses, and reaction with aromatic amines or phenolic reagents.
- How TLC, electrophoresis, GC, and HPLC techniques can be used to perform analyses for amino acids.
- How protein separations may be performed via precipitation, electrophoresis, immuno- or Western blotting, and chromatographic approaches.

- How proteins may be quantified using chemical methods including Kjeldahl, Biuret, Lowry, and bicinchoninic acid chemistries as well as dye binding approaches, spectroscopy and a number of physical-based determinations.
- The impact of proteomics and genomics on analytical science.
- Polymerase chain reaction (PCR) and how this may be used for applications such as DNA fingerprinting.

14.1 Biological chemistry meets analytical chemistry; an overview and introduction to bio-analytical chemistry

The biological sciences have enjoyed huge advances during the last few decades and have given rise to a whole new arsenal of bio-analytical tools. This chapter discusses many of the wet chemical techniques and other conventional chemistries that have for many years formed the mainstay of bio-analytical chemistry; these remain important and yet it is the areas of proteomics and genomic science that are causing the most profound changes to the way in which bio-analytical chemistry is being routinely practised. Existing techniques such as mass spectrometry are, for example, finding new application to protein sequencing, fluorescent dye labelling is central to DNA microarray technology, and the polymerase chain reaction allows many analyses to be based on DNA recognition that would otherwise not be possible. The reader is referred to Section 14.10 for a coverage of these rapidly expanding and immensely important areas.

It should not be forgotten that many target analytes of interest are either biologically derived (i.e. produced by an animal, plant, or microbe) or have biological significance (e.g. as a component within food, an atmospheric component, or environmental pollutant—to name but three examples). In many cases nature has developed chemistries that exhibit remarkable selectivity towards specific analytes that cannot yet be matched by conventional analytical approaches. Bio-analytical techniques attempt to harness this unique selectivity to allow analyses to be performed that would otherwise prove extremely difficult.

While the scope for bio-analytical chemistry is unquestionably vast, such approaches are not without their problems or limitations. In almost all cases bio-analytical determinations involve the use of biological molecules that are often labile (heat sensitive) as well as possessing poor shelf lives. Many biological reagents are, moreover, often notoriously difficult to harvest as well as being often extremely costly.

14.2 The scope of clinical chemistry and the modern hospital clinical biochemistry laboratory

Many bio-analytical techniques have originated from developments within the clinical and/or biomolecular sciences, although their applications now extend into many associated fields ranging from forensic science to environmental monitoring and food science. The largest area of application remains, however, in the field of medicine for use within hospital clinical biochemistry (pathology) laboratories, clinics, bedside settings, as well as for home monitoring.

Many mainstream analytical techniques are used for clinical applications including UV–visible absorption and flame photometry, although biochemical tests are becoming ever more widely used.

One of the mainstream workhorses found within many modern clinical biochemistry laboratories is the ***multi-analyser***, which essentially is an automated instrument based on wet chemistry tests such as titration-based analyses and UV–visible determinations. Multi-analysers are typically designed to perform 10 or more differing analyses (including blood glucose, O_2 and CO_2 partial pressures, K^+, Na^+ concentrations, heparin levels, as well as a number of urinary tests). Machines of this type do, however, require very high levels of maintenance due to the number of moving parts and the need, for example, to maintain the cleanliness of pipeworks. Many multi-analysers are supplied on a maintenance-leasing basis where the machine is owned and supplied by a pharmaceutical or clinical diagnostics company but maintained under a contract basis with an obligation that should a fault occur the company must restore the instrument to a working condition (or supply another) within a given time-period. Many larger hospitals with accident and emergency and/or critical care facilities often ensure that they have two or more multi-analysers to maintain essential analytical capability although such a policy, however, clearly adds to running costs.

14.2.1 Biological samples: blood, serum, and urine, and their use in clinical chemistry

Analyses of biological samples are amongst the most difficult to perform due to the complexity of the samples' chemistries associated with, for example, the blood clotting process or the immune system. Since the majority of biological analyses are performed within whole blood, blood serum, blood plasma, or urine, we shall briefly consider each of these first.

14.2.2 Whole blood (sample preparation and storage)

Blood drawn directly from the body (without any of its constituent components being removed) is known as **whole blood** and is used for many analyses. Blood contains a number of cellular components, comprising the erythrocytes (red cells), leucocytes (white cells), and platelets together with a number of colloidal macromolecular species and lower molecular weight solutes.

Whole blood will tend to clot when removed from the body and this will in most cases interfere with many clinical analyses unless steps are taken to prevent clotting by the addition of an anti-coagulant such as heparin or potassium oxalate.

Sodium fluoride is also normally added to blood samples taken for blood glucose determinations. The cellular components within blood are, of course, living and as such continually metabolize glucose via respiration. Sodium fluoride is a metabolic inhibitor and this helps maintain the glucose levels rather than them gradually falling with time due to the metabolic consumption of glucose. Sample tubes for blood glucose determinations can be obtained commercially and they contain small pre-prepared quantities of potassium oxalate and sodium fluoride and are normally used within clinical settings.

It is imperative that blood samples for CO_2 or O_2 determinations must be kept under anaerobic conditions and this can be achieved by introducing a small quantity of mineral oil to sample tubes; since the oil is less dense than the aqueous component, it floats and thereby covers the blood. Cork stoppers are normally used in preference to rubber bungs since the oil may cause rubber to swell.

Whole blood samples can typically be refrigerated for 48 h prior to analysis without deterioration but should always be allowed to reach room temperature prior to analysis. Whole blood samples unfortunately may not be frozen since this causes lysis of the cells.

Whole blood is not always suitable for all blood-based determinations and in these cases either plasma or serum samples should be used, the preparation of which are described in the sections that follow.

14.2.3 Clinical tests utilizing whole blood

A number of tests may be performed using whole blood, and one of the most important of these are for the determination of blood glucose levels. Blood glucose levels are often determined if a patient is either suffering from or suspected to be suffering from diabetes. It is hard to overemphasize the importance of blood glucose analyses ***since more blood glucose determinations are performed world-wide each day than any other single analytical test***. Sadly, the reason so many blood glucose determinations are required is due to the high and ever increasing incidence of diabetes in the western world and this, in turn, is related to dietary and other lifestyle factors. In some countries such as the United States and

United Kingdom more than 5% of the population now suffers from diabetes, with this figure increasing year on year.

There are a number of different approaches for determining blood glucose levels although the most commonly employed methods are based upon enzymatic biosensors. These can be laboratory-based sensors or portable devices and are described in subsequent sections.

14.2.4 Blood plasma (sample preparation and storage)

Blood plasma is formed by removing cellular components (normally by centrifugation). Blood plasma is again normally treated with either heparin or sodium oxalate as an anti-coagulant preservative. Plasma should be prepared soon after the blood sample has been taken. Blood plasma is straw coloured due to the absence of the red erythrocytes, with the remaining yellow coloration being due to the colloidal proteins. Samples may again be refrigerated for 48 h prior to analysis but may, unlike whole blood samples, be frozen for long-term storage. Care should be taken, however, to thoroughly mix samples upon thawing since fractionation into different layers may occur during the freezing process. Freezing samples for later analysis is common practice in hospitals for non-urgent tests. For these and similar analyses samples are typically collected and then a batch of tests performed together.

14.2.5 Blood serum (sample preparation and storage)

Serum is used for a number of clinical tests including calcium, magnesium, and chloride levels. Serum is again prepared via centrifugation but at a higher rotational rate than that used for the preparation of plasma to remove both the cellular components and fibrinogen. Samples can again be frozen for later use and again care should be taken to mix samples upon thawing to re-distribute evenly the solutes and/or colloidal components. Samples are often more stable if prepared as a protein-free filtrate (PFF). There are a number of methods for preparing PFF samples such as the use of trichloroacetic acid (TCA), which involves mixing nine volumes of TCA for every volume of serum and then removing the proteins via filtration following their precipitation.

Another approach known as the 'tungstic acid approach' involves adding one volume of 0.33 M H_2SO_4 together with seven volumes of water for every volume of serum and then allowing the mixture to turn brown (approximately 2 min) before adding a further one volume of sodium tungstate. The protein content of the serum will again precipitate allowing it to be removed via filtration or centrifugation.

14.2.6 Clinical tests utilizing serum

Blood electrolyte levels are normally quantified via analysis of serum with sodium, potassium, magnesium, chloride, and bicarbonate levels using atomic absorption or flame spectrophotometry.

Ion-selective electrodes are also sometimes used for the determination of calcium, potassium, and sodium within serum. Colorimetric methods may also be used for the determination of not only calcium and magnesium levels but also chloride levels.

Bicarbonate levels within serum are normally analysed via titrimetric approaches with standardized acids.

14.3 Blood gas analysers

Oxygen and carbon dioxide levels are both normally determined via dedicated O_2 and CO_2 electrodes incorporated into autoanalysers (see Section 14.2), examples of which may now be found in most chemical pathology laboratories world-wide. Whole blood samples are normally taken from the body and stored in sealed syringes or alternatively immediately placed in sample holders that are completely filled so as to exclude any air space that might otherwise allow gaseous diffusion to and from the sample. Glass syringes and/or sample containers made of glass (and not plastic) should be used to prevent any diffusion exchange with the atmosphere that can occur across most polymers to a limited extent. Samples should, moreover, always be analysed as soon as possible to prevent any gaseous interchange with the surroundings.

The oxygen electrode was developed by Leyland J. Clark in the early 1960s and consists of a platinum working electrode polarized at a potential of −600 mV vs. Ag/AgCl. The electrode is normally covered by a thin Teflon membrane that allows diffusion of O_2 to the underlying electrode whilst excluding almost all other molecules of greater molecular weights (and, therefore, possible interferents). Oxygen is electrochemically (amperometrically) reduced according to Eqn (14.1) and in this way the current observed may be related to the dissolved O_2 concentration via reference to a pre-determined calibration curve:

$$O_2 + 4H^+ + 4e^- \xrightarrow{-600\text{ mV vs. Ag/AgCl}} 2H_2O \qquad (14.1)$$

For clinical purposes, O_2 concentrations are normally expressed as partial pressures denoted as pO_2.

Dissolved CO_2 concentrations (pCO_2 levels) are determined by a modified glass potentiometric pH electrode covered with a teflon membrane to allow only the diffusion of CO_2 to the glass bulb of the electrode (for details of the operation of a *p*H electrode please refer to Chapter 10). For clinical determinations CO_2 levels are again normally expressed as partial pressure—pCO_2. CO_2 dissolves in water to form carbonic acid and this causes a lowering of pH values.

14.4 The determination of blood electrolyte levels

Blood electrolyte determinations are normally determined in blood serum and refer to the concentrations of Na^+, K^+, Cl^-, and CO_2 (HCO_3^-). A number of automated analysers are now marketed commercially, which rely on ion-selective potentiometric electrodes for the determination of each of these ions. Alternative methods for determining electrolyte levels are based upon flame photometry for the quantification of Na^+ and K^+ concentrations in blood serum, using modified pH electrodes (see Section 14.8), for the determination of dissolved CO_2 concentrations, and dedicated ion-selective Cl^- electrodes.

It is common practise to determine all of the electrolyte ion concentrations together as a suite of tests since an upset in the *relative concentration* of these ions has more clinical significance than the absolute concentration in any one of the ions. In practice, the *measured concentration of anions* in blood serum will always appear to be less than that of the *total cation concentration*. There are a number of reasons for this including the fact that there are a number of cations such as calcium and magnesium that are more difficult to determine in routine analysis. This measured imbalance is known as the ***anion gap*** and is expressed in terms of millimole concentration with a typical value for healthy subjects being approximately 12 mM.

It should be noted that the 'anion gap' is, in fact, a misnomer and corresponds to a 'measured anion gap' since the law of electroneutrality demands that in any solution the charge carried by the anions within solution *must always* equal that carried by the cations. Changes in the concentrations of each of the anion or cation concentrations do, however, give rise to changes in the measured anion gap, which has clinical significance since this is often indicative of a number of physiological disorders—particularly if the measured value is either <5 or >22 mM. Physiological conditions that can cause disruption of the measured anion gap include renal failure, raised blood pressure and atherosclerosis, ketoacidosis due to diabetes, alcohol poisoning, starvation, and/or the administration of a number of illicit or ethical drugs. Since so many conditions can give similar symptoms diagnosis normally requires a number of other tests to be performed.

14.5 Immuno-chemistry techniques

Immuno-chemistry techniques are now widely used for the clinical determination of extremely low concentrations of drugs, hormones, vitamins, and other compounds.

Immunological tests exploit the production of antibodies by many higher animals in response to foreign agents whether they be of biological (e.g. micro-organisms), or non-biological origin (such as airborne pollutants). The antibody produced by an organism is normally a gamma globular protein or immunoglobulin that is designed to bind or couple with the antigen to form an antigen–antibody complex.

Immunological analytical techniques are normally based upon competitive binding between an antibody and an antigen—and an antibody with a tagged antigen. In the test, the antigen is normally the analyte and a tagged form of the antigen is specially prepared in known concentrations. The tagged antigen is formed from the coupling of an antigen to some form of marker, such as an enzyme radioactive tracer, a fluoroprobe, or other molecule, that can be readily analytically quantified.

Antibodies for immunological tests are normally raised and harvested (produced) by injecting the antigen into an animal and recovering antibodies from blood serum.

14.5.1 Radio-immunological tests

The development of radio-immunological assays (RIAs) led to the award of a Nobel prize to Rose Yallow in 1977, and this approach is based on the ability of antibody proteins to bind to a radio-labelled antigen. Radio-labelled antigens must in all cases be first produced for use in the test although in many cases these can be purchased commercially. Antigens are normally labelled with ^{125}I, ^{131}I, ^{3}H, or ^{14}C. ^{3}H and ^{14}C are low-energy beta emitters, whereas ^{125}I and ^{131}I are both gamma emitters. Therefore the use of ^{125}I and ^{131}I has particular safety implications, and specific shielding requirements must be met.

The determinations are based upon a competitive binding of the antigen (the analyte) and a known molar quantity of antibody. The next stage is to separate the antigen–antibody complex so formed from the original analyte sample and any residual antibody and radio-labelled antigen. The radioactivity of the antibody–antigen complex may then be determined and related to the concentration of the analyte. It follows that the lower the radioactivity of the separated antibody–antigen complex, the greater will be the concentration of the analyte following binding of the unlabelled analyte to the antibody. In a similar manner, higher levels of recorded radioactivities for the separated antibody–antigen sample will correspond to lower concentrations of analyte, since less analyte will have bound with the antibody and this, in turn, will have allowed more of the radio-labelled antigen to bind with the antibody.

Radio-immunological tests of this type typically offer extremely favourable sensitivity and lower limits of detection down to nanomolar concentrations or better—as well as permitting analytical concentration ranges extending over many orders of magnitude.

The specificity offered by radio-immunological tests is extremely favourable in comparison to many other analytical approaches although this may be compromised by a number of processes such as the presence of and cross-reaction with other antibodies. Some of these processes can be minimized via the production and use of ***monoclonal antibodies***; that is, sources of antibodies produced with high purity towards a single antigen.

Antigens of molecular weights less than 1000–5000 will normally be too small to induce antibody formation and are known as ***haptens***. Haptens may, however, in some cases be linked to simple proteins to allow antibodies to be raised. An antibody preparation is known as an ***antiserum*** and these typically require freezing for long-term storage.

Incubation times that are required to reach equilibria for antigen–antibody binding vary from a few hours to a few days depending on the particular reactions.

The antigen–antibody complex must next be separated from the incubation mixture and this is often achieved by precipitation due to the addition of a solvent such as acetone or high concentrations of a salt such as $(NH_4)_2SO_4$. The sample may then be finally recovered following precipitation via centrifugation or filtration.

14.5.2 Fluorescence immunoassays

Antigens can also in some instances be labelled with fluorescent dyes for ***fluorescence immunoassay based assays***. There are two principle advantages that fluorescence-labelling techniques offer in comparison to radio-labelling approaches: first, fluorescent labels generally have fewer safety issues related to them; second, fluorescent labels do not decay with time and are, therefore, devoid of any of the problems associated with short half-life isotopes. Great care must be taken, however, to ensure that the analyte sample is free from proteins other than the analyte, since these may also become labelled and could lead to erroneous results.

In many cases the labelled antigen–antigen complex must be separated from the analyte sample and unreacted labelled antigen, for example, via the use of size exclusion chromatography approaches to remove any fluorescence-labelled antigen. In some cases, however, excess labelled antigen may be sufficiently quenched (see Chapter 5), thus avoiding the necessity for its removal.

14.5.3 ELISA tests

Enzyme-linked immunosorbent assays (***ELISAs***) have in recent years become one of the most popular forms of immunologically based tests.

There are a number of different ELISA formats, but all rely on the labelling or tagging of an antibody or antigen with an enzyme and determining the activity of the enzyme following reaction, or *incubation* of the antibody with its antigen. Enzyme activity can be determined by differing approaches depending on the enzyme in question, although many methods are based on colorimetric techniques.

Non-competitive ELISAs are based on the inhibition of the enzyme activity upon binding of a tagged antibody with an appropriate antigen. ***Competitive ELISAs*** are based on determining the activity of a known quantity of enzyme-tagged antigen following: (i) incubation with the analyte sample (containing the antigen/analyte), with the antibody immobilized on a suitable surface; and (ii) separation from the antibody. The enzyme activity remaining corresponds to the non-bound enzyme-tagged antigen and so the greater the concentration of analyte, within a sample, the lower will be remaining enzyme activity. Conversely, higher enzyme activities correspond to lower analyte concentrations.

Indirect ELISA tests are performed by first adsorbing the sample antigen (analyte) onto a suitable support and allowing this to incubate with a known quantity of an unlabelled (primary) antibody. The support is then washed and allowed to further incubate with a further (secondary) enzyme-tagged antibody. The enzyme activity of the excess antibody is again determined via a suitable approach and thus the concentration of the analyte (antigen) can be determined. One advantage that indirect-ELISA-based tests offer is that a common enzyme-linked secondary antibody can be raised against all primary antibodies of the same immunoglobulin class and this, in turn, removes the need to produce a whole series of enzyme-linked antibodies against each antigen (analyte) to be determined.

14.5.4 Simplified immunological test kits for use in clinics or at home

We have already seen how immunochemistry has given rise to a large number of analytical tests of biological and/or clinical significance. Tests such as these offer near unrivalled sensitivity for the determination of trace compounds in biological fluids. Immunological chemistry has, moreover, allowed the development of a number of highly sensitive and operationally simple tests for rapid analysis for use during surgery or even at home.

The most widely known example of such a test is the pregnancy urine test, many forms of which have been developed although all rely on the

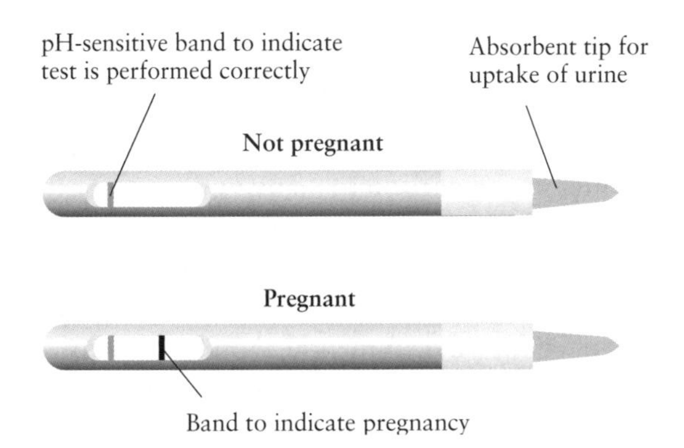

Figure 14.1 hCG-based pregnancy test.

determination of human chorionic gonadotrophin (hCG) levels. hCG is released into a woman's urine during pregnancy and may be used as an early indicator of pregnancy. There are a number of variations of this test currently being marketed although most operate essentially on similar principles. Coloured monoclonal anti-α-hCG antibodies are prepared and impregnated into a porous pad. An aliquot of urine is allowed to soak onto the pad and any hCG that is present will bind with the antibody to form a selenium-labelled anti-α-hCG complex that will continue to travel along the pad by capillary action till it reaches and travels past a region at which a polyclonal anti-hCG is immobilized.

If hCG is present in the sample, then the selenium-labelled anti-α-hCG formed with the monoclonal antibody will bind with the polyclonal antibody that is immobilized along a line to capture and thus concentrate the colour of the selenium complex. The remainder of the sample travels along the pad to interact with a pH-sensitive indicator to show that the test has been performed properly.

The test kits are available in a number of different forms but most are packaged in a pen-like housing with the absorbent pad protruding from one end of the device in the form of the 'pen tip'. A window along the 'pen barrel' (Fig. 14.1) permits viewing of the polyclonal antibody region, which becomes coloured if the test is positive. The pH-sensitive dye region at the end of the absorbent pad also becomes coloured to indicate that the test has been performed properly.

If used correctly, tests of this type can prove extremely sensitive and give results within typically 2 min of the introduction of the urine sample. Tests almost never give false positive results since a coloration in the positive window can only occur if hCG is present in the urine sample. False negative results can occur, however, if the levels of hCG are below those that are capable of giving rise to a discernible colour change to the human eye in the positive window. Commercial tests claim that results with >99% accuracy may be obtained on the first day after an expected period is due.

14.6 Introduction to biosensors

Biosensors are becoming ever more significant due to the simplified analyses they potentially offer in comparison to many more complicated and, therefore, costly analytical approaches. Biosensors in many cases allow very rapid analyses to be performed in situations that would otherwise prove to be totally inappropriate via other means, such as, for example, during surgery or by the side of a river for environmental monitoring. For these reasons biosensors are continuing to be the focus of much research although the greatest number of commercial sensors are for blood glucose determinations. Blood glucose determinations are likely to continue to dominate the biosensor market in the years to come, ***since more glucose determinations are performed than any other type of analytical measurement in the world.*** At the time of writing, in excess of 5% of the adult population of both the United States and United Kingdom suffer from diabetes and this figure is continuing to increase largely due to western lifestyle and dietary habits.

The determination of blood glucose levels has since the 1960s been greatly simplified due to the development of glucose biosensors, which have now become the most widely used approach for blood glucose determination.

There are number of differing definitions for what constitutes a biosensor, although probably the most widely accepted description is: *an analytical device in which a biological recognition entity is coupled in close proximity to a transducer*. In this context, the biological recognition entity may be an enzyme, antibody, cell, or even tissue slice. The transducer is a device that allows a quantifiable signal to be produced from a biological recognition event—that is, when the biological entity has recognized the presence of the analyte. There are again many types of transducers, including electrode, optical, mass determination, and/or thermal devices, to name but a few examples. There are many differing approaches for biosensor fabrication although these are normally characterized into groupings according to the type of transducer used. It is, therefore, common to find sensors being defined as, for example, 'amperometric electrochemical'-based devices or thermistor-based sensors, etc. Electrochemically-based sensors have, to date, enjoyed considerable attention as well as commercial success, and for these reasons we shall give considerable attention to these devices.

14.6.1 Electrochemically based biosensors

There have been many electrochemically based sensors developed since the first biosensor was described by Leyland J. Clark in the early 1960s. This first sensor utilized the enzyme glucose oxidase for the determination of glucose. This work lead to the commercialization of the first electrochemical blood glucose sensor by the Yellow Springs Instrument company in the United States. Updated versions of this instrument based on the original technology developed in the 1960s are still marketed today, being widely used in clinical biochemistry laboratories world-wide.

The Clark membrane-based glucose sensor (an example of a first-generation sensor)

The 'Clark' enzyme electrode is based on the enzymically catalysed oxidation of glucose by glucose oxidase, Eqn (14.2):

$$\text{Glucose} + O_2 \xrightarrow{\text{Glucose oxidase}} \text{gluconolactone} + H_2O_2 \tag{14.2}$$

The 'Clark' enzyme electrode is an example of what has, since its first description, become known as a 'first-generation' sensor format. The so-called 'first-generation' electrochemical sensors are based on the direct monitoring of: *either the depletion of one of the enzyme substrates or the accumulation of one of the enzyme products.*

Clark and co-workers described two glucose sensors based on first-generation principles utilizing the glucose oxidase reaction of Eqn (14.2).

The first of these relied on the amperometric reduction of O_2 at a cathodically polarized platinum working electrode, Eqn (14.3):

$$O_2 + 2H_2O + 4e^- \xrightarrow{-700\text{ mV vs. Ag/AgCl}} 4OH^- \tag{14.3}$$

This approach is, however, particularly susceptible to ambient fluctuations in O_2 levels that can, in unfavourable conditions, lead to unreliable results. This sensor comprises an oxygen electrode (Section 14.3) with glucose oxidase immobilized above the Teflon oxygen selective membrane.

The second approach Clark pioneered was based on the monitoring of H_2O_2 produced by the enzymically catalysed oxidation of glucose according to Eqn (14.4). In this arrangement, H_2O_2 may be amperometrically oxidized at an anodically polarized platinum electrode:

$$H_2O_2 \xrightarrow{+650\text{ mV vs. Ag/AgCl}} 2H^+ + O_2 + 2e^- \tag{14.4}$$

This approach gives responses that within a given concentration range may be directly related to the glucose concentration. This approach has proved to be generally far more reliable than those based on O_2 monitoring and indeed many laboratory-based commercial blood glucose analysers are still based on the amperometric oxidation of H_2O_2.

The majority of first-generation sensors employ two functional membranes between which the enzyme is immobilized to form an enzyme membrane laminate, Fig. 14.2. The underlying or inner membrane is ***permselective*** and serves to act as a screen to prevent electrochemical

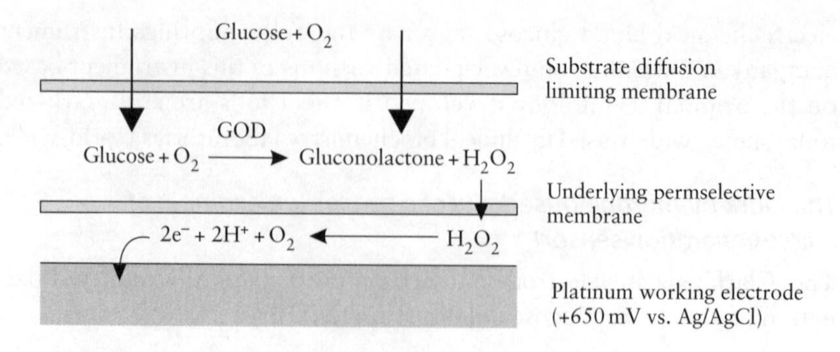

Figure 14.2 First-generation membrane-based sensor.

interferents from reaching the working electrode, while the outer membrane serves to act as both an outer biocompatible interface with the biological sample and a permselective barrier. We shall consider the function of these two membranes and the problems they seek to address in the following sections.

Chemical interferents for first-generation sensors and the use of permselective membranes Enzymes are uniquely specific catalysts and yet sensors are still susceptible to electrochemical interferents that can lead to erroneous responses. Although, for example, in the case of glucose oxidase based sensors, the production of H_2O_2 only occurs following the enzymic catalysed oxidation of glucose, this does not prevent the electrochemical oxidation of interferents such as ascorbic acid or paracetamol, which can be present in blood. The problems of electrochemical interferents are normally tackled using ***permselective membranes*** placed in between the enzyme layer and the working electrode.

Permselective membranes are normally formed from thin polymer films with intrinsic anionic charges within the polymer structure and may be fabricated from, for example, cellulose acetate, polyvinyl chloride (PVC) or the commercial perfluoro polymer Nafion®. Permselective membranes act to prevent the passage of most oxidizable interferents via a charge exclusion principle, Fig. 14.3, since most of these interferents dissociate upon dissolution in water to form anionic solutes together with the release of protons. For example, ascorbic acid or Vitamin C dissociates to form an ascorbate ion and uric acid gives rise to the urate ion in aqueous solution. H_2O_2, by contrast, is a neutral molecule in aqueous solution and is capable of diffusing across the permselective barrier with relative ease. The greatest problems are associated with electro-oxidizable interferents that are either neutral or extremely weakly dissociated in solution (such as paracetamol) and so can also diffuse across anionic permselective barriers and be oxidized at anodically polarized working electrodes. Since paracetamol is a

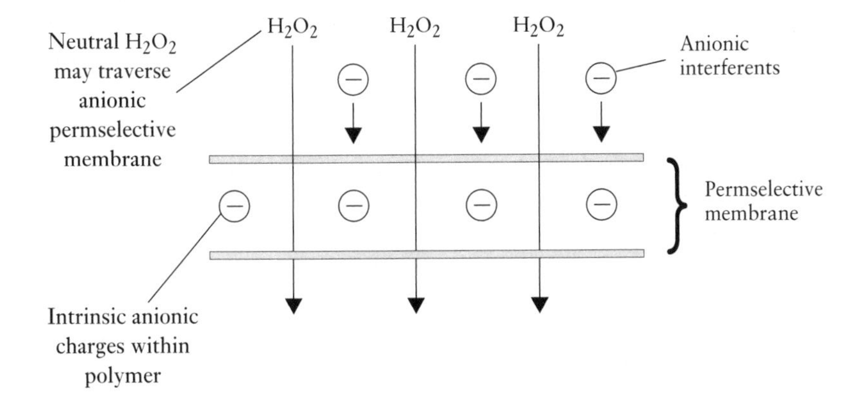

Figure 14.3 Anionic interferents unable to traverse a permselective barrier.

very widely used and easily obtainable analgesic that may be purchased without prescription, clinicians must take great care to instruct patients not to take paracetamol in the 24 h before a blood glucose determination. If an emergency blood glucose test is required then it must be established whether or not the patient has taken paracetamol during the previous 24 h or if this is not possible an alternative approach must be adopted.

The use of outer covering membranes for the linearization of sensor responses and the protection against the effects of biofouling

The outer covering membrane has to serve a number of functional roles. The first of these is to help extend (linearize) the sensor responses to cover a working concentration range over a clinically (or other) significant range. The problem is that the Michaelis–Menton constant (k_M) for many enzymes is significantly below a concentration range that may have to be determined for practical applications. The k_M value is defined as the concentration for a particular substrate that will allow the enzyme turn-over rate to be half of its maximal rate at a given temperature and pH provided that all other substrates are present in excess quantities, Fig. 14.4. Many

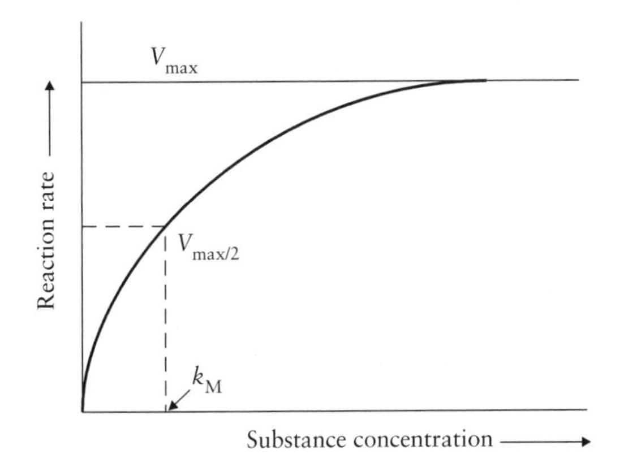

Figure 14.4 Plot of enzymatic reaction rate versus substrate concentration demonstrating Michaelis–Menton kinetics.

enzyme reactions (e.g. glucose oxidase) (Eqn 14.2) will have two or more Michaelis–Menton constants—one for each substrate. It follows that if an enzyme is exposed to a concentration for a given substrate in excess of the k_M value, then the enzyme activity will progressively tend towards a plateau and this will, in turn, limit the sensor response. If the sensor response tends towards a plateau at concentrations that may need to be determined, then some steps must be taken to allow higher concentrations to be quantified.

The simplest approach is to dilute the analyte sample by a known amount, although this can lead to complications such as the lysis of blood cells. Other approaches involve modifying the sensor in some way. Remember that the enzyme-catalysed reaction constantly consumes substrates. A membrane placed over the enzyme layer that can modulate the *rate* of substrate diffusion can, in turn, allow the concentration of the enzyme substrate (analyte) experienced by the enzyme to be lowered in proportion to the concentration of the substrate in the analyte sample. For example, the substrate diffusion and the limiting membrane may allow the concentration of the substrate experienced by the enzyme to be lowered by 50%; thus, if the concentration of glucose in the sample is 10 mM glucose, the enzyme will only experience a concentration of 5 mM. Membranes of this type are known as ***substrate diffusion limiting membranes*** and are often fabricated from microporous polymeric materials such as commercial microporous polycarbonate micro-filtration membranes.

The second problem that the outer membrane has to address is that of providing a *biocompatible outer barrier* that is capable of overcoming the problems associated with biofouling processes such protein deposition and blood clotting. *Biofouling* is a term used to describe many processes that can involve, for example, protein deposition and cellular attachment. The greatest surface biofouling effects are observed with whole blood and arise from the blood clotting cascade. It should be remembered that a blood clot is, of course, designed to form a waterproof seal over a wound. A blood clot forming over the surface of a sensor will, therefore, progressively coat and inhibit the diffusion of the analyte (enzyme substrate) to the enzyme and this, in turn, will lead to a loss of enzyme response until the sensor surface becomes completely sealed. While the biofouling of a surface may never be totally prevented, tailored biocompatible coatings at the surface of covering substrate diffusion limiting membranes, such as polysiloxanes (silicones), some polyurethanes, or an amorphous form of carbon known as diamond-like-carbon, can prevent most if not all surface cellular and/or platelet adhesion. It should be appreciated that there is no such thing as a totally biocompatible surface and that some protein will be deposited on almost any surface following exposure to whole blood. Proteins deposited at the outer surfaces of sensors will always act as an extra diffusional barrier to analytes gaining access to the enzyme. A slow but continual deposition of protein at the surface would lead to a slow but continual loss of sensor performance with time, which would render quantitative analysis impossible. In practice, covering membranes are, therefore, sought that can adsorb stable coatings of protein but do not increase in thickness with time; in this way a stable modified outer surface may be formed allowing calibration of the sensor following an initial exposure of the sensor to the biological fluid. Many sensors are, for this reason, designed so as to be exposed to the biological fluid sample as a pre-conditioning treatment prior to calibration.

Another process known as ***electrode passivation*** often leads to a further loss of sensor response on exposure to biological samples. Passivation occurs as a result of low molecular weight solutes (such as phenolics) being able to pass through both the outer biocompatible/substrate diffusion limiting and inner permselective membranes. Upon reaching the working electrode these solutes can, in some instances, irreversibly coat and thereby passivate or partially insulate the electrode surface. These problems are harder to overcome although the use of pre-conditioning periods often helps to overcome the worst of these effects since a near steady-state coating on the electrode is often formed after a given time exposure to the biological analyte.

The ferrocene-mediator-based blood glucose sensor (an example of a second-generation biosensor)

The so-called 'second-generation' of electrochemical biosensors are those that utilize charge transfer ***mediators*** to facilitate charge transfer from the enzyme to a working electrode. One practical example of a sensor utilizing a mediator is the ExacTech range of portable pocket-sized blood glucose sensors marketed by Medisense®. This sensor, at the time of writing, represents the most widely used and commercially successful biosensor to date. This device revolutionized the treatment of diabetes for many sufferers since it, for the first time, allowed diabetics to monitor their blood glucose levels at home or at work during the day via a highly simplified and reagentless approach.

The sensor again utilizes the enzymatic oxidation of glucose, but the mediator ferrocene, Fig. 14.5, now acts as the electron acceptor in place of molecular oxygen, Fig. 14.6. Ferrocene is a iron-containing bi-pentyl carbon ring compound that may be easily and reversibly oxidized or

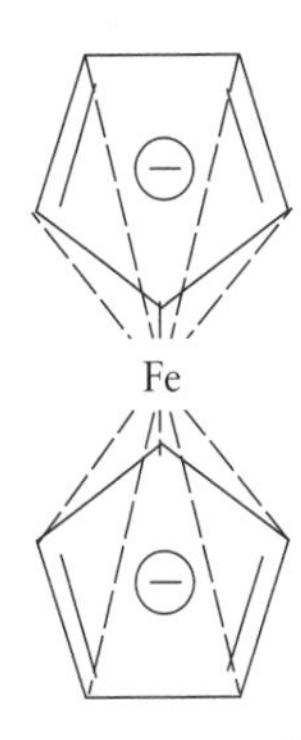

Figure 14.5 Structure of ferrocene.

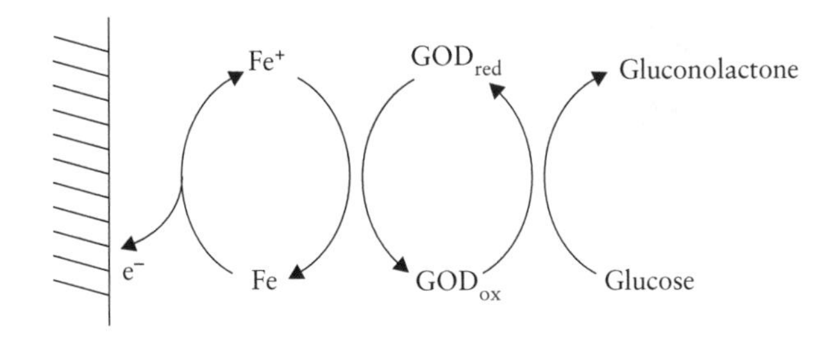

Figure 14.6 Ferrocene acting as artificial electron acceptor in place of O_2 for glucose oxidase.

reduced between the Fe/Fe^+ redox states. The ferrocene once reduced by the enzyme glucose oxidase may be re-oxidized by the electrode, thus giving up the electron it acquired from the reduced form of the enzyme. The mediator can once more acquire an electron from another enzyme molecule and in this way can constantly shuttle charge from the enzyme to the working electrode. The mediator is not consumed in the process but rather is constantly re-cycled, ready for further use.

One of the benefits of using a mediator is that the enzyme process now becomes independent of localized O_2 concentrations and it will be recalled that this is one of the principle drawbacks associated with the 'Clark'-type enzyme electrodes. The mediator may, moreover, be oxidized at the working electrode at a lower overpotential of approximately +300 mV vs. Ag/AgCl in comparison to the +650 mV vs. Ag/AgCl required for the oxidation of H_2O_2. Many anionic biological interferent solutes such as ascorbate (ascorbic acid) and urate (uric acid) are not capable of being oxidized at such a low potential and their effects may, therefore, be eliminated. In certain cases, some interferent species such as paracetamol may be oxidized at a working potential of +300 mV (vs. Ag/AgCl) and, therefore, the effects of chemical interferents cannot be completely discounted.

It should be remembered that the mediator diffuses between the active site of the enzyme and the surface of the working electrode. Unfortunately, the inclusion of a permselective membrane as used in 'Clark'-type biosensors would prevent the diffusional movement of the mediator and for this reason permselective membranes cannot be used in mediator-based sensors as a further protection for preventing the access of inteferents to the working electrode.

The Medisense blood glucose sensor utilizes a disposable strip in conjunction with a small potentiostat with an LCD for easy display of blood glucose readings. Various forms of the portable instrument have been developed for use with the same strips, the most popular of which is a pen-shaped sensor, Fig 14.7, which can be easily kept in a jacket, handbag, or purse.

Blood glucose tests are performed by first inserting the strip into the end of the instrument, which activates the unit from its 'sleep' mode. A finger

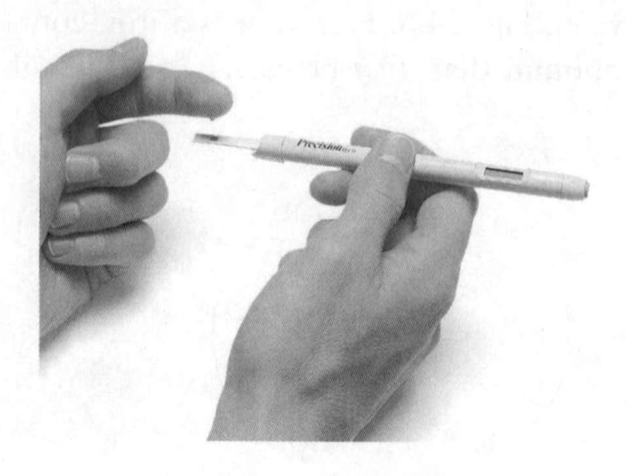

Figure 14.7 Medisense pen.

lance, again with a disposable tip, is then used to obtain a drop of blood at the end of a finger, which may be placed upon the sensor tip thus bridging the working, secondary, and reference electrodes. The drop of blood completes the circuit by allowing an electrical contact to be made between the electrodes and this, in turn, initiates a 20-s countdown on the LCD during which time the current due to the enzymatic oxidation of glucose is recorded. The current recorded is then referred to a calibration curve (look up table) in the electronic memory of the instrument and the concentration displayed on the LCD.

Figure 14.8 Medisense companion.

One of the most commonly encountered symptoms associated with diabetes, especially with more elderly suffers, *is diabetic retinopathy*, in which eyesight impairment occurs due to the death of retinal cells in the eye following irregular blood glucose levels, normally for many years. Such sufferers often find reading difficult and for these patients the 'credit card'-shaped sensors, Fig. 14.8, with larger displays often prove more suitable than the pen-shaped devices.

14.6.2 Thermometric, optical, and mass determination based biosensors

Despite the inherent sensitivity and widespread applicability that electrochemistry offers, a number of other biosensors based on a number of different transduction approaches exist.

Thermometric sensors are based on the determination of heat generated via exothermic enthalpy changes that accompany enzymatically catalysed reactions. Glucose oxidase, for example, as we have seen, catalyses the oxidation of glucose to H_2O_2 and gluconic acid, and this is accompanied by the release of 80 kJ mol^{-1} of glucose oxidized as an exothermic reaction. In devices such as these, the enzyme is immobilized at or close to the surface of a transducer to allow a quantitative monitoring of the rate of the enzymatically catalysed reaction. The greatest problem associated with devices of this type is the temperature fluctuations in the ambient environment, which can give rise to erroneous results.

Optical transduction approaches are also widely used in biosensor designs. Optical techniques, as we have seen, are very widely used for enzyme assays and a number of other biological determinations. The use of optically based transduction approaches within biosensors can therefore be thought of as a logical progression building upon existing biological techniques in the developing field of biosensor technology.

A good example of how optical techniques may be used is in biosensors that utilize nicotinamide dinucleotide (NAD) or nicotinamide dinucleotide diphosphate (NADP) dependent enzymes. The dehydrogenase enzymes utilize either NAD/H or NADP/H as co-factors that are re-cycled during the enzyme reaction. Two examples are given below for the

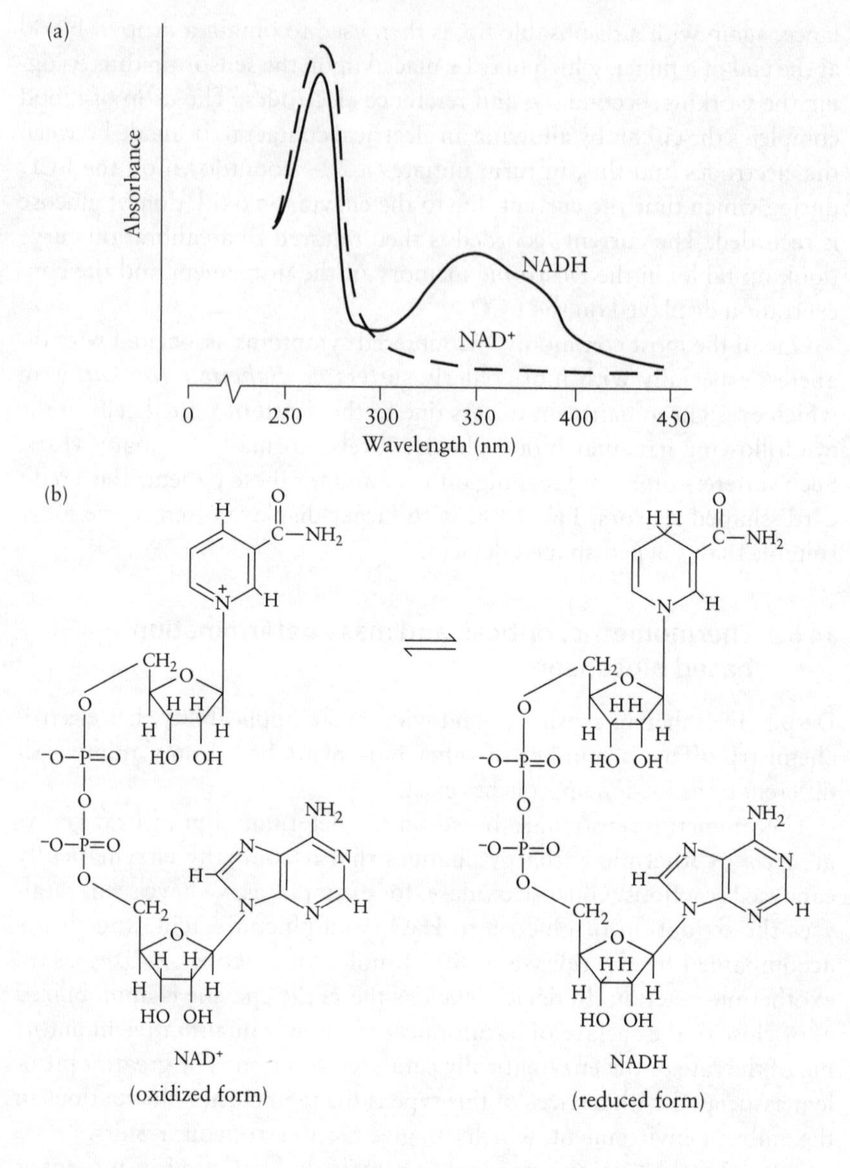

Figure 14.9 (a) UV–visible spectra for NAD^+ and NADH. (b) Structures of NAD^+ and NADH.

enzymes glucose dehydrogenase (Eqn (14.5)) and lactate dehydrogenase (Eqn (14.6)).

$$\text{Glucose} + H^+ + NAD^+ \xrightarrow{\text{glucose dehydrogenase}} \text{gluco-6}'\text{-lactonephosphate} + NADH + H^+ \quad (14.5)$$

$$CH_3CHOHCO_2^- + NAD^+ \xrightarrow{\text{lactate dehydrogenase}} CH_3COCO_2^- + NADH + H^+ \quad (14.6)$$

The UV–visible absorption spectrum of NAD or NADP, Fig. 14.9(a), changes significantly upon reduction with the generation of a new absorption peak at approximately 350 nm, Fig. 14.9(b). Monitoring the absorption of a NAD/H or NADPH enzyme dependent reaction, therefore, allows us to monitor the rate of the enzymic reaction and the concentration of the substrate (analyte) in question. This may be incorporated into the format of a biosensor relatively easily by immobilizing the enzymes and other biological reagents close to or at the surface of an optical fibre (often termed an *optrode*). The optrode is then often coupled to a suitable absorption spectrometer or colorimeter that now may often be miniaturized into the form of a hand-held unit for ease of use.

14.6.3 DNA sensors

DNA sensing techniques based on different modes of nucleic acid interaction offer great promise for clinical monitoring. Applications are numerous and include the identification of nucleic acid sequences from pathogenic (disease causing) micro-organisms for the facilitation of the diagnosis of microbial based diseases. DNA-based sensors also hold promise for the detection of toxins such as those capable of acting as mutants and/or carcinogens, due to the interaction and subsequent damage they can cause to DNA.

There are many different approaches that are currently being explored for the development of DNA-based sensors. However, one arrangement that is receiving particular interest involves coupling nucleic acid recognition layers with electrochemical transducers. Many devices of this type rely on nucleic acid hybridization via DNA base pair matching and involve immobilization of a short (20–40mer) synthetic oligomer (the single-stranded DNA probe), which is complementary to the sought for target.

The sensor upon exposure to the sample in this way allows base pairing hybridization of the complementary nucleic acid sequences to allow duplex formation, which may be measured by either optical or electrochemical approaches. In this context the hybrid may often be detected by exposure to a solution of either an electroactive or coloured indicator such as a redox-active cationic metal complex that can bind strongly to the hybrid to allow signal transduction.

14.7 Carbohydrate determination

There are many different approaches available for both the qualitative and quantitative determination of carbohydrates and the approach chosen will depend on both the composition of the mixture and the specificity required towards each or any of the components.

Historically, many determinations were based upon the reaction of monosaccharides with phenylhydrazine and then recovering the osazone derivative that may be identified by melting point determination and observation of the characteristic crystalline structure. One of the problems with this approach, however, is a lack of specificity since the reaction involves two carbon atoms of the sugar. Also, the three hexoses glucose, mannose, and fructose all share the same enediol form and so yield identical osazones.

14.7.1 Separation of mixtures of carbohydrates

In practice, many samples can contain mixtures of different carbohydrates and will need some separation as part of the analytical process. Qualitative and semi-quantitative determinations may often be performed by means of either paper or thin-layer chromatography using a solvent system chosen to facilitate the separation of the suspected constituent within the sample. There are no simple rules available for the choice of the solvent system although Table 14.1 gives some possible suggestions for solvent mixture systems if some knowledge of the carbohydrate mixture is known.

Table 14.1 Choice of solvent mixture for resolution of carbohydrate mixtures by TLC

Solvent mixture v/v and % composition	Uses
n-Butanol—45% Pyridine—30% Water—25%	Used for a wide range of separations and when composition not known Optimal support: cellulose
Tertiary butanol—40% Methyl ethyl ketone—30% Formic acid—15% Water—15%	Separation of monosaccharides and disaccharides Optimal support: cellulose
Ethyl acetate—55% Pyridine—30% Water—15%	Separation of pentoses and hexoses. Will separate glucose and galactose Optimal support: cellulose
Ethyl acetate—60% Ethanol—10% Pyridine—10% Acetic acid—10% Water—10%	Separation of pentoses and hexoses Optimal support: silica gel
n-Butanol—50% Acetic acid—25% Water—25%	Separation of monosaccharides and disaccharides. Also useful for separation of sugar acids Optimal support: silica gel
n-Butanol—50% Acetic acid—30% Diethyl ether—15% Water—5%	Separation of monosaccharides and disaccharides and also mixtures of mono-, di-, tri-, and oligiosaccharides Optimal support: silica gel

14.7.2 Determination of carbohydrates via HPLC

Carbohydrate mixtures may normally be resolved and determined quantitatively using cation exchange columns with water as the mobile phase at 70°C. Reverse phase HPLC (Section 14.8) using an acetonitrile:water solvent mixture, for example, may also be used to resolve many carbohydrate mixtures. Chemically bonded stationary phases of the quaternary ammonium cation-exchange type together with a mobile phase of 50:50 v/v acetonitrile–0.1 M acetic acid may also prove beneficial in some instances.

Separation of carbohydrates can also be facilitated via the use of alkaline borate buffers since this promotes formation of anionic carbohydrate complexes that are more easily resolved via ion-exchange chromatography.

14.7.3 Determination of carbohydrates via gas chromatography

Gas chromatography may in some situations be the most appropriate approach for the quantification of two or more carbohydrates within a mixture, especially when present in trace quantities. GC allows the resolution of carbohydrates of very similar structure and will yield separate peaks for the enantiomers for some monosaccharides.

One of the most significant disadvantages with using GC for carbohydrate analysis is that the carbohydrates must be first derivatized to form volatiles such as trimethylsilyl-*O*-methyl oximes, *O*-methyl ethers, *O*-acetyl ethers, or *O*-trimethylsilyl ethers. The trimethylsilyl (TMS) derivatives that are most widely used may be prepared using a 2 : 1 : 10 mixture of hexamethyldisilazane (HMDS), trimethylchlorosilane (TMCS), and pyridine. In many circumstances, it is important to use only weak silylating conditions such as these since this will prevent random isomerization that can, in turn, give many spurious chromatographic peaks. Conversely, if the mixture contains carbohydrates combined with nucleic acids or containing amino, carboxylic, or phosphate moieties, then a stronger silyating agent will be required.

The optimal choice of stationary phase has again to be chosen depending on the mixture to be resolved, although a methylpolysiloxy gum (OV-1) column will suffice for many carbohydrate mixtures. Further resolution may sometimes be achieved via the use of two-dimensional chromatographic runs at 90° to each other with different solvent systems (see Chapter 8).

14.7.4 Determination of carbohydrates via chemical methods

Many of the early chemical approaches for determining carbohydrates offer advantages in terms of simplicity and also cost but are unfortunately

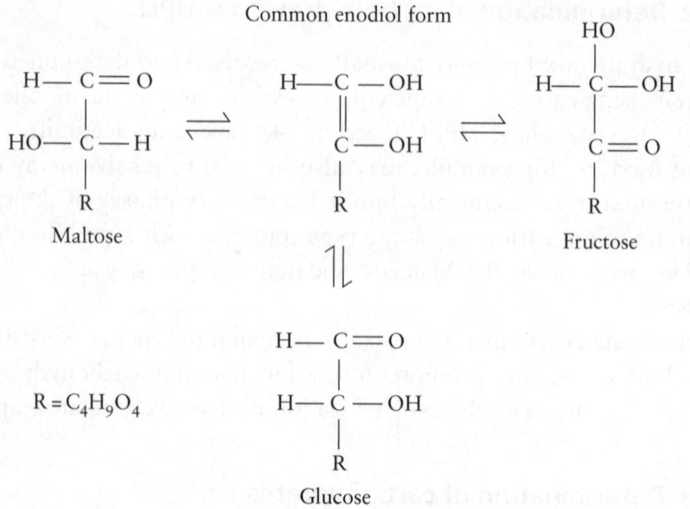

Common enodiol form

Maltose

Fructose

$R = C_4H_9O_4$

Glucose

Figure 14.10 Carbohydrates in enediol form.

inherently non-specific and this limits their usefulness. Chemical tests typically offer qualitative information, although in some cases may be used for semi-quantitative determinations.

Reduction approaches

Carbohydrates containing potentially free aldehyde or ketone groups are known as reducing carbohydrates, since in basic conditions they reside in the enediol form (Fig. 14.10), which can act as an effective reducing agent. Tests based on the reducing power of a saccharide may be used provided that the aldehyde or ketone group of at least one of the monosaccharides has not been eliminated in the glycosidic bond. Not all carbohydrate sugars are reducing and we consider sucrose as an example; the anomeric carbon atoms of both monosaccharides are involved in the glycosidic bond and it is for this reason that sucrose is not a reducing sugar.

It follows that tests based on the reducing power of certain carbohydrates may also be used to distinguish between reducing and non-reducing sugars.

A group of tests (including those using Fehling's or Benedict's reagents) are based on the reduction of cupric (Cu^{2+}) ions to the cuprous state (Cu^{+}), which forms yellow cuprous hydroxide in basic solutions and upon heating forms insoluble red cuprous oxide (Cu_2O). It is necessary to keep the cupric salts in solution and prevent precipitation. In the case of Benedict's reagent this is achieved by the inclusion of sodium citrate in the reaction mixture while in the case of Fehling's solution sodium potassium tartrate is used.

Reaction with o-toluidine and other aromatic amines

Aldoses and ketoses will react with o-toluidine and a number of other aromatic amines in glacial acetic acid to form coloured products whose λ_{max}

may allow identification of a particular sugar although confirmation by other tests should also be obtained. A number of aromatic amines that will undergo reactions of this type have now been banned in most western countries as well as other parts of the world due to their carcinogenic properties.

The formation of coloured compounds with *o*-toluidine is also often useful for the visualization and possible identification of sugars following separation by thin-layer or paper chromatography.

Reactions with phenol and strong acids

Pentoses and hexoses will react on heating with a strong acid and phenol to form coloured products and although this reaction may only be used for qualitative tests to identify the presence of either a pentose or hexose, both of these react in a similar manner. There are a number of variations based on this type of analysis, the most popular of which is the ***Molisch test*** in which concentrated sulphuric acid and α-naphthol are used to form red–violet compounds in the presence of a carbohydrate.

The basis of this type of test is the dehydration by the acid of the pentose or hexose to form furfural or hydroxyfurfural derivatives, the aldehydes moieties of which will then condense with the phenol to form a coloured product.

14.8 Analysis of amino acids

The quantization and identification of amino acids in a mixture is often required, for example, for metabolic studies or alternatively to help with the elucidation of a protein structure. For simple qualitative analyses, paper or thin-layer chromotographic approaches often suffice, although quantitative determinations for the resolution of more complex mixtures often require the use of electrophoresis, gas-chromatography, HPLC, or a dedicated amino acid analyser.

14.8.1 Paper and thin-layer chromatographic determination of amino acids

Paper and TLC may be chosen for simplicity when only a qualitative identification is needed for components in a mixture through reference to known standards by comparison of R_f values (see Chapter 8).

Some samples may require the removal of interferents such as carbohydrates, proteins, or salts via the use of an ion-exchange resin column prior to running a chromatograph.

There are many solvent systems that may be chosen (Table 14.2) depending on the composition of the mixture being analysed.

Table 14.2 Choice of solvent mixture for resolution of amino acid mixtures by TLC

Solvent mixture v/w and % composition	Uses
n-Butanol—60% Water—25% Glacial acetic acid—15%	Suitable for the resolution of a wide range of amino acid mixtures
n-Butanol—35% Acetone—35% Water—20% Glacial acetic acid—10%	Suitable for the resolution of a wide range of amino acid mixtures
Isopropanol—80% Water—15% Formic acid—5%	Suitable for the resolution of a wide range of amino acid mixtures
Phenol—80% Water—20%	Useful for the separation of amino acids that are difficult to resolve via other solvent mixtures—gives wide ranging R_f values
Phenol—80% Water—19.5% Ammonia—0.5%	May be chosen for the separation of basic amino acids
n-Butanol—37% Acetone—37% Water—18% Diethylamine—8%	Possible choice for resolution of mixtures that prove difficult to separate via other solvent mixtures

The choice of the solvent mixture is often largely a case of trial and error especially if little is known concerning the components in the sample—although a few guidelines can often help. Increasing the proportion of water generally leads to the increase in all R_f values, while the inclusion of ammonia will increase the R_f for basic amino acids. Further resolution may again sometimes be achieved via the use of two-dimensional chromatographic runs at 90° to each other with different solvent systems (see Chapter 8).

Locating reagents are often applied either by spraying or dipping to help visualize the amino acid components following separation. A 2 g dm^{-3} solution of ninhydrin in acetone is the most popularly used locating reagent, sometimes with the addition of acetic acid and 2,4,6-collidine (both 5–10% v/v) so as to form intensely coloured bands of different colours with different amino acids. All amino acids will form coloured bands within a few hours at room temperature. This process can often be accelerated if the TLC strip or paper strip is heated in an oven although this can cause the formation of further coloured bands if other compounds possessing a primary or secondary amino group are present. Again, this reaction chemistry may provide further information since if an analyte fails to form coloured bands unless the strip is heated, the analyte is almost certainly not an amino acid.

Other reagents specific for certain amino acids may also be used but are beyond the scope of discussion for this book. The interested reader is referred to Plummer (1987).

14.8.2 Electrophoresis for the separation of amino acids

Amino acid mixtures may often be separated and subsequently identified by electrophoresis. Different amino acids carry different charges at a given pH and so can be induced to travel along a thin-layer support medium such as paper, silica gel, or cellulose under the influence of an electric field. Identification of components in a mixture is again achieved with reference to known samples and via the use of visualization reagents as already described for paper chromatographic and TLC analyses. The use of higher voltages will help accelerate the rate of separations as well as sometimes helping separating amino acids from possible interferents such as salts, carbohydrates, or proteins.

Electrophoretic separations are normally performed at pH 2.0 or 5.3, since at pH 2.0 all amino acids will carry a positive charge and the basic amino acids will migrate towards the cathode at the fastest rate. In contrast, at pH 5.3, migration will occur towards either electrode depending on the charge carried on an amino acid. This process is extremely useful for the determination of the acidic or basic nature of an unknown amino acid or dipeptide.

14.8.3 Gas–liquid chromatography based determinations of amino acids

The single greatest problem with attempting to analyse amino acids via GC is that they first require derivatization to render them sufficiently volatile.

Trimethylsilyl derivatives are the simplest to prepare via the addition of *N*,*O*-bis(trimethylsilyl)trifluoroacetamide (BSTFA) in acetonitrile and heating at 150°C under anhydrous conditions for approximately 2 h in a sealed tube.

Another approach for forming suitable derivatives can be achieved by acetylation of the methyl, propyl, or butyl esters of the amino acids to give the trifluoroacetyl or heptafluorobutyryl derivatives.

Difficulty can sometimes be experienced in separating all of the amino acid derivatives from a mixture and, to achieve the optimal separation, consideration needs to be given to the choice of the stationary phase, the derivatization route employed, and the nature of the amino acid mixture.

In some cases it may be necessary to use two columns simultaneously with different stationary phases.

The choice of detector is also sometimes problematic and indeed it may be necessary to split the gas stream and use two detectors simultaneously. The flame ionization detector is commonly chosen since it will detect all of the amino acid derivatives formed via the routes described although its relative molar response will vary from one amino acid to another and for this reason it is necessary to prepare separate calibration curves for each target analyte in the mixture.

14.8.4 HPLC analyses for amino acids

Reverse-phase HPLC offers another alternative approach for the determination of amino acids although derivatization again needs to be performed to form either fluorescent or coloured complexes. Buffered mobile phases containing a polar solvent such as methanol or tetrahydrofuran (with the proportions depending on the mixture and derivatization approach adopted) should be used possibly with gradient elution for the resolution of more complex mixtures.

Derivatization to form coloured complexes with all amino acids allows detection at 436 nm and may be achieved following reaction of the amino acids with dabsyl chloride (4-dimethyl-aminoazobenzene-4′-sulphonyl chloride). One of the greatest disadvantages with this approach is that the excess reagent causes serious deterioration of columns and can limit the lifetime of columns to less than 100 injections. Optimal sensitivity via this approach cannot, moreover, match that achievable via fluorometric approaches.

There are a number of approaches for the formation of fluorometric derivatives although some disadvantages can be experienced. One simple approach involves derivatization of primary amino acids with *o*-phthalaldehyde (OPA) in the presence of ethanethiol under pH conditions of 9–11 to form fluorescent products with an excitation wavelength of 340 nm and an emission maximum at 455 nm. The fluorescent emission yield will vary from compound to compound in a mixture and for this reason separate calibration curves must be first determined using standardized samples for quantitative determinations.

An alternative approach that allows derivatization of both primary and secondary amino acids to allow a fluorometric HPLC based determination involves reaction with 9-fluorenylmethyl chloroformate, although excess reagent must be removed prior to analysis. This approach, moreover, allows sensitivity comparable to *o*-phthalaldehyde derivatization with lower limits of detection down to pico- or femtomolar concentrations, although reproducibility can sometimes be poor.

14.9 Analysis of proteins

Proteins are polymers of amino acids. Smaller amino acid polymer chains are known as polypeptide chains while longer and normally folded chains form proteins. Some analyses may allow quantification of the total protein content in a sample, some a qualitative identification of specific proteins, and others the quantification of individual proteins. Since there is such vast variety of proteins we shall briefly look at protein structures. There are 22 different amino acids and the sequence of amino acids within a protein is known as its ***primary structure***. The amino acids within proteins are linked by peptide bonds formed between an amino group of one amino acid and a carboxyl group of another. The formation of a peptide bond results in the loss of the amino and carboxyl groups of each amino acid, although the terminal amino acids at each end of the so-called polypeptide chains retain in one case an amino (N-terminal) and in the other a carboxyl (C-terminal) group. The N-terminal amino acid is always designated the first amino acid if determining the sequence of a polypeptide chain or protein. The three-dimensional shape or folding of a polypeptide chain is known as the ***secondary structure*** of a protein. A helical structure is often adopted, which is stabilized by intra-chain hydrogen bonds formed between the amide nitrogen of one peptide and the oxygen of a carbonyl group of another. Interchange hydrogen bonding between parallel extended chains via interaction of hydrogen and oxygen atoms can also give rise to pleated sheet structures.

The ***tertiary structure*** of a protein describes the three-dimensional shape of the protein. These three-dimensional shapes are often determined by folding of the protein chain. Some form spherical-like or globular shapes and these are known as globular proteins. Such proteins are typically only semi-soluble in water (so forming colloidal solutions) and normally become crystalline upon purification and separation. Globular proteins normally play functional roles within the cell such as enzymes, or immunoglobulin proteins. The fibrous proteins, by contrast, are all linear polypeptide chains and fulfil roles as structural proteins.

The polypeptide chains of some fibrous proteins intertwine to form helical structures and these typically possess mechanical elasticity, an example in this context being keratin. The elasticity of helical structures may, however, be lost if there is a high degree of bonding between individual helices. Other fibrous proteins adopt a pleated sheet structure such as in silk and are typically non-elastic.

The ***quaternary structure*** of a protein describes a further level of organizational structure found in some globular proteins that relates to the association of protein units to produce an aggregate protein with clearly defined functional properties. Aggregate proteins of this type are typically held together via non-polar bonds between the non-polar regions of the molecules concerned. Conjugated proteins are aggregate proteins that further contain non-polypeptide components. Haemoglobin is an example of such a protein that is formed from four polypeptide chains together with an iron porphyrin moiety for the binding of oxygen.

14.9.1 Separation of proteins

It is often necessary to either qualitatively identify the presence of a particular protein or even quantitatively determine a protein within a mixture; in such cases it is normally necessary to first separate the mixture into its corresponding consituents. The high molecular weight of many proteins often renders simpler approaches such as paper chromatography inappropriate and to further complicate the issue many proteins may be irreversibly denatured (so destroying the tertiary and quaternary structures) by even relatively mild experimental conditions such as gentle heating.

14.9.2 Precipitation approaches

Precipitation approaches are based on the addition of a solute (more soluble than the component(s) to be separated) to a mixture of proteins to exceed the total solubility product for the solvent and so force the least soluble components to precipitate from the mixture.

High concentration of salts such as sulphites or sulphates may be used to precipitate a range of proteins. The most widely used approach is to introduce salt(s) in a step-wise addition to give a series of fractions each containing a mixture of proteins. Fractions are removed after each addition of salt either by centrifugation or filtration.

Another approach involves the addition of alcohols, although care should be taken with proteins that are known to be susceptible to denaturation and this may be helped, for example, by using lower temperatures.

14.9.3 Electrophoresis

Electrophoresis permits the separation of proteins on the basis of charge and is a popular technique due to its simplicity even though complete resolution of more complex mixtures is sometimes not achievable. There are several approaches for the electrophoretic separation of proteins; one of the most widely used technique is *zone electrophoresis*. Zone electrophoresis employs a solid support onto which a sample is applied as a streak. Following separation, the bands of protein may be precipitated into the pores of the support medium using trichloroacetic acid and then stained via a dye such as nigrosin. There are a number of different ways for providing a semi-quantitative determination, the simplest of which involves determining the intensity of coloration by the dye for each band; if the total protein concentration is known, then the ratio of the different constituent bands may allow a concentration for each component to be estimated. Another approach involves first cutting the strips to isolate bands and then eluting the dyed proteins into fixed volumes of solvent. In this way, it is possible to colorimetrically determine the intensity of individual component bands and thus permit the quantification of the components in a mixture—via comparison with the total protein content in a sample.

A third approach involves scanning the strip band by band with a densitometer. The area of peaks recorded for each band may then be correlated with the total protein content.

All the electrophoretic approaches described for identifying and quantifying proteins rely, of course, on effective separations and this is critically affected by the choice of supporting media and the pH of the buffer solution. Filter paper is often used as a support although it also incurs a number of disadvantages including complications associated with the adsorption of proteins. Alternative choices to minimize protein adsorption include media such as cellulose acetate, starch, or polyacrylamide supports. Although almost any pH can be used, values above the isoelectric pH for all of the proteins within the mixture will facilitate separations and in practice pH values of between 8 and 9 almost always suffice.

Sodium dodecyl sulphate (SDS) electrophoresis

SDS electrophoresis operates on the principle that proteins may be broken down to their polypeptides and then separated via electrophoresis on polyacrylamide gels. Samples are first dissolved in buffer containing SDS together with β-mercaptoethanol to reduce any disulphide bonds and stabilize the constituent polypeptides. Samples are normally first boiled for a few minutes to denature the protein, allowing exposure of the full length of the polypeptide chains. The anionic detergent masks the charges held upon the polypeptides and so the mobility of each polypeptide will depend essentially upon molecular weight. Polypeptide chains can then be typically visualized using a suitable dye following separation. The mobility of polypeptide chains will be lowered with increasing molecular weight and in practice a plot of the mobility with respect to the logarithm of the molecular mass will often allow identification of constituents if comparisons are made with reference to a series of known polypeptides and/or proteins.

14.9.4 **Immuno- or Western blotting**

Immuno- or Western blotting is another approach that can be sometimes used for the identification of individual proteins within mixtures, as long as an antibody (either mono- or polyclonal) can be obtained for a specific protein.

Proteins are first electrophorectically separated on a medium chosen to offer optimal separation. The proteins are then transferred or 'blotted' electrophorectically onto a robust adsorptive membrane for subsequent identification of the proteins via antibody binding. Transfer is achieved by sandwiching the gel and the membrane soaked in electrophoretic buffer between two electrodes that are polarized with a potential difference of typically 100 V. This typically permits transfer in approximately 1 h or so. The membrane and adsorbed proteins are then incubated with suitable antibodies for a time period of typically not less than 1 h. Excess antibody is then removed via washing and the remaining bound antibody is detected via incubation with a further antibody against the first, which is labelled either with an enzyme, colloidal gold, or suitable isotope such as ^{125}I.

14.9.5 Chromatographic separations

The principles of chromatography are described in Chapter 8; however, brief discussion will be provided in this chapter due to the importance of chromatographic approaches for bio-analytical determinations.

The most frequently used approaches are based upon column-based techniques since these lend themselves for the collection of fractions that may then be subsequently quantified (as described below).

Ion-exchange chromatographic columns utilizing ion-exchange cellulose based packing materials such as diethylaminoethyl cellulose or carboxymethyl cellulose are often chosen for the separation of mixtures of unknown composition. Gel permeation approaches, however, often offer cleaner separations of protein mixtures than are obtainable using ion exchange columns, although some knowledge of the mixture is normally needed to allow selection of the most appropriate gel.

Reverse-phase HPLC may also be used for the separation of some mixtures of peptides and/or proteins with octadecylsilane (C_{18}) coated columns and are typically most appropriate for the separation of peptides with 50 amino acid residues or less whereas tetryl (C_4) or octylsilane (C_8) columns are more suitable for the separation of larger molecules.

Affinity chromatography exploits the highly specific binding of an antibody to its corresponding antigen to collect fractions of individual proteins via the use of antibodies raised towards specific proteins. Affinity-based chromatographic separations are increasingly being used due to: (i) to the unrivalled separational performance they offer; and (ii) the ease by which antibodies may be raised and harvested via monoclonal techniques.

14.9.6 Quantification of proteins

The nature of the sample will always affect the choice of technique selected with fluid samples typically being simpler to quantify than solid samples. The presence of interferents can also affect the choice of determination and may necessitate purification and/or separation by, for example, chromatography (see Section 14.9.5). Another approach for the removal of interferents is to induce the precipitation of the soluble protein, which may then be subsequently re-dissolved and analysed as required.

Chemical methods

Kjeldahl method The ***Kjeldahl method*** is based on the determination of the nitrogen content in a compound and since all proteins contain nitrogen, they lend themselves well for Kjeldahl determinations. One of the major problems associated with the Kjeldahl approach is that protein mixtures must be first separated since the amino acid (and, therefore, the nitrogen) content of different proteins differ. Clearly the nitrogen content for a particular protein must be known for a quantitative determination to be undertaken.

The protein must normally be precipitated to remove any nitrogen-containing interferents that otherwise could lead to erroneous results. The precipitate is collected and then heated with concentrated sulphuric acid under reflux together with a catalyst (e.g. via the addition of cupric or mercuric ions) for the oxidation (digestion) of all of the nitrogen-containing compounds. This procedure takes several hours during which sulphur dioxide fumes will be produced and, therefore, care should be taken to always perform this reaction in a fume cupboard. The solution will first turn brown or black before turning clear although the heating process should be continued for a further 2 h or so.

Upon cooling the mixture should be made basic via the addition of an excess of sodium hydroxide and then transferred to a steam distillation flask. The mixture should then be distilled and collected into a receiver flask to collect the ammonia that may be trapped by bubbling into a solution of boric acid. The ammonium ion content can then be quantified via titration with hydrochloric acid. The calculation for the nitrogen content is simple, since 1 mol of nitrogen will give rise to 1 mol of ammonia and this, in turn, will require 1 mol of HCl for the end-point reaction (see Chapter 3).

The Kjeldahl method although being labour intensive is, however, still recognized as one of the standard reference methods for the determination of proteins since it offers extremely good reproducible and accurate determinations if due experimental care is taken.

Biuret method The ***Biuret method*** is a simple and robust method for the quantification that is based on the formation of a purple-coloured complex formed between the protein and copper when a protein is introduced to a basic solution of copper sulphate. Potassium tartrate or sodium citrate are often added to prevent the precipitation of cupric ions as the hydroxide.

The naming of the Biuret method is historical and now rather misleading in nature; the test was named following the discovery that Biuret reacts with copper sulphate in alkaline solution to yield a purple-coloured complex. Similar complexes are, in fact, formed with any compound containing $-CONH_2$, $-CH_2NH_2$, $C(HN)NH_2$, or $CSNH_2$ groups (linked through C or N atoms), and this explains why this technique can be used for the quantification of proteins.

Complexes of this type exhibit λ_{max} at approximately 545 nm and hence the absorption at this wavelength can be followed spectrophotometrically, used for quantification with reference to a calibration plot. The Beer–Lambert law (Chapter 4) is typically followed up to concentrations of 2 $g\,dm^{-3}$. The reaction should be allowed to proceed for approximately 15 min prior to quantification although the complex should remain stable for a number of hours to allow plenty of time for the spectrophotometric determination. The single biggest drawback associated with the Biuret method is its lack of both sensitivity and specificity. All proteins react in a similar manner with Biuret, so quantification of individual proteins requires prior separation.

Lowry method The ***Lowry method*** again is based upon a spectrophotometric determination and offers greater sensitivity than the Biuret approach, with a working analytical concentration range of approximately 20 mg to 2 $g\,dm^{-3}$. The reaction is based on the formation of the copper–protein complex (via use of the Biuret reaction), which, in turn, reduces phosphotungstic and phosphomolybdic acids to tungsten blue and molybdenum blue, respectively, and this may then be determined spectrophotometrically within the wavelength range of 600–800 nm—since these two compounds give rise to broad and overlapping absorption peaks.

The reagent is difficult to prepare but may, however, be readily purchased, which simplifies the experimental procedure greatly. The reaction does not follow the Beer–Lambert law well and for this reason a pre-determined calibration curve will always be needed for quantification.

Bicinchoninic acid based methods Bicinchoninic acid based methods are based on the Biuret and Lowry methods, although in this instance the copper–protein complex is chelated by bicinchoninic acid to form very stable complexes with λ_{max} at around 562 nm. This approach offers the further advantage of offering several orders of magnitude greater sensitivity in comparison to the Biuret and Lowry approaches. Commercial reagents can be purchased, which greatly simplify experimental protocols and hence ease of use.

Spectroscopic methods

UV–visible spectroscopic determinations are normally the simplest of approaches available for the determination of proteins although quantitative analyses can normally be undertaken in samples that are known to contain only a single protein. Protein mixtures must normally be separated prior to analysis since most proteins absorb with absorption maxima λ_{max} that overlap, although the molar absorption coefficients vary from protein to protein. Total protein quantification of protein mixtures may, therefore, not be undertaken by spectroscopic approaches unless separated into fractions with the individual determinations being summed.

Almost all proteins absorb in the UV region with λ_{max} values typically being found at around 280 nm due to the presence of aromatic amino acid residues. The molar absorption coefficients vary from one protein to another due to the variable content of amino acids from one protein to another and this explains the need to first separate mixtures prior to analysis it quantification is to be acheived.

Dye binding (absorbance shift) methods

Proteins can in some instances give rise to absorbance shifts (colour changes) in some dyes and this has led to the development of a number of spectrophotometric-based determinations for proteins.

Coomassie Brilliant Blue is widely used for the general determination of proteins since complexation of proteins with the dye causes a shift in absorbance from 464 to 595 nm, which can be followed spectrophotometrically. The quantity of dye binding to different proteins does vary and so does the observed colour change. Therefore, standard calibrations must first be determined for a particular protein, and protein mixtures must first be separated prior to any quantitative determination being attempted.

Bromocresol Green is another dye that is regularly used for the spectrophometric determination of albumin. At pH 4.2, Bromocresol Green binds with a high selectivity towards albumin and is accompanied by a shift from a yellow to a blue colour. The exact shift in absorbance differs from one source of albumin (e.g. egg or serum) to another and does not follow the Beer–Lambert law well, and this necessitates the empirical determination of calibration curves for quantitative analyses.

Physical methods

A number of physical techniques are routinely used for the determination of protein in fluid samples, most of which rely on prior precipitation either via the use of organic acids such as trichloroacetic or picric acid—or alternatively via the use of antibodies. The precipitation of the protein lowers the turbidity of the solution and this may be followed either within a simple spectrophometer or nepheleometer for enhanced sensitivity.

14.10 A brief introduction to bioinformatics, genomics, and proteomics in analytical science

It is not an understatement to say that analytical biochemistry has undergone a revolution during the 1990s to the present day, with the evolution of a new broadband suite of techniques that have emerged from molecular biology while also using modern computing techniques—and this has given rise to a whole new field known as ***Bioinformatics***. Although the term bioinformatics has been applied to widely differing applications, bioinformatics has particular relevance to analytical science due to the unrivalled specificity capable of identifying sources of very complex analytes of biological origin by exploiting understandings in genomic and proteomic science. The area of molecular biology and bioinformatics is a rapidly changing area that is already having major impacts on the way analytical science is being performed in many widely differing areas.

We cannot in the context of this book attempt to give a thorough treatment of this area, but we can provide an introduction to the areas of genomics and proteomics as applied to analytical science.

The term ***genome*** describes the entire collection of genes within an organism that provides all of the information required for organisms that respire, synthesize materials, and reproduce—as well as all of the other functions to maintain life.

The term ***genomics*** describes the study of the genome of a particular organism. The genome constitutes the total DNA (deoxyribonucleic acid) that is found within each cell of a living organism (with the exception of specific cell types, such as red blood cells). Genomics involves identifying and sequencing acid–base pairs that code all the information for a gene.

Not all of the genes encoded in a cell's DNA are expressed. Different cells will express different combinations of genes depending on their function (for example, a cell from the kidney will express a different combination of genes to a skin cell), and so will produce different combinations of proteins. The ***proteome*** is the entire collection of proteins produced by a specific cell type at a particular point in time: it is a snapshot of the genes that are being expressed within a cell at that time point. Whereas an organism carries the same genome in each of its cells, the proteome will vary from cell to cell. Just as genomics describes the study of an organism's genome, so ***proteomics*** is the study of an organism's proteome.

Proteomics is not just the study of protein chemistry, but rather the study of complex multi-protein systems. Many proteomic studies are therefore directed towards complex mixtures; separational science plays a major role in enabling these complex systems to be characterized, through the identification of the individual components of a system. The identification of individual components does not, however, provide all of the information about the system that is available. The primary focus of proteomics is to characterize the behaviour of the system as a whole—which proteins are being expressed where and when, and how this varies with time—rather than just trying to identify all of the individual components. Proteomics can also help to identify why certain genes are expressed in some situations while others are not, and this can be applied to further biomedical and therapeutic research.

Genomic and proteomic techniques have many applications in analytical science ranging from medical diagnostics to forensic science. For example, DNA fingerprinting (see Section 14.10.3) allows law enforcement agencies to identify individuals on the basis of tissue or body fluid samples. It can also be used to identify the source of a diverse range of biological materials, from food sources (for example, to enable the characterization of a food as organic or genetically modified) to leather and wool (to help prevent smuggling of premium goods such as cashmere garments).

A number of techniques have emerged in recent years that have allowed biological information—at the level of both the genome and the proteome—to be tapped. These techniques include:

- the use of ***DNA microarrays*** in combination with protein fluorescence to identify the expression of specific genes within a genome (Section 14.10.2);
- the matching of DNA base pair sequences from two different DNA samples using ***DNA fingerprinting*** (Section 14.10.3);
- the very rapid identification and sequencing of many different proteins in complex mixtures by ***mass spectrometry*** (see Section 9.10.1 for more details of the application of mass spectrometry for peptide sequencing).

Many excellent treatments of the applications of genomics and proteomic analytical science are provided for the interested reader in Gibson and Muse (2002) and Liebler (2002). We now look at three techniques that are gaining widespread application and acceptance for mainstream analytical use; these are the polymerase chain reaction (PCR), DNA microarrays, and the DNA fingerprinting technique.

Full details of these books are given in the further reading section at the end of this chapter.

DNA fingerprinting essentially involves the matching of DNA base pair sequences from two different sources of DNA that have originated from the same organism. If large quantities of DNA are present, matching is a relatively simple process. Minute quantities of bodily fluids, e.g. saliva, blood, will contain DNA that may be uniquely linked to, for example, an individual human following a crime. Analysis of such samples is, however, extremely difficult. PCR techniques, however, allow the amplification of a tiny quantity of DNA to produce large quantities of DNA that may be readily analysed. We shall, therefore, consider the PCR technique and then give a brief description of the DNA fingerprinting technique. With PCR techniques it is possible to rapidly replicate or multiply DNA segments that are present in extremely small amounts and within a few hours millions of copies of a particular DNA sequence can be made in order to produce specific DNA for DNA fingerprinting, sequencing, or other analytical applications.

14.10.1 Polymerase chain reaction (PCR) techniques

Introduction and significance

The ***polymerase chain reaction*** (PCR) provides a fast and inexpensive approach for amplifying or copying specific DNA sequences. The importance of PCR stems from its ability to provide sufficient copies of DNA segments to allow their analysis or exploitation by other techniques. In this sense PCR can be thought of as a 'molecular photocopier' for DNA. Since PCR operates by copying specific segments of DNA, these targeted regions may be focused or restricted towards, for example, specific genes or other regions of interest or significance. For PCR to be used, the exact nucleotide sequence that borders either side of a given region of DNA must be known. This is seldom a problem since the nucleotide sequences of many genes and regions bordering many genes are known and well documented. These border regions act as target sequences for primer attachment.

The importance of PCR was recognized by the awarding of the 1993 Nobel prize for chemistry to Kary Mullis, the inventor of the PCR technique. This technique operates by exploiting a particular ***polymerase*** enzyme in conjunction with synthetic primers that together initiate a chain reaction that produces a rapidly growing population of specific DNA sequences. If the nucleotide sequence of target regions is known, short single-stranded DNA primers may be chemically synthesized and should ideally range from 20 to 50 nucleotides in length. DNA polymerase is then added to catalyse the synthesis of the production of complementary DNA strands using primers as starting points. The DNA polymerase routinely used for the PCR technique was first isolated from the bacterium *Thermus aquaticus* and may withstand temperatures up to 95°C, and in fact operates at an optimal turnover rate at a temperature of 72°C, which as we shall see is important for the PCR technique.

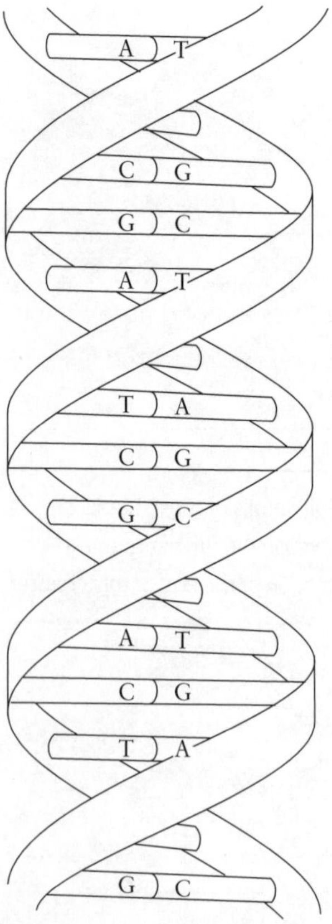

Figure 14.11 DNA double helix.

The technique

The PCR procedures involve three steps:

Step 1: The DNA in its double helix format, Fig. 14.11, is first heated at 95°C to denature the DNA causing the paired strands to separate, Fig. 14.12.

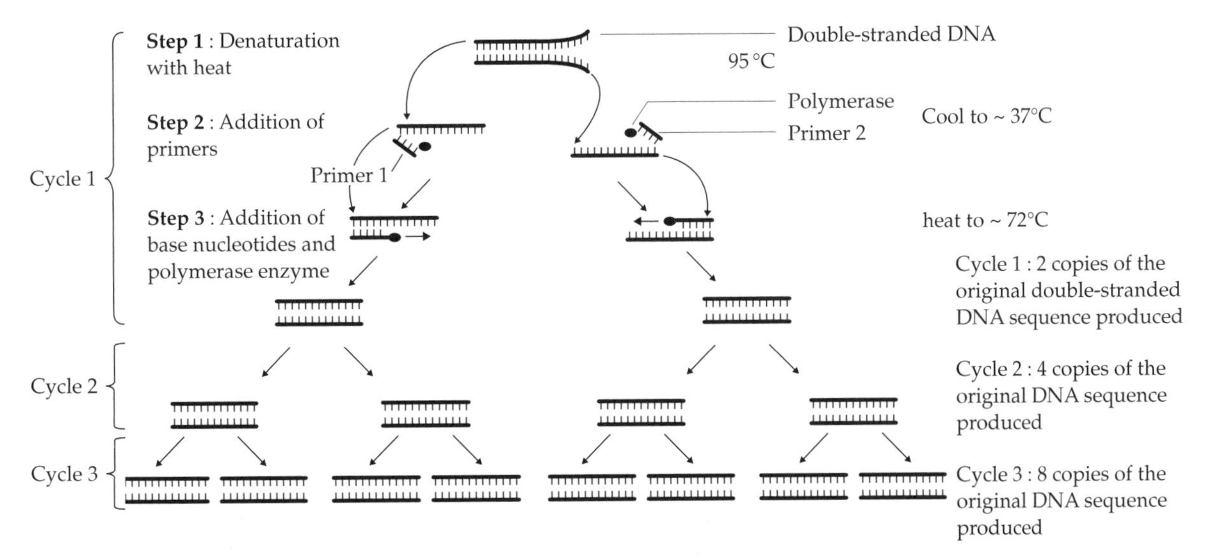

Figure 14.12 The Polymerase chain reaction.

Step 2: The DNA is then cooled to ~37°C and specific primers are added to bind to targeted complementary DNA target sequences. The primers are normally added in vast excess to help ensure that the two strands will always bind to the primers, and not to each other. The primers may be synthesized in the laboratory using a DNA synthesizer or may be purchased commercially. Ideally their base pair sequence they should be unique within the genome of the organism, so that they will only match the area chosen, so limiting and defining the sequence to be copied.

Step 3: The temperature is raised to 72°C (the optimal temperature for the DNA polymerase from the thermophilic bacterium *Thermus aquaticus*).

A mixture of adenine, thymine, cytosine, and guanine together with the polymerase enzyme is added to the single-stranded primer-linked DNA. The DNA polymerase now synthesizes a complementary strand to the first strand of DNA. In a few minutes an entirely complementary copy of the targeted DNA sequence is produced, thereby doubling the initial starting quantity of DNA.

The nucleotide sequence of a section of DNA is the precise order of occurrence of the four bases adenine, thymine, cytosine, and guanine as they occur within a given section of DNA.

The whole process may now be repeated to form a second copying cycle. The reaction mixture is once again heated to denature the two new helices, more primer is added and this binds to the DNA. The quantity of DNA is again doubled. This process is repeated many times until sufficient quantity of DNA is produced. With each process the DNA quantity is doubled, and in the majority of situations 20–30 reaction cycles will produce sufficient DNA for analysis. After 30 PCR cycles approximately 1×10^9 target copies are produced. PCR techniques therefore make it possible to identify an individual person from the minuscule amount of DNA present, or indeed identify the source of any biological sample.

Complementary lengths of DNA hybridize by complementary base pair association. Cytosine always associates with guanine. Adenine always associates with thymine.

The PCR cycle can be repeated many times within a commercial ***thermal cycler*** to allow complete automation of the PCR process. The PCR cycle (typically involving 20–35 cycles) may be completed in a few hours. The use of polymerase enzymes from thermophilic bacteria such as *Thermus aquaticus* means that these are not destroyed by the 95°C heating step required to denature the DNA and so they retain their activity through the many cycles required in a typical application procedure.

14.10.2 DNA microarrays

DNA microarrays provide information that allows us to identify which genes are being expressed in a cell under a given set of conditions. The information from DNA microarrays can, for example, help to identify an individual who may be susceptible to a particular disease – or alternatively can help to compare gene expression in healthy and cancerous cells.

Microarray technology relies on and exploits the binding of single-stranded DNA to complementary single-stranded DNA sequences. Microarray chips comprise glass slides onto which DNA is immobilized at defined spots (Fig. 14.13). One DNA microarray chip may possess tens of thousands of spots. Spots are either printed robotically by inkjet-type

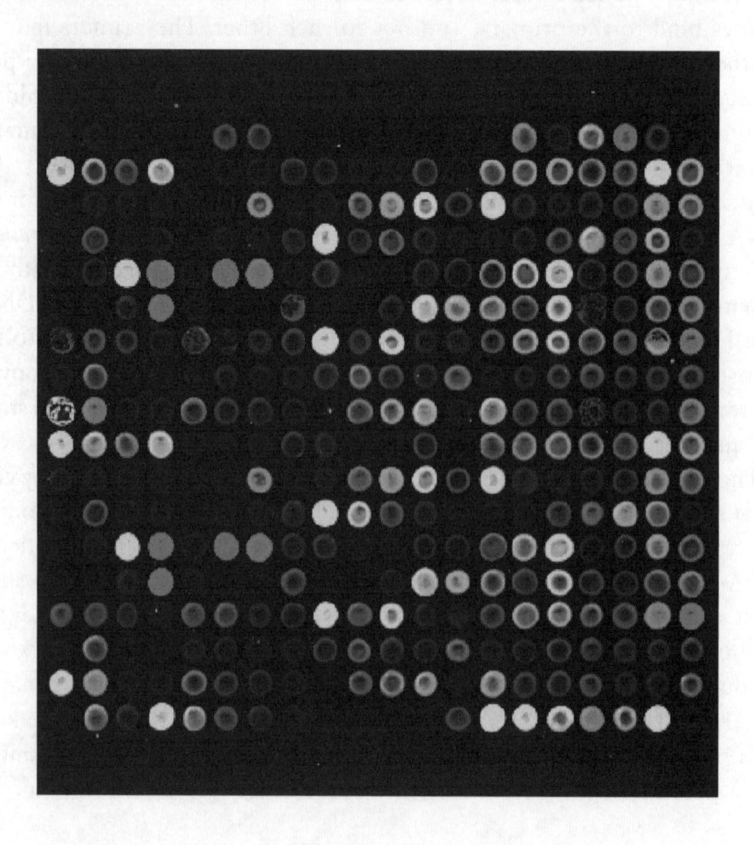

Figure 14.13 Comparison of gene expression patterns in liver (red) and brain (green). The liver RNA is tagged with a red fluorophore, the brain RNA with a green one, then both are exposed to the aray. Red spots correspond to genes active in the liver but not in the brain. Green spots correspond to genes active in the brain but not in the liver. Yellow spots correspond to genes active in both brain and liver. (Courtesy Dr P. A. Lyons.)

printing or formed by photolithographic means. Dimensions of DNA microchips vary but are often ~2.5 by 2.5 cm, with each spot diameter being of the order of 0.1 mm or less. Each spot in turn comprises tens of thousands of identical DNA fragments ranging in length from 20 through to hundreds of nucleotides. Ideally each spot should allow the identification of one gene or (another) specific DNA sequence.

Let us consider how DNA microarrays may be used. Imagine that we have two cells that are identical except for the fact that one has been taken from healthy tissue, and one has been taken from diseased tissue. We want to determine any differences in the genomes of the two cells that might determine why one cell is healthy and the other cell diseased. A procedure similar to that illustrated in Fig. 14.14 might be used. First, the genomic DNA from the two cell types is isolated, and single-stranded copies of the DNA (the ***probe*** DNA) are labelled with contrasting fluorescent tags (called ***fluorophores***). Usually, one tag (representing one cell type) is green, and the other (representing the second cell type) is red.

Second, genomic DNA representing all the target genes that may be present in the two cell types is fixed to the microarray chip. The microarray is exposed to both of the tagged probe mixtures to allow complementary hybridization to occur. Hybridization will occur at spots containing a gene that is also present in the probe mixture from one or both of the two cell types. Loose probe mixture (i.e. those DNA probes that have not undergone hybridization with DNA fixed to the microarray chip) is then washed off.

The spots can then be 'visualized' by exciting the fluorescent tags with a laser of a suitable wavelength. Fluorescence can be detected by simple optical means.

Let us assume that the probe DNA from the healthy cell is tagged with a red fluorophore, and that the probe DNA from the diseased cell is tagged with a green fluorophore. If probes from both cell types are present at a spot in approximately equal quantities (indicating the presence of a particular gene in the genomes of both cell types), the spot will appear yellow. If probes from neither sample are present the spot will not show fluorescence and so will appear black (indicating that the gene represented by that spot is absent from the genomes of both cell types). By contrast, a green spot indicates a gene that is present in the healthy cell but not the diseased cell, while a red spot indicates a gene that is present in the diseased cell but not the healthy cell.

Therefore, the colours of the spots can be used to characterize the relative genomic composition of the two cell types, and to identify specific genes whose presence or absence may make the difference between a cell being healthy or diseased.

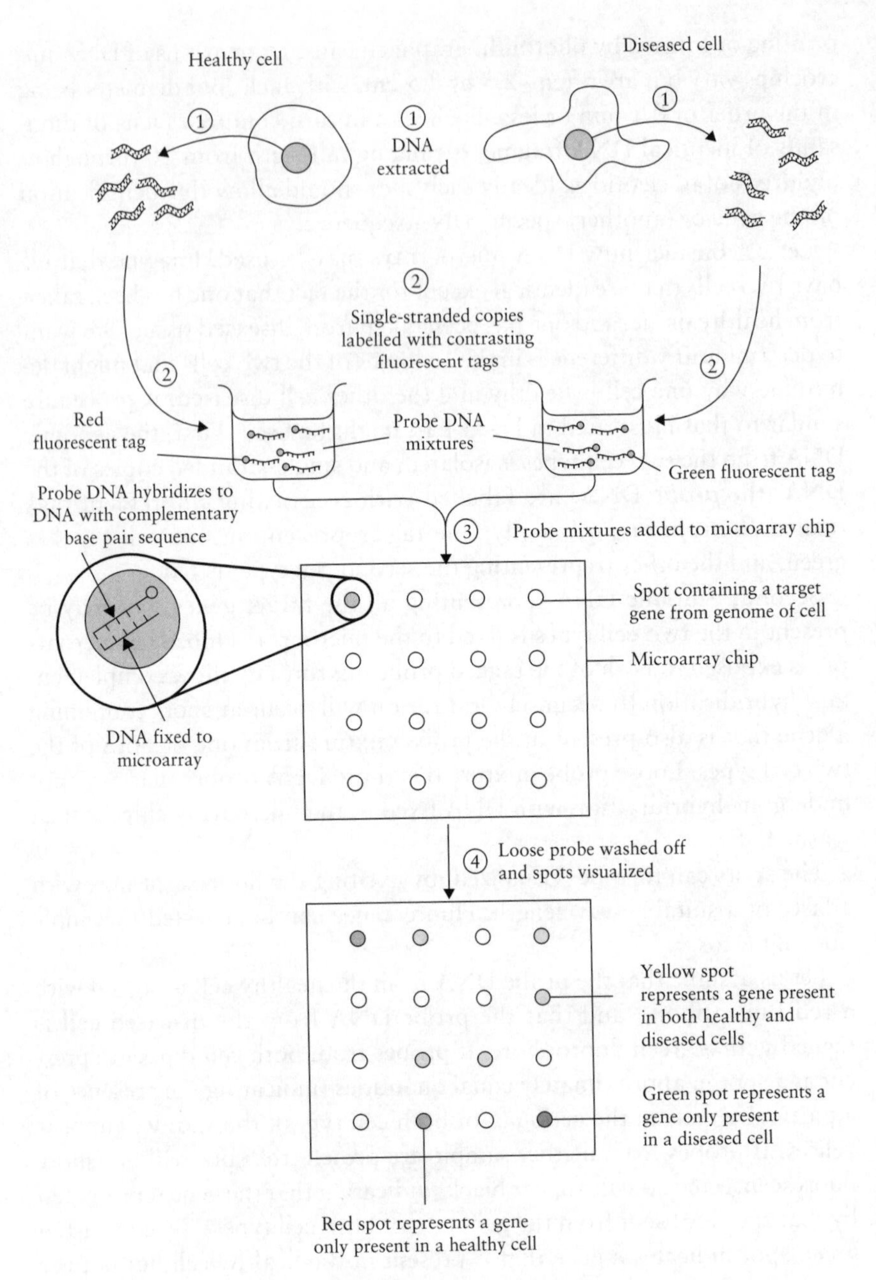

Figure 14.14 Schematic representation of the preparation of a DNA microarray.

The use of DNA microarrays described above allows the presence or absence of genes to be investigated. Similarly, the relative level of gene *expression* in two cell types (whether genes are 'switched on' or 'switched off') can be determined by fluorescently tagging mRNA molecules from the two cell types, and using these RNA probes to hybridize with DNA representing the genes of interest on the microarray chip. Again, such microarray studies can be used to characterize differences between different cell types – for example, they may reveal a specific gene that is switched on in a healthy cell but (inappropriately) switched off in a diseased cell, causing the disease state.

14.10.3 DNA fingerprinting

DNA fingerprinting is a powerful technique that allows identification of DNA from one sample to be matched with DNA from another sample and this is used in a variety of forensic techniques to help, for example, with the solving of a crime, providing evidence within paternity court cases, or confirming the source of biological materials such as foodstuffs. For DNA fingerprinting techniques, DNA is first extracted and so-called ***restriction enzymes*** added to produce ***restriction fragments***. Restriction enzymes are enzymes that cut DNA at specific points and leave shorter fragments, which may be analysed more readily. The mixture of restriction fragments may be separated by electrophoresis techniques. Each sample forms a characteristic set of bands; following electrophoresis DNA on the gel is denatured first by heating and the single strands are transferred onto special paper by a blotting technique. A radioactive probe is then added to each blot paper.

The DNA probe is a radioactive single-stranded DNA complementary to the DNA to be matched. The radioactive probe will only attach to bands containing complementary DNA by base pairing. The excess probe is rinsed off and a sheet of photographic film placed over the plate blot. The radioactivity within the radioactive probe will expose the film to form an image corresponding to the DNA bands. Each developed band on the photographic film then corresponds to bands on the DNA electrophoresis blot. DNA fingerprinting, for example, may be used to identify the person from whom bodily samples such as serum, semen, or saliva originated when found on a victim's clothing following a crime. Some biologically derived materials with commercial premium such as leathers or wools (e.g. cashmere) can also be analysed and verified for their authenticity to help prevent counterfeiting or smuggling.

Exercises and problems

14.1 What is meant by primary, secondary, tertiary, and quaternary structures of a protein?

14.2 How are whole blood, serum, and plasma samples prepared and stored; what analyses may these be used for?

14.3 How may blood electrolyte levels be determined?

14.4 What is meant by and what is the significance of the 'anion gap' within serum samples.

14.5 Explain the principle of immunoassays.

14.6 Describe the operation of competitive and non-competative ELISAs.

14.7 What is meant by a first-generation biosensor and how does this differ from a second-generation biosensor?

14.8 Why is sodium fluoride often added to whole blood?

14.9 What are the main problems associated with O_2-based glucose sensors?

14.10 Why may permselective membranes not be used in second-generation biosensors?

14.11 Compare and contrast the Kjeldahl, Biuret, Lowry, and bicinchoninic acid approaches for determining protein concentrations. Describe the basis for the polymerase chain reaction—how can this be used in forensic science?

14.12. Compare and contrast two approaches for performing amino acid analysis.

Summary

1. Multi-analysers form one of the mainstream workhorses in many clinical biochemistry laboratories and are essentially automated instruments for wet-chemical techniques such as titrations and UV–visible determinations.

2. Many clinical tests are performed using urine, whole blood, plasma, or serum samples.

3. Whole blood is often mixed with an anticoagulant such as heparin to prevent blood clotting prior to analysis.

4. Sodium fluoride is often added to whole blood for glucose determination to prevent glucose consumption via blood cell metabolism.

5. Blood samples taken for CO_2 and O_2 determinations should be kept under aerobic conditions (under mineral oil) to prevent gaseous mixing with the atmosphere.

6. Blood glucose determinations are the most commonly performed analyses world-wide as a consequence of diabetes.

7. Blood plasma is blood with its cellular components removed.

8. Blood serum is blood with both its cellular components and fibrinogen removed.

9. O_2 and CO_2 may both be determined electrochemically.

10. Blood 'electrolyte' levels, Na^+, K^+, Cl^-, and CO_2 (HCO_3^{2-}) are of great clinical significance and may be related to a number of clinical conditions.

11. The measured concentrations of anions in blood serum will always appear to be less than that of the total cation concentration and this is known as the 'anion gap'; disturbance of the anion gap holds great clinical significance.

12. Immunochemistry techniques involve utilizing antibodies and may be used in enzyme radioactive tracer approaches, immunofluro analyses, and enzyme-linked immunosorbent 'ELISA'-based tests.

13. Biosensors are devices in which biological entities (e.g. antibodies, enzymes) are coupled in close proximity to transducers (e.g. electrodes) to permit reagent-less analyses. Glucose biosensors for diabetes are the most widely used biosensors.

14. Glucose biosensors exploit the enzymatic oxidation of glucose by glucose oxidase.

15. First-generation sensors are based on the monitoring of an enzyme reactant or product (e.g. H_2O_2-based glucose sensors).

16. Second-generation sensors are based on the use of a mediator (e.g. sensors using ferrocene as a mediator and an artificial electron acceptor in the place of O_2 for glucose oxidase).

17. Other biosensor transduction approaches involve thermometric, optical, and mass-based determination approaches.

18. Carbohydrates are of great clinical significance and may be determined by many different means. Some carbohydrate mixtures require separation by HPLC, TLC, or other chromatographic approaches.

19. Chemical approaches may be determined via reaction with phenol and strong acids to form coloured products although these approaches do not lend themselves to quantitative tests.

20. Amino acid mixtures sometimes require separation by paper, TLC, GC, HPLC, or electrophoretic approaches.

21. Proteins are formed from 22 different amino acids. The sequence of amino acids that forms a protein is known as its primary structure. Chains of amino acids joined by peptide linkages are known as polypeptides.

22. The three-dimensional shape or folding of polypeptide chains is known as the secondary structure of a protein.

23. The tertiary structure of a protein describes the three-dimensional shape of the protein.

24. The quaternary structure describes a further level of organization structure in some globular proteins, relating to the association of protein units to produce an aggregate protein.

25. Proteins may be separated via both precipitation approaches and electrophoresis.

26. Analytical approaches for protein identification include immuno- and Western blotting techniques and chromatographic analyses.

27. The quantification of proteins may be classified by both chemical and physical approaches.

28. Chemical approaches for the quantification of proteins include the Kjeldahl method based on determining the nitrogen content of the protein, the Biuret method based on the formation of a purple complex with copper, the Lowry approach also based on the formation of a copper–protein complex, and the bicinchoninic acid based approach, which is a derivative of the Lowry and Biuret techniques.

29. The majority of physical based approaches for quantifying proteins involve their precipitation.

30. Genomics involves the study and exploitation of the genome of an organism, that is, the genome profile of an organism.

31. Proteomics is the study of multi-protein systems and the roles they play in a biological system.

32. Genomics and proteomics collectively may be termed bioinformatics.

33. Bioinformatics finds application in DNA fingerprinting and a number of biological and forensic sciences.

34. Many DNA-based techniques have been made possible by exploitation of the polymerase chain reaction, which allows very small segments of DNA to be replicated many times over so that sufficient DNA is made available for other analytical techniques.

Further reading

Brown, T. A. (1999). *Genetics: a molecular approach*. Chapman & Hall, London.

Durbin, R., Reddy, S. R., Krogh, A., and Mitchison, G. (1998). *Biological sequence analysis: probabilistic models of proteins and nucleic acids*. Cambridge University Press.

Gibson, G. and Muse, S. V. (2002). *A primer of genome science*. Sinauer Associates Inc., MA.

Liebler, D. C. (2002). *Introduction to proteomics: tools for the new biology*. Humana Press.

Manz, A. (2004). *Bioanalytical chemistry*. Imperial College Press.

Mikkelson, S. R. (2004). *Bioanalytical chemistry*. Wiley.

Plummer, D. T. (1987). *An introduction to practical biochemistry* (3rd edn). McGraw-Hill, New York.

Watson, J. D. (1980). *The double helix*. W.W. Norton, New York, London.

Winter, P. C., Hickey, I., and Fletcher, H. L. (1998). *Instant notes in genetics*. Garland Bios.

Environmental analyses and assays

15

Skills and concepts

This chapter will help you to understand:

- The reasons for choosing intermittent or continual air sampling approaches.
- How rotamers operate and are used for intermittent air-sampling approaches.
- The operation and suitable application of sampling devices, including rotamers, wet and dry test meters, filtration based devices, impingers, impactors, and solid absorbent based sampling devices.
- The importance of water-based approaches and what is known by the biological oxygen demand (BOD) for environmental samples.
- Suitable approaches for determining inorganics within environmental samples including the use of UV–visible based spectroscopy.
- How a number of organic compounds, including organo-, nitrogen, or phosphorous based pesticides as well as polychlorinated biphenyls (PCBs), may be determined in environmental samples.

15.1 Introduction to environmental analyses

Environmental issues such as maintaining air and water quality have received much attention in recent years. Environmental protection is a very complex area and at the heart of this subject, environmental analysis provides the information to help maintain the quality of our environment as well as how to assess and deal with pollution events. Within this chapter, we will discuss some of the analytical techniques used for environmental analyses. It should be remembered that many of the techniques routinely used for environmental analyses are common to other areas of analytical chemistry, so cross-references to appropriate chapters will be provided in each of these cases.

At the heart of any environmental protection policy is the need to perform analyses so that pollution may be controlled and monitored. The term 'pollution' embraces a complex concept, and many different definitions are used although a widely accepted definition for pollution from the ***Organization for Economic Co-operation and Development*** **(OECD)** states that *pollution means the introduction by man, either indirectly or directly, of substances or energy to the environment resulting in deleterious effects of such a nature to endanger human health, harm living resources or to interfere with other legitimate uses of the environment.*

It follows that many chemicals or other environmental factors that damage the environment can be considered as pollutants. Non-chemical factors can of course include, for example, noise, although in this book we shall restrict ourselves to solely chemical forms of pollution.

Pollution is often associated with the introduction of chemical compounds into the environment. Many of these compounds may be made artificially by man via industrial processes too numerous to mention, although the production of chloro-fluorocarbons (CFCs) represents a good example of this type of pollution. CFCs, although now banned in many parts of the world, were for several decades widely used as refrigerants and aerosol propellants and their effects in helping to destroy the ozone layer in the stratosphere are well known. Ozone is a natural compound that is found in the lower stratosphere but may also be considered as a pollutant itself when produced by a number of man-made processes, for example, as a component within motor vehicle emissions. Carbon dioxide is similarly produced as a result of the respiration of all living organisms, however, its percentage increase in the atmosphere as a by-product of fuel combustion may be contributing to global warming and, in this context, may also be considered as a further pollutant in some circumstances. As a third example, nitrates form an essential component within the nitrogen cycle; Fig. 15.1. However, the over use of nitrate based fertilizers can ultimately lead to an increase in the nitrate concentration within natural waters due to run-off from farmland. Nitrate pollution of rivers and lakes is a matter of serious environmental concern since this can lead to damage of the fauna and flora of surrounding rivers or lakes.

For environmental applications, samples for analysis are normally collected to: (a) help understand the chemistry of the localized environment in question; (b) establish hazardous levels of pollutants in the environment for efficiency of environmental protection; or (c) determine the source of an environmental pollutant.

15.2 Air analysis

Air quality is fundamentally important to human life since we maintain respiration by breathing in oxygen and expiring carbon dioxide. A frequent

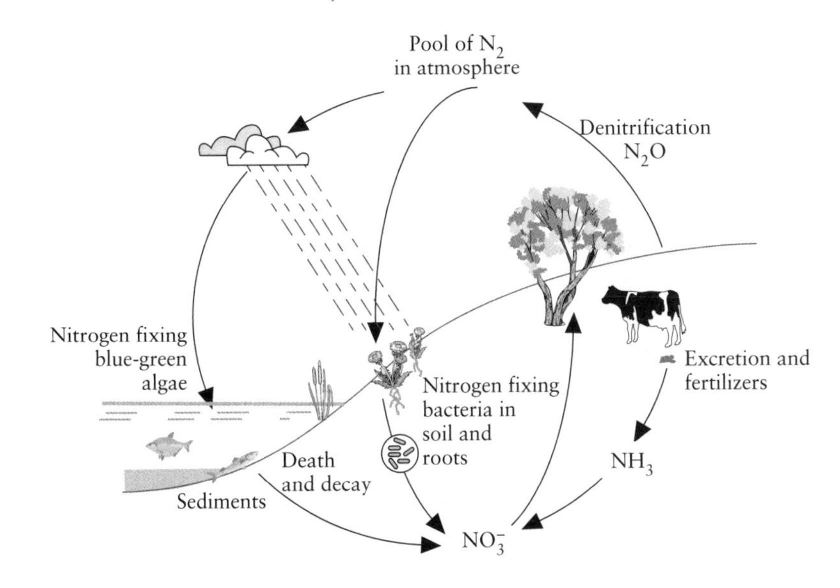

Figure 15.1 The nitrogen cycle.

interchange of air with our lungs, however, allows an easy route for pollution to enter and accumulate within our bodies. At one time it was thought that when a pollutant entered the atmosphere, air-borne analysis of the pollutant would provide an accurate indication of the degree of contamination. However, it is now understood that many chemicals undergo reactions (e.g. photochemical) with other components of the atmosphere and that this can lead to the formation of further pollutants. A good example in this context is the photochemical formation of 'smog' following the interaction of hydrocarbons, carbon dioxide, and UV light.

Analysis of air samples first involves the collection of air samples. How these samples are collected is often dictated by the nature of the pollutant, the lowest limit of detection required, and whether or not, for example, information is required relating to whether the amounts of pollution increase or decrease on a seasonal or daily basis. Great care must be taken to establish the volume of sample to be collected, as well as the sampling interval. In some cases, the time of day that a sample is taken may also be relevant and as a final point, issues relating to sample storage should not be overlooked.

15.2.1 Intermittent air sampling

If intermittent air sampling is chosen for the collection of air samples, then the central components of the air collection apparatus must include a vacuum source, the means for accurately determining the volume of air sampled and a series of collection vessels (sampling device or collector). In

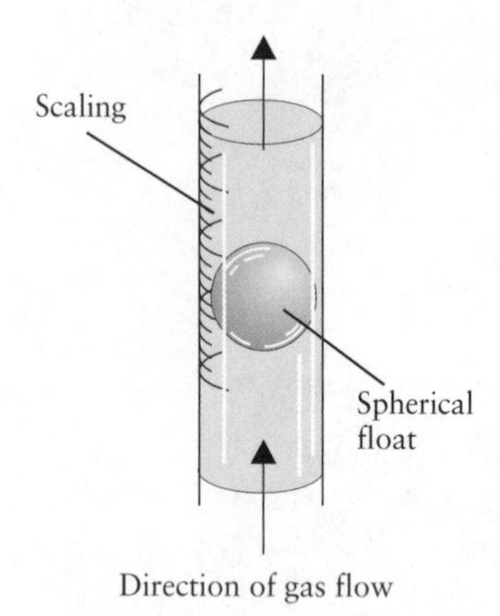

Figure 15.2 Schematic of a rotamer.

many arrangements the sampling device must be used in conjunction with a vacuum pump and metering device. Vacuum sources are required to draw the sample through the collection device and may be either motor or hand driven. Measuring devices are typically designed to measure either (i) the *rate* of air flow, or (ii) the volume of air collected. Measuring devices that determine the rate of air flow include, for example, ***rotameters***, a simple design of which comprises a spherical float within a tube that has a scaling as shown in Fig. 15.2. The bore of the tube is tapered, becoming wider towards the top; gas enters the bottom of the tube and the float rises in direct proportion to the rate of gas flow. The scale must be graduated for the gas being collected although in most practical cases this will be air and therefore may only need to be done once. Volume measuring devices are various and include so called 'dry test' and 'wet test' meters. Dry test meters typically operate via the alternate filling and emptying of plastic bellows thereby driving a system of bell-cranks and so the meter gauge.

Wet test meters involve the gas driving some form of paddle that is coupled to a meter. The instrument is normally partially filled with water and must be calibrated each time it is used. When using either dry or wet meters, measurements of temperature and environmental factors such as environmental air pressure (thermometer and manometer), are normally required to follow fluctuations in atmospheric temperature and pressure.

15.2.2 Sampling devices for aerosol constituents

Aerosol constituents are normally calibrated either via ***filtration*** or by use of an ***impinger***. Following collection by filtration, aerosol constituents can be detected via: (i) chemical analysis; (ii) weighing; or (iii) particle sizing. Filters are typically made from paper, glass fibre or granular materials such as fritted glass, porous ceramic, or sand. For particle sizing membrane based filters fabricated from, for example, polycarbonates or cellulose esters are normally used since they can selectively entrap particles of a given minimum size. Particle filters made from polymers can normally not be used at elevated temperatures. However, glass filters can be used for temperature up to 800°C or even greater. Impinger devices collect both solid particles and aerosols. ***Dry impingers*** are also sometimes known as ***impactors*** and, as the name suggests, impingers involve air streams impinging on surfaces. The surface upon which the air stream impacts normally takes the form of a slide, for example of glass. A number of impactors are arranged so that the air stream travels through a series of jets with diminishing diameters to selectively collect particles down to a few micrometres in diameter. In practice, impingers of this type are normally used for the collection of aerosols and microscopic examination.

Wet impinger air streams impinge on a surface immersed within a liquid; Fig. 15.3. In these devices, a glass tube is directed towards a surface at the base of a vessel to allow particles to be retained within the liquid. Collectors of this type are particularly useful for the collection of extremely small particles of diameters down to sub-micrometre levels. It is important, however, that the chosen fluid *cannot* act as a a solvent for the particles for collection so that the particles do not dissolve. A number of more sophisticated devices also exist for the collection of particles down to one-thousandth of a micrometre in diameter and these include electrostatic precipitators and thermal precipitators.

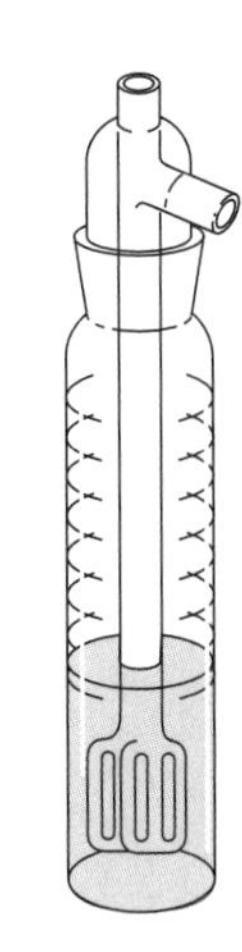

Figure 15.3 Schematic of a graduated wet impinger.

Gaseous constituents may also be collected by absorption within liquids and/or adsorption onto solid surfaces by condensation, freezing, or the filling of evacuated chambers. Vapours, by contrast, may be easily condensed at normal atmospheric temperatures and pressures, and, for this reason, may be collected rather more readily.

15.2.3 Solid adsorbents

Solid absorbents are normally used for the collection of low concentration organic compounds and permit analyses by techniques such as gas chromatography and/or mass spectrometry. Sampling may be performed either via ***passive*** or ***active*** sampling approaches. ***Passive samplers*** (sometimes also known as ***diffusion samplers***), contain adsorbents such as an activated charcoal or Tenax® porous polymer and these are often fabricated in the form of a lapel badge to be pinned to one's clothing for personal monitoring; Fig. 15.4. ***Active sampling*** methods involve drawing air through a sampling tube using a pump for, for example, pollution monitoring in remote locations; Fig. 15.5. Active sampling typically offers greater sensitivity and lower limits of detection due to the continued

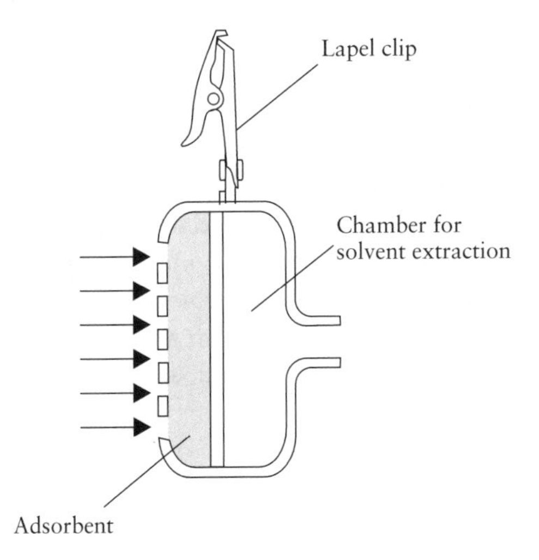

Figure 15.4 Schematic of personal passive diffusion sampler.

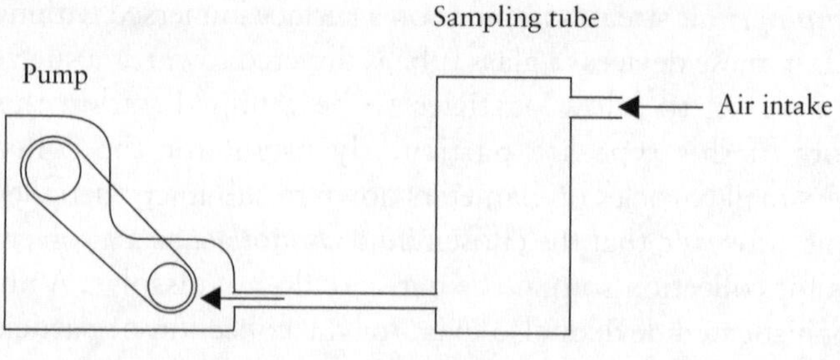

Sampling tube
Pump
Air intake

Figure 15.5 Schematic showing active sampling with sampling tube and pump.

passage of air through the device, itself giving rise to pre-concentration within the sample tube. Analytes are typically detected following thermal desorption or solvent extraction approaches.

Air sample analyses commonly used include determinations for nitrogen dioxide, sulphur dioxide, total hydrocarbon content, and also the total organic content of samples, each of which are considered in the following sections.

Nitrogen dioxide

Nitrogen dioxide (NO_2) concentrations within the atmosphere can be determined following absorption of nitrogen dioxide into a solution of sulphanilic acid containing an azo dye. A pink coloration is formed upon reaction with hydrogen peroxide and allows concentrations of NO_2 as low as 5 ppm to be determined.

Analyses for total oxides of nitrogen (excluding nitrous oxide), for example, from gaseous effluents from combustion processes, are performed by first collecting gas within an evacuated flask containing a hydrogen peroxide solution in dilute sulphuric acid. The oxides of nitrogen are converted into nitric acid upon dissolution, and the nitrate ion is allowed to react with phenol disulphonic acid to produce a yellow product that can be determined colorimetrically. This approach typically allows lower limits of detection down to concentrations of 5 ppm in the atmosphere.

Sulphur dioxide

Air-borne sulphur dioxide may be determined by first passing the air sample through a solution of sodium tetrachloromercurate. Sulphur dioxide is retained by the formation of dichlorosulphursitomercurate(II) ion, $HgCl_2SO_3^{2-}$. This complex is then allowed to react with formaldehyde and pararosaniline to form a pararosaniline methylsulphonic acid which is highly coloured with a λ_{max} at 560 nm, which permits spectrophotometric lower limits of around 0.003 ppm SO_2 in the atmosphere. This form of analysis has been automated within a number of commercial instruments.

Total hydrocarbon determinations

Total hydrocarbon concentrations may be determined using IR spectroscopy since they absorb in the wavelength range of 3–4 μm. Analyses are performed by collecting samples within a liquid oxygen condensation traps.

Total organic content

Total organic content (TOC) determinations can be performed following extraction with an organic solvent, and the components may then be determined by two approaches. The first of these involves oxidation with potassium peroxodisulphate. The CO_2 produced is determined in aqueous solution either via absorption in the infrared region or by measurement of the conductivity of solutions (prepared with distilled water). A second approach involves reduction of the organic content to methane and the determination of the methane content either by flame ionization techniques or IR spectroscopy.

15.3 Water analysis

The maintenance and monitoring of water quality is crucial for environmental protection as well as the maintenance of human life. Water pollution can be caused by many different industrial processes, including the production of steel, coal, paper as well as the disposal of chemicals from industry and the home. Water is naturally one of the best solvents that exists. However, this unfortunately means that many contaminants can dissolve and be transported by water. It should not be forgotten that water can also help transport solid particulate pollutants.

The maintenance of water quality can involve sampling from river streams, collection of rainwater, waste water, potable (drinking) water and in many cases analytical protocols have to be designed that are capable of coping with fairly complicated mixtures. The first stage of this process is the sampling regime that may have to vary considerably for different samples. Sampling at a riverside, for example, may be relatively trivial; depth samplers are often required, however, to collect water from large bodies of water at specific depths.

Water chemistry is frequently complex and the pH and temperature of the water as well as the presence of dissolved gases such as oxygen and/or carbon dioxide must often be measured if the analytical information (viz. analyte concentration and speciation) is to be trusted or indeed be relevant to the environmental purpose for which it is sought. Water samples can contain single-celled organisms (protozoa) through to higher animals such as fish as well as course plant life. All of these living organisms impart a ***biological oxygen demand*** (BOD) to the environment and for this reason the BOD is a widely used criterion for assessing the environmental well-being or otherwise of ecosystems. For some analyses it is necessary to *fix* gases via the use of reagents to prevent the gaseous content of samples being altered with time as a consequence of continued biological activity and this is especially important if samples are to be transported some distance to a laboratory.

For potable (drinking) water distribution, remote sensors are becoming ever more widely used in conjunction with telemetry to allow remote monitoring to provide real-time information throughout a distribution network.

In a similar way, remote sensing can be used for the maintenance of rivers and may permit real time monitoring of water quality and/or possible pollution events. It should be remembered that the analyses of water samples are often similar to those described in earlier chapters of this book, and include atomic absorption techniques (Chapter 7), electrochemical determinations for heavy metal ions in water (Chapter 10), or flame photometry for the determination of inorganic salts (Chapter 7).

One of the greatest problems associated with environmental water monitoring is the complicated mixture of different compounds or particulates, which often demands some pre-filtration and/or treatment prior to analysis. There are a number of comprehensive laboratory manuals describing in detail analyses that the interested reader can refer to. The concentrations at which different compounds may have biological significance can be considerably lower than would otherwise be expected, since many organic compounds are capable of bio-accumulating within organisms to concentrations that are considerably higher than those found in their immediate surroundings.

15.3.1 Gas chromatography for water analyses

Gas chromatography is widely used for water analyses since it typically offers high sensitivity and excellent separation efficiencies. Many separations can be achieved using just a few stationary phases and this widens the applicability and ease of application of GC for the analysis of different water samples. Commonly used detectors for use with water samples include approaches based on flame ionization, electron capture, thermal conductivity, flame photometric detectors and mass spectrometry. For further details of gas chromatography, please refer to Chapter 8.

Extraction procedures

Analytical determinations for organic compounds often necessitate extraction from water prior to chromatographic analysis and subsequent determination. Since GC columns are incompatible with water, many analyses require transfer of the analyte from an aqueous to an organic phase. One of the simplest approaches involves solvent extraction. This is performed by shaking the sample with an immiscible organic solvent in which the analyte(s) are soluble. Light petroleum and hexane are often particularly suitable although chlorinated solvents are sometimes used.

The organic layer is separated and dried prior to being passed through the chromatographic column. The pH of the extractions can sometimes be altered to facilitate the extraction of either acidic or basic compounds, as required. Acidified media are often used to help extract acidic components such as carboxylic acids, while basic media are often used to facilitate the extraction of compounds such as amines. When considering the choice of extraction solvents, the detector should also be considered since, for example, hexane will give rise to a prominent chromatographic peak if a flame ionization detector is used.

A second approach involves a so-called ***head-space analysis technique*** in which a water sample is paced within a container with a septum seal to allow a pre-defined volume of air space to be entrapped. Following equilibriation of the air space with water, an air sample is extracted, which should contain some of the organic components. These organic compounds are then allowed to equilibrate between the aqueous and gaseous phases, which permits the gaseous sample to be injected directly into the gas chromatograph.

A third approach is known as a ***purge trap*** technique, which involves the extraction of volatile organic solvents from the sample via purging with a gas stream. After a pre-defined time, the collection tube is flash heated to release organic analytes directly into the chromatographic column. Another related approach involves collecting the volatile organic compounds into a liquid nitrogen cold trap as a secondary solvent and then rapidly heating the liquid nitrogen for the release of the organic compounds into the chromatographic column.

A fourth approach based on ***solid-phase extraction*** involves passing water through a short disposable column of absorbent material of the reverse phase type such as octadecylsilane (ODS) (see Chapter 8) bonded to a silica support. Organic compounds retained on the column may elute via the use of solvents such as hexane that can then be directly injected into the GC column.

15.3.2 Applications

Fingerprinting of oil spills

Oil spillages at sea can cause widespread environmental damage to both fauna and flora. Analysis of oil samples by chromatography (following separation from water) can often allow the source of the spillage to be matched or 'fingerprinted' to allow comparison with libraries of known reference samples. It is often necessary to use capillary columns to separate the components, although it is more usual to identify the source of an oil for legal reasons if, for example, compensation claims are being sought. The overall profile of the chromatogram will often provide information relating to the nature of the oil contamination. Hydrocarbon oils, for example, often exhibit chromatograms with regularly spaced peaks whereas lubricating oils yield fewer peaks, and vegetable oils fewer still.

The determination of pesticides such as DDT represents a further example of water based analyses that require analysis via gas chromatography. Analysis for DDT typically involves the extraction of the organic component into hexane. The solvent must be first be dried using, for example, anhydrous sodium sulphate powder and in some cases the concentration may be increased by partial evaporation of the solvent. The sample will often require cleaning prior to analysis using, for example, a silver nitrate column to remove polar compounds. The use of silica gel columns can be used to help separate non-polar interferents from the sample and in this situation, the hexane-based sample can be passed through the column. In this situation, DDT is retained on the column and may be subsequently eluted using a more polar solvent, such as a mixture of diethyl ether in hexane.

15.4 Analysis of inorganics within water based samples

The largest grouping of inorganic analytes of environmental significance is the metal ions and their salts. The most abundant ions found within environmental samples include sodium, calcium, potassium, and magnesium and often occur at concentrations ranging from $\mu g\ dm^{-3}$ to $mg\ dm^{-3}$. Magnesium and zinc are sometimes found at concentrations up to $\mu g\ dm^{-3}$ ranges although most other metal ions are likely to be found at trace concentrations at the $\mu g\ dm^{-3}$ range or lower. Metal ions sometimes originate from naturally occurring ores as well as anthropogenic or man-made sources, including, for example, industrial processes such as metal purification and refining, solid waste disposal (and subsequent leaching from, e.g. land fill sites) and/or industrial effluents.

The analytical approaches most widely used for the analysis of metal ions within the environment include: (a) ***flame atomic absorption spectroscopy***; (b) ***flameless atomic absorption spectroscopy***; (c) ***inductively coupled plasma–optical emission spectroscopy (ICP–OES)***; and (d) ***inductively coupled plasma–mass spectrometry (ICP–MS)***. Samples should normally be stored within polyethylene bottles since these are less likely than glass bottles to contaminate samples with metal ions. Samples should normally be first acidified by, for example, the addition of dilute HNO_3 to prevent precipitation of metal salts. The majority of analyses are for the total metal ion contents for a given element irrespective of the chemical state or speciation of the metal in question. ICP–OES and ICP–MS approaches lend themselves particularly well for total metal ion analyses of this type, since with these techniques sample pre-treatment is normally not required and analysis can be performed irrespective of the chemical state of the metal ions in mixed samples.

Flame absorption spectroscopy offers favourable sensitivity towards sodium, potassium, calcium, and magnesium. The greatest consideration that should be given is that some samples may require dilution to allow quantitative determinations whereas others require pre-concentration. The simplest approach for pre-concentration involves partial evaporation of an acidified sample and this is most likely to be necessary for the determination of zinc, iron, or magnesium. Another approach involves solvent extraction for the determination of trace metals. Flameless atomic absorption spectroscopy involves the replacement of the flame for atomizing the sample and in this way the sensitivity may be greatly increased. Flameless atomic absorption approaches negate the need for sample preconcentration, and this can be especially useful for the determination of trace metals. The most popular approach for atomization is via the use of a graphite furnace although the approach adopted depends on the metal ion to be analysed. Mercury salts, for example, can be chemically reduced using tin(II) chloride or sodium borohydride. The elemental metal mercury that is produced may then be carried by a stream of nitrogen for either collection or direct injection into a modified spectrophotometer.

Tin and lead may also be reduced, for example, by sodium borohydride to evolve some hydrides which may then be swept from a sample by a gas stream. In this instance, mild heating can be used to break down the hydrides to their constituent elements in their ground states. A third approach involves the use of ICP–OES and the related technique of ICP–MS—the reader is referred to Chapter 7 for further details.

In both of the ICP-based approaches, the sample is first atomized in the plasma flame at temperatures ranging from 6000 to 10 000 K. With ICP–OES techniques, the emission spectrum is monitored directly, which permits simultaneous determination of up to sixty elements at pre-set wavelengths. Some instruments perform sequential analyses at each of the required wavelengths for specific elements but often allow for rapid determinations, (typically <5 s per element). One of the greatest problems associated with ICP is interference between different elements, since different elements produce many more lines within the emission spectra. These problems may be overcome via ICP–MS—in which the inductively coupled plasma is essentially used as an ion source for the mass spectrometer which then quantifies individual metal ions. A more comprehensive discussion of ICP–MS is provided in Chapter 9. However, it should be remembered that separation and identification of ions occur on the basis of mass to charge ratios. Since the mass spectra of inorganic mixtures are typically simpler than those associated with organic compounds, spectral resolution is easier than is encountered with many organic compound containing samples. The sensitivity offered by ICP–MS is unfortunately somewhat lower than that obtainable by graphite furnace atomic absorption techniques but may still permit determinations down to 1 μg dm^{-3} concentrations and for these reasons is becoming established as the bench-mark approach for the determination of many metal ions in environmental samples.

15.4.1 UV–visible spectrometry based analyses of metal ions in environmental samples

UV–visible techniques for the determination of metal ions has largely been replaced by atomic absorption approaches although UV–visible approaches are still used for some of the simpler analyses. UV–visible spectroscopy is normally only possible for determination within relatively unpolluted water samples such as lakes, whereas atomic absorption spectroscopy lends itself more readily for the analysis of, for example, sewage effluents. Many samples may be analysed via the formation of a coloured complex. One example is the determination of iron, as described in Chapter 3, although this technique may suffer from interference from sulphate, cadmium, or lead. Alternative approaches include the use of solvent extraction although these approaches have largely been surpassed by the use of atomic absorption spectroscopy since the advantages offered in terms of simplicity are lost if a pre-treatment step is required.

15.4.2 Adsorptive stripping voltammetry of metal ions in environmental samples

Another approach that is widely used is adsorptive stripping voltammetry—the reader is referred to Chapter 10 for a more thorough treatment of electroanalytical techniques. In this technique a working electrode consisting of a mercury drop suspended at the end of a glass mercury filled capillary (known as a hanging mercury drop electrode) is used for the pre-concentration of metal ions which are subsequently voltametrically stripped from the surface and the current that flows during this process may be related to the concentration of metal ions within the analyte sample. The hanging mercury drop working electrode is polarized for the reduction of metal ions to the free metal and this may be described by the general reaction shown in Eqn (15.1):

$$M^{n+} + ne^- \rightarrow M \qquad (15.1)$$

The current measured may then be related via the use of pre-determined calibration plots to the concentration of the metal ion within a sample.

The metal ion is drawn to the cathodically polarized (negative) electrode where it is reduced. Further metal ions are drawn to the electrode and accumulate over a period of time. After a pre-determined time, the potential of the working electrode is swept in the reverse (positive) or anodic direction.

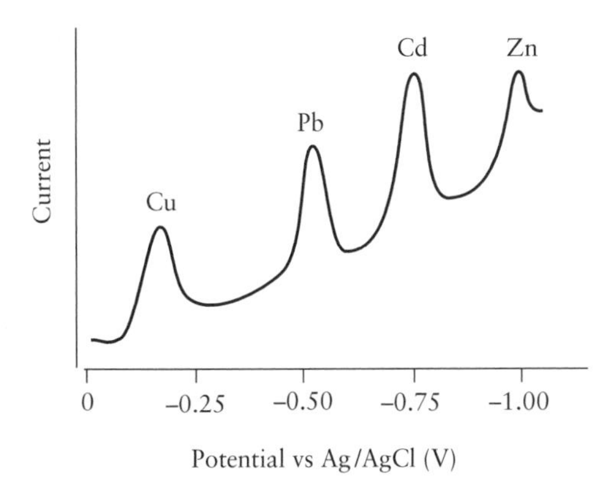

Figure 15.6 A representative anodic stripping voltammogram.

A number of metal ions may be deposited at the working electrode surface during the cathodic sweep and as the potential is progressively swept back in the anodic direction each metal ion is returned back to the solution. The resulting current/potential profile is known as a stripping voltammogram; Fig. 15.6. The general equation describing such a process takes the form as shown in Eqn (15.2):

$$M \rightarrow M^{n+} + ne^- \qquad (15.2)$$

Adsorptive stripping voltammetry can allow for the determination of metal ions down to concentrations of parts per million and, in some cases, even parts per billion concentrations. Typical analyses require accumulation times of several minutes. The instrumentation, which involves the use of a hanging mercury drop electrode, potentiostat, and associated recording equipment is, however, relatively simple, inexpensive, and robust.

15.5 Organics in water-based samples

There are many organic compounds of environmental significance, although two have been of particular cause for concern in the protection of the environment; these are pesticides and ***polychlorobiphenyls*** (PCBs).

15.5.1 Organo-, nitrogen, and phosphorus pesticides

Pesticides are of great environmental concern due to their widespread use in agriculture. DDT was the first synthetic insecticide to be widely used following the Second World War, and although it is now banned in many areas of the world DDT is still found in the environment due to its previous widespread use and wash-off from farmland. DDT samples contain a mixture of closely related compounds, although the major active component is D,P′-DDT, accounting for approximately 70–80% of the total content of a typical sample. Much of the remaining 20% of the sample is p,p′-DDT, which is similar in structure but with a $CHCl_2$ side chain replacing CCl_3. Since samples contain varying mixtures of compounds, analyses require calibrating with respect to samples for each source of pesticide. Analyses are normally based upon some form of chromatographic approach. With most DDT containing samples, extraction must first be performed using a solvent such as hexane to allow a pre-concentration of the analyte. The solvent is then typically dried using anhydrous sodium sulphate powder. This may then be further pre-concentrated via partial evaporation. Samples can sometimes require pre-clean up using either, for example, alumina/silver nitrate columns that retain polar compounds but allow less polar compounds such as DDT to be eluted in hexane. An alternative approach involves the use of silica gel columns that are less polar than silver nitrate columns. On passing a sample dissolved in hexane through the column, less polar interferents such as PCBs are eluted allowing the DDT component to be retained; the DDT can subsequently be eluted using a more polar solvent mixture such as hexane and benzyl ether. Samples can then be introduced to a gas chromatograph for final separation and quantification.

15.5.2 **Polychlorinated biphenyls**

Polychlorinated biphenyls such as 2,4,5,2′,5′–PCB (Fig. 15.7) are formed as by-products of a number of industrial processes. The exact analytical procedure adopted may vary depending on the nature of the sample (e.g. in water or a soil sample). Most determinations are performed using either GC or GC–MS, see Chapter 8. Sample preparation for either approach is essentially similar, using, for example, either hexane or acetone. The solvent in either case must normally be dried using anhydrous sodium sulphate. Concentration steps may be undertaken either via partial solvent evaporation or solid phase extraction and possibly further sample clean up for particularly problematic samples. Final separation and/or identification is often achieved using either GC or GC–MS. Identification of individual analytes (especially when using GC–MS) is normally performed by matching spectra with those obtained using known samples.

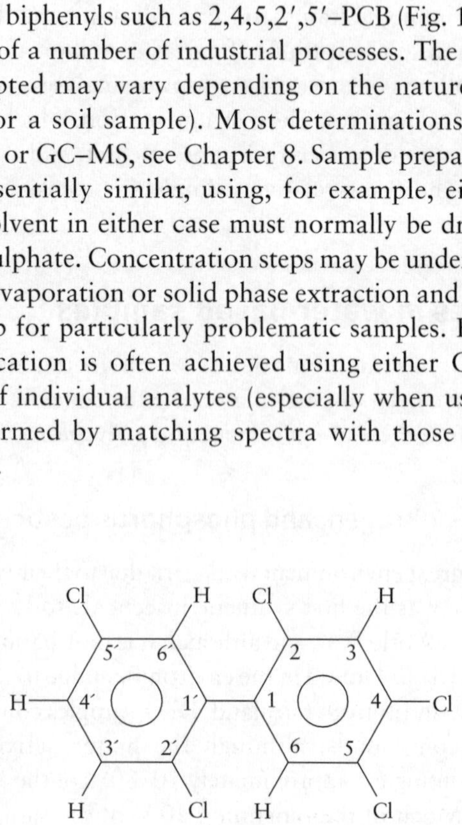

Figure 15.7 Structure of 2,4,5,2′,5′-pentachlorobiphenyl.

Exercise and problems

15.1. Compare and contrast the use of impactors and impingers.

15.2. Why are personal monitors sometimes used in place of impinger or impactor based sampling devices?

15.3. What is meant by the biological oxygen demand (BOD) in environmental monitoring and what is its significance?

15.4. Describe a practical approach for determing NO_2 concentrations in air samples.

15.5. Describe how sulphur dioxide concentrations may be determined in air samples.

15.6. Describe a practical approach for determining the concentration of polychlorinated biphenyls (PCBs) in environmental samples.

Summary

1. Environmental analysis and protection is a complicated subject that involves the study of many interrelated chemistries; this is seen, for example, in the nitrogen or carbon cycles.

2. Air analyses can either be performed using intermittent air sampling or continuous real-time monitoring approaches.

3. Intermittent sampling approaches must include a vacuum source, the accurate determining of the volume of air sampled and a series of collection vessels (sampling devices or collector).

4. Measuring devices are typically designed to measure either (i) the range of air flow, or (ii) the volumes of air collected.

5. Measuring devices for air-flow rates include, for example, rotamers.

6. Volume measuring devices for air samples include both 'dry test' and 'wet test' meters.

7. Dry test meters typically employ plastic bellows that may be alternately filled and emptied, thereby driving a system of bell cranks and ultimately a gauge.

8. Wet test meters involve the gas driving a meter.

9. Sampling devices for aerosol constituents include filtration and impinger devices.

10. Dry impingers are also sometimes known as impactors.

11. Wet impingers involve air impinging on a surface immersed within a liquid.

12. Solid adsorbers are normally used for the collection of low concentration organic compounds and permit analyses, for example, via gas chromatography or mass spectrometry; sampling with solid adsorbers may be performed by passive or active approaches.

13. Passive (or diffusion) samplers contain adsorbers such as charcoal or porous polymers and, for example, are sometimes used in the form of lapel badges for personal monitoring.

14. Active sampling involves continually passing air through the device so allowing pre-concentration of an analyte.

15. Passive and active sampling devices may be used for air-borne total hydrocarbon, total organic, or sulphur dioxide determinations.

16. One of the most frequently used environmental analyses is for the determination of biological oxygen demand.

17. Gas chromatography is used for many water analyses. Some analyses require extraction of organics via, for example, head space analyses, purge trap, or solid phase extraction approaches.

18. Oil spillages, for example, may be fingerprinted via chromatographic approaches to identify the source of a spillage. Other examples of chromatographic based analyses are for the environmental determination of pesticides and herbicides.

19. Inorganics such as metallic ion concentrations within environmental samples are normally determined via approaches described in other chapters, for example, ICP atomic absorption or UV–visible spectroscopy.

20. Polychlorinated biphenyls (PCBs) are of great significance for environmental pollution and may be determined by GC or GC–MS based approaches.

Further reading

Ahmad, R., Cartwright, M., and Taylor, F. (2001). *Analytical methods for environmental monitoring*. Prentice Hall.

Kekkebus, B. B. and Mitra, S. (eds) (1997). *Environmental chemical analysis*. CRC Press.

Natusch, D. F. S. and Hopke, P. K. (1983). *Analytical aspects of environmental chemistry*. Chemical Analysis Series, Wiley.

Reeve, R. N. (2002). *An introduction to environmental analysis*. Wiley.

Critical choice of technique, good laboratory practice, and safety in the laboratory

16

Skills and concepts

This chapter will help you to understand:

- The importance of consulting the academic literature prior to undertaking a chemical analysis to gain experience of prior knowledge and the best practice.
- The financial considerations and constraints as well as the importance of choosing the most suitable analytical method for a given set of requirements.
- The importance of validating analytical protocols and how this can be achieved via the involvement of third parties.
- The importance of good laboratory practice and how this can be achieved, how to keep a good laboratory notebook, and how to help avoiding the loss of electronic data.
- The importance of safe working practices in the laboratory.

16.1 Choice of technique

16.1.1 Consideration of the problem

One of the primary concerns that must be considered before any analysis is undertaken is to carefully identify and assess the problem, although this may not be as trivial as it first appears. For example, there is no need for greater levels of sensitivity or accuracy than are required for a particular purpose, and excess complexity can simply represent wasted effort and/or money that might be put to better use. Equally, analyses by techniques that cannot offer the required sensitivity or lower limits of detection needed are useless. Erroneous data for whatever reason (due to the presence of interferents, poor experimental procedure and/or the introduction of other errors) are at best useless and may often result in inappropriate action being taken. They may lead to dangerous practices.

It should be recognized that any analytical procedure includes all of the stages from sample collection, sample pre-treatment (if appropriate), sample storage and handling, the analytical methodology itself, and validating the analytical technique and finally reporting the data.

16.1.2 Considering the literature, validating analytical techniques, and considering finance

One of the first stages in identifying an appropriate analytical technique is to consult the literature and research fully the methods that others have used in similar circumstances. As we have seen in many instances throughout this book, there may be several different approaches for undertaking a particular analysis. Differing techniques often offer different advantages, for example, in terms of simplicity, lower limits of detection, and/or detection *per se*. There are often considerations that render one technique more suitable than another for any particular determination. For example, if it is known that particular interferents affect one technique, another may be chosen in preference. It is for reasons such as these that it is important to consult as wide a range of literature in books and/or journals as possible that may prove useful and only then should consideration be given to a choice of technique.

Any technique should first involve some experimentation to verify that the laboratory is capable of reproducing results as reported by other workers. Validation will often require analyses to be performed with samples containing potential interferents that might be present within the sample and the effect that these might have on the results that would be obtained. To help in this process, professional bodies such as the Royal Society of Chemistry and the American Institute of Standards produce many standard reference materials. Some central or national laboratories may also be ready to undertake third-party validation of techniques. The reader is referred to Chapter 2 where similar factors are considered. The use of certified reference materials helps identify experimental errors and the identification of, for example, determinant and indeterminant errors and their possible sources. Many instrumental techniques will require careful maintenance of the instrument which will often include regular calibration by third parties, again often using certified reference samples or materials. Many companies offer certification based on third-party accreditation schemes operated by, for example, the International Standards Organization (ISO). It is clear that any method must be evaluated and assessed for its precision, accuracy, and reproducibility and this normally dictates that analyses are performed within certain conditions and parameters. If third-party reference materials are not available then reference materials must first be prepared.

One approach is to analyse a pre-prepared sample into which known quantities of analyte are introduced via the process of spiking. These tests operate by the principle that if essentially 100% of the spiked analyte is recovered (plus or minus the appropriate uncertainty that is allowed) then a level of confidence can be assigned relating to the appropriateness and reliability of the technique chosen. Great care must be taken in such cases, however, to ensure that the effect of any potential interferent is fully understood and that their presence does not cause results to fall outside pre-defined acceptable levels of tolerance.

The availability of equipment and/or financial restraints should also not be overlooked. If many analyses of a similar type need to be undertaken at regular intervals then it may well be in the interests of a company to purchase the appropriate equipment and to train staff as necessary. If, however, a few analyses need to be undertaken at intermittent time intervals, then it may be more cost effective to employ a third party, thereby saving money whilst also possibly making use of the expertise of others who are familiar with the techniques to be used. Another approach may be to access facilities within, for example, a university or national laboratory via the payment of 'access charges'.

Alternatively, specialized equipment may in some cases be hired on a long-term leasing basis, for example, to health care laboratories where instrumentation for critical care applications must always be maintained to ensure critical cover. Contracts of this type normally provide for 24 h call-out for maintenance of the machines and/or replacement as required although these types of contract are typically more expensive than simply purchasing the equipment since payment is being made for immediate back-up, assistance, and/or maintenance as required.

As a final point of consideration, it is widely accepted that many analytical determinations performed are either inappropriate or not required at all. Estimations for the number of inappropriate determinations that are undertaken vary but extend to 20% or more of all analyses world-wide being either wholly inappropriate, or duplicated when not required; this represents, at a conservative estimate, 5% of the gross national product of most developed countries. These figures do not, moreover, take into account those analyses that are performed incorrectly or the wastage that occurs as a result of inaccurate or erroneous analytical data. It is hard therefore to overestimate the importance of ensuring that appropriate analyses are performed and that all analytical methodologies are regularly and critically evaluated.

16.2 Good laboratory practice and safety when performing wet chemical analyses

Good laboratory practice should always be followed to ensure the quality of data. It should always be remembered that good laboratory practice and safety are intrinsically linked, and so we shall consider safety first.

16.2.1 Safety

The importance of safety cannot be overstated. We only have one life and it is extremely easy to irreparably damage our eyes or skin with less than careful behaviour that seems to offer little danger at the time. Good safety takes only a little effort—which is a price well worth paying to avoid serious injury.

The following properties should always be followed:

- ***Safety glasses with side protection*** should always be worn within the laboratory. It is extremely easy for solutions to come into contact with the eye.
- ***Fire alarms, extinguishers suitable for chemical and electrical fires and fire exits should all be located prior to any work being performed.*** Nobody expects a fire. The time to start looking for a fire extinguisher, the fire alarm or a fire exit is not when you are faced with dealing with an emergency.
- ***Identify where the nearest safety shower is***—and ensure you know how it operates.
- ***Identify where the eye wash station and first aid boxes are.*** Again make sure you are familiar with the operation of eye wash bottles and/or taps.
- ***Laboratory coats should always be worn and be buttoned up.*** A loose flapping laboratory coat may act as a further hazard by being caught around equipment or within handles and may moreover be easily be set alight by Bunsen burners; an unbuttoned lab-coat also does little to protect our clothing at the front, yet this is after all its primary job.
- ***Long hair should be tied up and kept out of the way*** to prevent catching or snagging in equipment, as well as to minimize becoming contaminated with reagents or being set alight.
- ***Food and drink (and this includes chewing gum) should never be brought into the laboratory (and certainly never consumed within the laboratory).*** It is extremely easy for food to become contaminated—even if it is to be consumed away from the laboratory at a later stage.
- ***A careful written assessment of the hazards associated with any experimental procedure should be made before it is attempted.*** This

indeed forms the basis of statutory regulations in many countries. Many accidents can be avoided by careful thought prior to the experiment.

- ***Great care should be taken with glassware.*** Many people have suffered horrific injuries by exerting excessive force to push pipette stems into rubber filling bulbs which have broken; ***it is extremely easy to push a broken pipette stem straight through the palm of one's hand and out the other side as the pipette stem breaks!*** Do not pick up broken glass with fingers—you will invariably cut yourself and then possibly contaminate the wound with acid or base etc which was on the glass. Use a dust-pan and brush, and if necessary, a mop.
- ***Great care should be taken when handling wet reagents in the vicinity of electrical instrumentation.*** If possible, keep flasks and other glassware away from instruments to avoid damaging spillages. If an accident does occur isolate the mains supply preferably at the mains plug or if this is wet at the nearest isolation switch. ***Once again identify where prior to working within the laboratory—do not wait till an accident occurs. If someone does receive an electric shock and is unconscious, turn off the power supply before approaching them otherwise you may receive a shock yourself. If this is not practically possible, separate the person from the instrument with an insulating object such as a wooden broom or stool.***
- There should be a number of ***registered first-aiders working*** in your area. Once again ***identify who these people are prior to working in the laboratory*** and ensure you know how to contact them quickly (e.g. via internal phone numbers) and what procedures should be taken to summon help. Identify where there are telephones to summon help and ensure that emergency numbers are clearly displayed.
- ***Never work in the laboratory alone.*** Always ensure someone is close at hand to help if something goes wrong.

16.3 Importance of keeping and maintaining a hard-backed laboratory notebook and the backing up of electronic data

A good-quality hard-backed laboratory notebook should always be kept. The maintenance of a laboratory notebook is an essential activity for any analytical chemist working either in industry or in a university/research environment. It is common practice in many laboratories for a supervisor to check and sign every notebook before the close of each day's work.

The notebook should always be hard-backed since this offers some protection against spillages and generally adds to its durability in a rather hostile environment. ***The notebook should be written in with ink or ball-***

point pen, and not pencil. A permanent record of activities being undertaken forms an essential part of many quality assurance and accreditation processes (Chapter 2).

All work, data, and observations and experimental procedures should be clearly entered within the notebook, together with the date and time.

Mistakes should be neatly crossed out, leaving the original material still visible. If you decide later to reject data due to an identified problem that has been recorded in the laboratory notebook, it is easier to justify your actions with hard evidence.

Remember, a laboratory notebook facilitates the writing of reports either in the university or industrial environment; pieces of paper are easy to lose and all too easily become illegible when they are subject to the inevitable spillages that occur in any laboratory.

Analytical data are increasingly being obtained in electronic form—it is absolutely essential that a regular and systematic back-up procedure is put into place; again the time to think about doing this is not after the first loss of data. Electronic data can be very easily overwritten, corrupted, or damaged by equipment (e.g. the magnetic fields of NMR machines) or by human errors as well as computer viruses/worms.

Data back-up processes should be performed on a weekly basis as an absolute minimum and within industrial environments data back-up should be performed daily.

Summary

1. Before embarking on a new analytical approach, always consult the literature to exploit as much prior knowledge as possible. This also facilitates best practice.

2. Safety is paramount and involves, for example, the use of safety glasses, familiarization with fire fighting equipment, as well as the location and the use of safety showers.

3. Food and drink should not be brought into or consumed in the laboratory.

4. Care should be taken when handling glassware.

5. Care should be taken when using electrical equipment in the laboratory, especially in the vicinity of water and aqueous solutions.

Further reading

Adams, K. (2002). *Laboratory management: principles and processes*. Prentice Hall.

Cold Spring Harbor Laboratory (2001). *Safety sense: a laboratory guide*. Cold Spring Harbor Laboratory Press, 2001.

Picot, A. (1994). *Safety in the chemistry and biochemistry laboratory*. Wiley.

Index